1st International Conference on Upstream Engineering and Flow Assurance 2012

Topical Conference at the 2012 AIChE Spring Meeting and 8th Global Congress on Process Safety

Houston, Texas, USA
1-5 April 2012

ISBN: 978-1-62276-569-0

Printed from e-media with permission by:

Curran Associates, Inc.
57 Morehouse Lane
Red Hook, NY 12571

CURRAN ASSOCIATES INC.
proceedings.com

Some format issues inherent in the e-media version may also appear in this print version.

Printed by Curran Associates, Inc. (2012)

AIChE
3 Park Avenue
New York, NY 10016-5991

Phone: (203) 702-7660
Fax: (203) 775-5177

www.aiche.org

Additional copies of this publication are available from:

Curran Associates, Inc.
57 Morehouse Lane
Red Hook, NY 12571 USA
Phone: 845-758-0400
Fax: 845-758-2634
Email: curran@proceedings.com
Web: www.proceedings.com

TABLE OF CONTENTS

Process Safety and Corporate Responsibility .... N/A
Michael Dolan

Horizons in Multiphase Flow 1
Geoffrey F. Hewitt

Transient Multiphase Flow - Past, Present and Future with Flow Assurance Perspective 2
Tom Danielson

"Steady-State Multiphase Flow Modeling " Past, Present and Future with a Flow Assurance Perspective 3
Mack Shippen, William Bailey

Characterization of Confidence in Multiphase Flow Predictions 4
G. Kouba, Hariprasad J. Subramani, Selen Cremaschi

High-Viscosity Oil Multiphase Pipe Flow 5
Hong-Quan Zhang, Cem Sarica, Eduardo Pereyra

Dynamic Models in Multiphase Flow 6
Ole J. Nydal

R& D Efforts to Control, Monitor and Identify Drilling Fluid Invasion in Reservoir Rocks 7
Andre Leibsohn Martins, Alex Tadeu A. Waldmann, Elessandre Souza, Alex Rodrigues De Andrade, C. M. Scheid, L. A. Calçada, R. Z. Moreno, C. E. Dannenhauer, S. A. Loureiro

Comparative Analysis of Gelled Fluids Through Large Amplitude Oscillatory Shear Rheology (LAOS) 18
Jason Maxey

Using Polymer Elasticity to Scale up Lab Characteristics to Field Applications of Friction Reducers 31
Ahmed M. Gomaa, Jing Zhou, Hong Sun, Qi Qu

Efficient Displacement of Synthetic or Oil-Based Mud and Transitional Phase Inversion 50
Lirio Quintero, Jean-Louis Salager, Ana Forgiarini, Gianna Pietrangeli, Jonathan Brege

Mass Transfer Effects in Bore-Hole Filter-Cake Cleanup 62
Jeffrey Harwell, David Sabatini

Advances in Produced Water Treatment 63
Timothy P. Daigle, Savanna Hantz, Brian Phillips, Rafique Janjua

Advanced Pipeline Monitoring for Flow Assurance with Fiber Optics 64
John D. Hedengren, David Brower

Real-time Monitoring of a Subsea Boosting System 66
Phaneendra B. Kondapi, Anita Sengebusch

Quantifying Flow Assurance Risks During Front End Engineering Design 70
Allan Rydahl, Richard Shea

Multiphase Flow - The Flowing Phases Know What They Are Doing, But Do We? 71
Sam Kashou

Integrating Life Cycle Dynamic Simulation and Training for Efficient Design, Startup, and Operation: Insuring Against Future Disasters 72
Norm Stewart

Quantifying Asphaltene Impairment Risks During Front End Engineering Design 78
Stefan Smuk, Ian Roberts

Designing and Delivering Integrated Control, Monitoring and Optimization Systems for Subsea or Topside Systems 79
Espen Storkaas, Olav Slupphaug, Bjorn Bringedal

Interfacially-Active Nanohybrids for Improved Oil Recovery 80
Daniel E. Resasco, Jeffrey Harwell, Santiago Drexler, Jimmy A. Faria, M. Pilar Ruiz, J. Baez, Bor-Jier Shiau

Effect of Nanoparticle As Interfacial Modifiers in EP Operations 81
Ramanan Krishnamoorti

Rock-Based Micromodel and CFD Studies of Nanomaterial Transport for Reservoir Applications 82
Karsten E. Thompson, Saade Bou-Mikael, Daniel S. Park, Dimitris Nikitopoulos, Clinton S. Willson

Stability of Nanoparticles At Reservoir Conditions and Applications for Conformance Control .... N/A
Quoc P. Nguyen

New Nano-Particle Based Drilling Fluids 83
Mukul M. Sharma, Martin Chenevert, Chung-Min Jang, James Friedheim

Designing Brine-Stable, Transportable Nanoparticles for Downhole Sensing 84
Michael S. Wong, James M. Tour, Mason B. Tomson

Flow Assurance: Modeling Wax Deposition in Subsea Pipelines 85
H. Scott Fogler

Asphaltenes Deposition 86
Jill Buckley

Review of Paraffin Deposition Under Multiphase Flowing Conditions 87
Cem Sarica, Ekarit Panachareonsawad

Asphaltene Solubility and Fluid Compatibility 88
Irv Wiehe

Wax Deposition and Control in North Sea Oil and Gas Production Systems 89
Hans P. Ronningsen

New Asphaltene Nanoscience and Its Impact on Reservoir Characterization 90
Oliver Mullins

Available Know-how in Transforming an Emulsified Drilling Fluid to Be Removed from Unwanted Location Into a Low-viscosity Single Phase System 91
Jean-Louis Salager, Ana Forgiarini, Lirio Quintero

Aqueous Polymer-Surfactant Mixtures for Wellbore Treatment and Completion 106
Mario R. Rojas, Jeffrey Harwell, Bor-Jier Shiau

New Method for Removing Formation Damage with Nanoparticle Dispersions 113
Paul M. McElfresh, David Holcomb, Daniel Ector

Temperature Control of Drilling Fluid with Phase-Change Materials 122
Othon R. Monteiro, Lirio Quintero, Martin Bates, Duncan Bruce, Elizabeth Wilcock, Antonina I. Smirnova, Alaistair J. Logan

Phase Behavior of Reservoir Fluids: Comparisons of PC-SAFT and Cubic Equation of State Simulations 131
Sukit Leekumjorn, Kristian Krejbjerg

Volume-Translated Cubic Eos and PC-SAFT Density Models and a Free Volume-Based Viscosity Model for Hydrocarbons At Extreme Temperature and Pressure Conditions 143
Mark A. McHugh, Babatunde Bamgbade, Yue Wu, Ward A. Burgess, Deepak Tapriyal, Bryan Morreale, Yee Soong, Hseen Baled, Robert M. Enick

SAFT Model for Upstream Applications 163
Walter G. Chapman, Sai R. Panuganti, Francisco M. Vargas, Doris L. Gonzalez

Prediction of Gas Injection Effect on Asphaltenes Precipitation Using Different Thermodynamic Models 220
Doris L. Gonzalez, Frank Lim, Nikhil Joshi, Wilson A. Cañas, Raul Osorio

A Generalized Partial Molar Algorithm Provides Fast Estimates of CO_2 Storage Capacity in Depleted Oil and Gas Reservoirs 256
Maria Barrufet, Ernesto Valbuena

Integrated Equation of State Modelling for Flow Assurance 270
Richard Szczepanski, Nuno Pedrosa, Xiaohong Zhang

Unconventional Natural Gas How Much At What Cost? 286
Stephen A. Holditch

Natural Gas Hydrates: An Emerging Massive Energy Source 287
Michael J. Economides, Xiuli Wang, Harold Brannon

Production Strategy for Marine Hydrates Reservoirs 288
Jyoti Phirani, Kishore K. Mohanty

Massively Parallel Simulation of Field-Scale Oceanic Gas Hydrate Deposits 289
Matthew T. Reagan, George J. Moridis, Jeffrey N. Johnson, Lehua Pan, Katie L. Boyle, Craig M. Freeman

Hydraulic Fracture Height Growth - Real 290
Kevin Fisher, Norman Warpinski

Optimization of Marine CNG Transport for Stranded Gas 291
M. Nikolaou

Adsorption of Ethoxylated Alkyl Phenols from High Salinity Brines 292
David Goodwin Jr.

CFD Modeling of Drop Coalescence - Improving Separator Efficiency 304
Santosh Appathurai

CFD Modeling of Electrically-Enhanced Oil-Gas Separators 305
Krishnaraj Sambath

On-Line Monitoring of Multiphase Composition Using Differential Dielectric Sensor (DDS) ..... N/A
Hui Li, Ram S. Mohan, Jack D. Marrelli, Shoubo Wang, Ovadia Shoham

Asphaltenes From Tahe and Liaohe Oils: Structural Analysis and Their Interaction with Comb-Type Copolymers 306
Tao Yang, Jun Xu, Xuhong Guo

Effect of Process Parameters on Dispersion Characterization of Crude Oil-Water/Brine Emulsions .... N/A
Prem Bikkina, Yi Zhang, Ram S. Mohan, Luis Gomez, Ovadia Shoham, Carlos Avila

Multiphase Flow in Downcomers and Vertical Separators .... N/A
Jose Lopez, Ram S. Mohan, Ovadia Shoham, Gene Kouba, Karl Anderson

Scaling in Oil and Gas: Understanding Surface and Bulk Processes 307
Anne Neville

The Role of Scale in Flow Assurance 308
Mason B. Tomson, Amy T. Kan, Wei Shi, Wei Wang, Essmaiil Djamali, Chao Yan, Nan Zhang, Haiping Lu, Lu Wang, Jonathan A. Pennington, Sarah N. Work

Emulsion Stabilization, Breaking and Inversion: Advantage or Disaster in Flow Assurance 309
Jean-Louis Salager, Ana Forgiarini

An Overview of Emulsion Stabilization Mechanisms in Petroleum-Water Mixtures 310
Peter Kilpatrick

Effects of Multiphase Flow on Corrosion 311
Srdjan Nesic

Estimation of Parameters for Simulation of Steady State Foam Flow in Porous Media 312
Kun Ma, Sibani Lisa Biswal, George J. Hirasaki

Chemically Enhanced Oil Recovery: Novel Surfactants - The Concept of Large Hydrophobe Alkoxy Carboxylate Surfactants 324
Upali Weerasooriya, Gary A. Pope, Sriram Solairaj, Jun Lu, Stephanie S. Adkins, Gayani P. Arachchilage, Do Hong Kim, Christopher Britton

A New Acid System for Matrix Stimulation of Deep Oil and Gas Wells 344
Hisham A. Nasr-El-Din

Chemical EOR by Wettability Alteration 345
Gaurav Sharma, Robin Gupta, Kshitij Mohan, Kishore K. Mohanty

Polyelectrolyte Nanoparticles for Flow Assurance 353
Stephen Johnson, Jenn-Tai Liang

Smart Surfactants for CO_2 Foams in Enhanced Oil Recovery .... N/A
Keith P. Johnston

For a Better Well - Pump More Proppant .... N/A
Stephen A. Holditch

Consideration in Production of Liquids From Shale 360
Milind D. Deo

Well Test Analysis of Shale Producing Wells .... N/A
Mohan Kelkar

Stress Interference Effects in Hydraulic Fractures in Horizontal Wells in Shales .... N/A
Mukul M. Sharma

Multi-Scale Discussions on Mass Transport in Gas Shale Reservoirs 368
I. Yucel Akkutlu

Supercritical Fluid Interaction Between Kerogen and Gas in Shale 369
James Holste, Kenneth R. Hall

Methodology for Characterization of Fluid Properties 370
Matt Honarpor, Nagi Nagarajan

Hydrate Blockage Formation 371
Jefferson Creek

Field Results for Slurry Flow 372
Michael Eaton, Jason W. Lachance, Don P. Shatto, Larry D. Talley, Doug J. Turner

Do We Have New Solutions to the Old Problem of Gas Hydrates? 373
Bahman Tohidi, Ross Anderson, Antonin Chapoy, Jinhai Yang, Rod Burgass

Developing a Comprehensive Transient Hydrate Formation Model: From Laboratory to Field Applications 374
Amadeu K. Sum

Hydrates in the Ocean--Beneath, Around, and Above Production Equipment 375
Gregory Hatton

Understanding and Modeling Hydrate Formation in High Water Producing Oil Pipelines 376
Sanjeev Joshi

Design and Optimize Field Architecture for Deepwater Production Wells Based On Flow Assurance Criteria 377
Sridhar Appanapalli, Manjunath Terwad

Lithologic Heterogeneity and Focused Fluid Flow Governing Gas Hydrate Distribution in Marine Sediments 385
Sayantan Chatterjee

Droplet Transport in near-Horizontal Low Liquid Loading Flows 386
Kiran Gawas
Understanding Foam Flows - Application to Gas-Dominated Flow Systems 387
Haijing Gao
Author Index

(32a) Horizons in Multiphase Flow

Geoffrey F. Hewitt, Department of Chemical Engineering, Imperial College London, London, United Kingdom

This presentation reviews the current situation on understanding and predictilon of systems involving multiphase flow. Such systems are typified by hydrocarbon recovery systems where the flows are commonly gas-liquid, liquid-liquid and gas-liquid-liquid flows. The focus will be on the prediction of*flow regime*or*flow pattern*as a first step. Once the flow pattern is esatblished, then*phenomenological models*can be used for the specific regime. The presentation will cover the principal regimes in both vertical and horizontal flws.

(32b) Transient Multiphase Flow - Past, Present and Future with Flow Assurance Perspective

Tom Danielson, ConocoPhillips, Houston, TX

Historically, transient multiphase flow codes grew out of nuclear safety codes developed specifically to model nuclear plants during loss-of-coolant accidents (i.e., steam-water). These codes were heavily augmented by the Upstream industry to deal with the more complex fluids, geometries, and thermodynamics associated with oil and gas production.

Although several attempts were made to develop transient mixture models with a single momentum equation, development settled on two-fluid formulations, employing separate momentum equations for each phase (developed out of a force balance), as the best way forward. While such models were quite successful in simulating hold-up and pressure drop during both steady-state and transient operations, many associated flow assurance issues (e.g., hydrates, hydrodynamic slugs) remain quite rudimentary. Also, the simulators themselves are terrifically slow by general CFD standards.

In the future, simple, powerful mixture models may well make a comeback, supplanting the more complex two-fluid models. The rationale is three fold: First, the simplicity of these models will allow much greater computational speed, resulting in the possibility to run much finer grids (on the order of a pipeline diameter) than currently used by transient pipeline simulators. Second, compositional tracking could be routinely used, allowing for much simpler simulations of, for example, complex networks of differing fluids. Third, flow assurance phenomena such as hydrate formation could be folded directly into the multiphase model on a fundamental level, rather than as an*ad hoc*addition.

(32c) "Steady-State Multiphase Flow Modeling - Past, Present and Future with a Flow Assurance Perspective"

Mack Shippen, Schlumberger, Houston, TX and William Bailey, Schlumberger, Cambridge, MA

A monumental amount of research and development work has been invested in multiphase flow modeling over the past fifty years. Yet, many challenges remain as we incorporate additional phases, account for exotic fluids and push our simulation tools to the brink in the desire to optimize ever more complex production systems fraught with difficult flow assurance issues.

A visual history of the evolution of steady-state multiphase flow models will be presented leading to the current "state-of-the-art". Looking towards the future, it is important that the models be advanced to address areas of greatest uncertainty and align with trends in field development strategies. The authors' views of the top 5 areas of R& D necessary for these purposes will be presented.

(32d) : Characterization of Confidence in Multiphase Flow Predictions

G. Kouba, Chevron Energy Technology Company, Houston, TX,**Hariprasad J. Subramani**, Flow Assurance, Chevron Energy Technology Company, Houston, TX and Selen Cremaschi, Department of Chemical Engineering, University of Tulsa, Tulsa, OK

Confidence in multiphase flow predictions can influence key decisions on design, location, and operations of pipelines and separation systems. These decisions in turn can profoundly affect system reliability, operability, safety, and overall project economics. There may be no application where lack of confidence in predictions is felt more strongly than for long distance tiebacks of pipelines to subsea processing in deep water. The uncertainties and resultant confidence in simulator outputs of flow regime, pressure drop, entrainment, and insitu velocities dictate the estimated confidence levels for opposing and complimentary systems and are key contributors in a Decision Analysis exercise.

Uncertainties in benchmarking data, input data to simulator and of internal models for various phenomena propagate to the simulator outputs of flow regime, pressure drop, entrainment, and insitu velocities. How these discrepancies are treated will affect the scaling of the simulator to different conditions where benchmarking data may not exist.

This paper aims to raise the awareness of the importance and value of uncertainty analysis in multiphase flow systems, provide a methodology with examples, and identify some of key gaps in the uncertainty analysis.

(32e) High-Viscosity Oil Multiphase Pipe Flow

Hong-Quan Zhang[1], Cem Sarica[1] and Eduardo Pereyra[2], (1)Petroleum Engineering, University of Tulsa, Tulsa, OK, (2)Petroleum Engineering, The University of Tulsa, Tulsa

High-Viscosity Oil Multiphase Pipe Flow

Hong-Quan Zhang, Cem Sarica and Eduardo Pereyra

The University of Tulsa

Heavy oil, extra heavy oil and bitumen make up about 70% of the world's total oil resources of 9 to 13 trillion bbls. Many recent large development projects have been targeting heavy oil. The high viscosity of heavy oil poses great challenges for its production and transportation through wells and pipelines. Experimental results show that the high-viscosity oil multiphase flow behavior is significantly different from the low viscosity oils. Most of the available mechanistic multiphase flow models were developed based on low-viscosity oil experimental results. When compared with high-viscosity oil data, these models displayed significant differences.

In this paper, previous experimental findings on high-viscosity oil multiphase pipe flow behavior are reviewed and compared with low-viscosity oil experimental results. These include flow pattern, pressure gradient, holdups, slug characteristics, oil/water mixing, core annular flow, gas-lift effect, and film distribution. Modeling approaches, such as that of Zhang *et al.* (2003) model, are summarized. Individual closure relationships are analyzed based on experimental measurements and observations. These closure relationships include slug length, slug liquid holdup, slug translational velocity, interfacial shear, wetted wall fraction, liquid entrainment fraction in gas core, oil/water dispersion and inversion, emulsion viscosity, etc. Finally, future experimental needs and modeling issues are discussed.

(32f) Dynamic Models in Multiphase Flow

Ole J. Nydal, Energy and Process Engineering, Norwegian University of Science and Technology, Trondheim, Norway

Transport of oil and gas mixtures in pipelines involves flow dynamics on a wide range of time and length scales. Liquid slugs and waves occur at scales extending from diameter scales ("hydrodynamic slug flow") and up to riser lengths ("severe slugging"). Dynamic flow models are used for design and for operational studies of multiphase pipelines. The approach in such 1D dynamic multiphase flow modeling depends on the scale to be resolved. The small scale hydrodynamic slugging is normally of minor importance compared with severe slugging, and can often be treated as averaged flow. Some cases, however, show a coupling between the small scale hydrodynamic slug flow and the large scale severe slugging phenomenon.

A hybrid two fluid model and a slug tracking model is described. A two fluid model is applied on a stationary grid in the gas-liquid stratified flow region until a slug is formed (or initiated), when a slug tracking method with a moving a grid takes over. The potentials of the model are demonstrated in relation to cases of different time and length scales. The first cases relates to small scale dynamics, wavy flows, and slug flow after bend initiation (slug length distributions). The second concerns a rapid acceleration of a pig (model systems) and the third case is a severe slugging case where slug flow initiation upstream the riser base affects the severe slugging characteristics.

R&D Efforts to Control, Monitor and Identify Drilling Fluid Invasion Into Reservoir Rocks

A. T. A. Waldmann[1], A. L. Martins[1], E. A. Souza[1], S. A. Loureiro[1], A. R. de Andrade[2], C. M. Scheid[3], L. A. Calçada[3], R. Z. Moreno[4] and C. E. Dannenhauer[5]

[1]Petrobras, S.A., RJ, Brazil, [2]Baker Hughes, RJ, Brazil, [3]Federal Rural University of Rio de Janeiro, Brazil, [4]State University of Campinas, SP, Brazil and [5]ESSS, SC, Brasil

This invited paper was prepared for presentation in the 1st International Conference on Upstream Engineering and Flow Assurance, a part of the 2012 AIChE Spring Meeting, held in Houston, Texas, USA, 1-5 April 2012.

Abstract

In exploratory wells, during sampling operations, the drilling fluid filtrate invasion into reservoir increases operation time and the possibility of contaminating samples. Inadequate fluid samples can lead to inaccurate information about the economic feasibility of the reservoir development. This article aims to summarize R&D efforts to control, monitor and diagnose drilling fluid invasion in reservoir rocks. The main focus include: Modeling drilling fluid invasion in two phase radial flows considering: fluid and filter cake compressibility, relative permeabilities, anisotropic rocks, capillary effects, fines migration, break out pressures, inversion log analysis; Experimental methodology to evaluate and optimize filter cake properties, including: shape and size distribution of bridging agents, static and dynamic filtration processes, quantitative evaluation of filter cake compressibilities, filtration in anisotropic and fractured cores; Software development for bridging agent optimization and the necessary requirements for field implementation.

Introduction

Offshore oil well drilling jobs are time consuming and expensive operations in which to minimize drilling fluid filtrate invasion is a requirement to guarantee the original reservoir rock properties. During a conventional drilling job, the bottom hole pressure must be kept between pore and fracture pressures in order to prevent undesirable influxes and formation fracture. Such influxes would cause serious security problems for the rig team, since they are usually highly flammable hydrocarbons. Consequently, as the pressure differential from the well to formation is always positive, the fluid will tend to invade the reservoir rock (**Figure 1**).

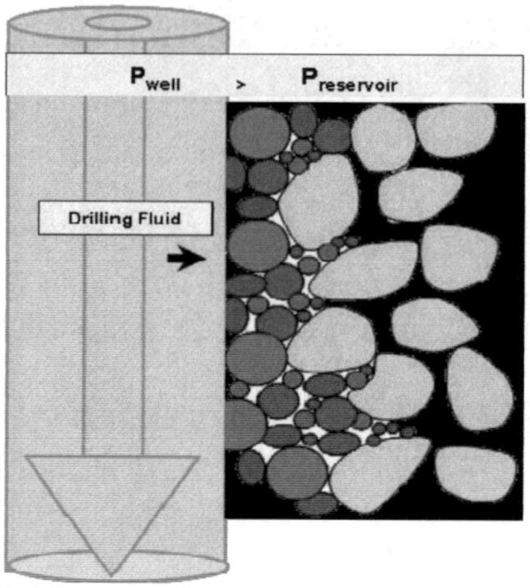

Figure 1. Drilling fluid invasion through the reservoir rock

Drilling fluid filtrate invasion can cause irreversible damage to the reservoir, reducing its original permeability and affecting the well and/or field production profile (Moreno et al., 2009). In development wells, the quantification of drilling fluid invasion can help to define the volume of acid to be pumped in routine stimulation jobs, reducing time and operational costs (Pereira et al., 2007). In exploratory wells, during the sampling operation, the drilling fluid filtrate invasion into reservoir increases the operation time and the possibility of contaminating samples. Inadequate fluid sampling, in exploratory wells, can lead to inaccurate information about the economic viability of the reservoir development (Coelho, 2005). In heavy oil reservoirs, oil and filtrate interaction may generate stable emulsions, reducing its initial and/or its long term productivity. Invasion in light oil reservoirs is less critical due to its good mobility properties. Other critical scenario is the low permeability gas reservoirs where imbibition effects may result in deep invasion, causing a phenomenon known as water blocking, dramatically reducing production (Ding, 2006).

This paper details research efforts in minimizing drilling fluid invasion considering: experimental work (dynamic filtration tests, static filtration tests through non isotropic porous media, injection experiments in consolidated sandstones and evaluation of imbibition mechanisms in low permeability porous media) and numerical simulation (invasion profile evaluation considering: non Newtonian flow; non isotropic flows; compressible flows; bridging agents optimization and dynamic filtration modeling).

Experimental Work

Dynamic Filtration

This topic describes a set of dynamic filtration tests obtained from an experimental filtration loop. The main idea is to investigate the filtration mechanisms that occur in the well, filtercake build up mechanisms and determination of filtercake permoporous properties (Scheid et al., 2010). The dynamic filtration loop consists of a pipe formed by a permeable wall which filtrates an aqueous suspension of different sized calcium carbonate particles. In a 0.5 m^3 tank a 1.5 horsepower mixer homogenizes the suspension. A positive displacement 1.5 horsepower pump is responsible for the fluid movement. A computer system continuously records, as a function of time, the weight of the filtrate squeezed out during the test into a tank atop a balance. **Figure 2** illustrates dynamic filtration flow loop.

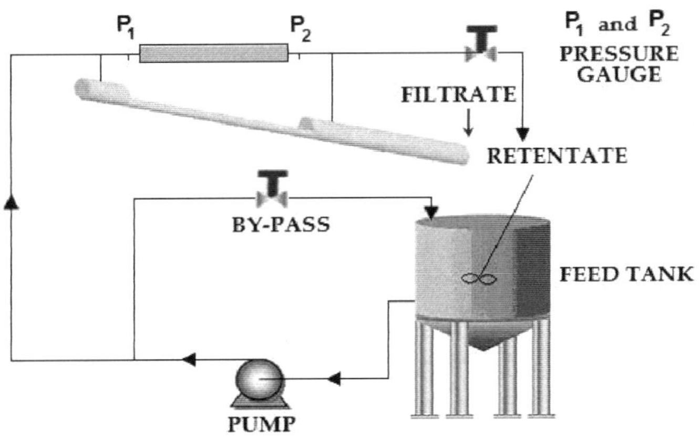

Figure 2. Dynamic cross flow filtration loop diagram.

By using the experimental crossflow filtration loop, it was possible to calculate: the filter cake properties and the filtration flow rate. In these experiments, aqueous suspensions of calcium carbonate ranging in concentration from 2 to 6% (w/w) and pressures ranging from 1 to 2 Kgf/cm^2 were used. The porous medium made out of polyester is substituted after each experiment. The calcium carbonate suspension rheology follows Newtonian laws. The experiments developed in the dynamic filtration loop consisted in the direct measurement of filtrate volume as a function of the filtration time. The experiments were run with controlled feed pressure and flow rate. The filter cake parameter such as porosity, thickness and permeability were determined after each experiment.

Erosion of filter cake will occur when the shear stress imposed by circulating mud at the cake surface exceeds the shear strength of the filter cake at that level. Two components define the shear strength of the filter cake (Dewan and Chenevert, 2001),

$$\tau_{th} = \tau_{th_0} + f \cdot \tau_g \quad (\text{dynes/cm}^2) \qquad (1)$$

The first term on the right side of equation 1 is the shear strength with zero pressure across the filter cake and the second term is the additional strength due to the applied pressure. The second term tends to hold the particles together. The parameter f is the coefficient of friction and the parameter τ_g is the grain-grain stress resulting from the applied pressure. Outmans (1963) gave one approximation for the grain-grain stress:

$$\tau_g = 68950 \cdot P_c \cdot \left(\frac{d}{e(z,t)}\right)^{\frac{1}{1-\nu}} \quad (\text{dynes/cm}^2) \qquad (2)$$

where Pc is given in psi, d is the mean particle diameter (10 microns) and ν is the compressibility exponent. The filter cake parameters were determined by non linear regression using the experimental data obtained under tests conditions presented in **Table 1**. In this experimental setup, the average value obtained for filtercake thickness was 1.5 mm.

The results obtained through the filtration curves analysis present a consistent and coherent physical behavior. Increasing pressure provides an increase in the volume of filtrate. An increase on the solids concentration, for the same differential pressure, causes a reduction in the filtrate volume (**Figure 3**). **Table 1** shows filtercake and porous media properties obtained by filtration curves analysis.

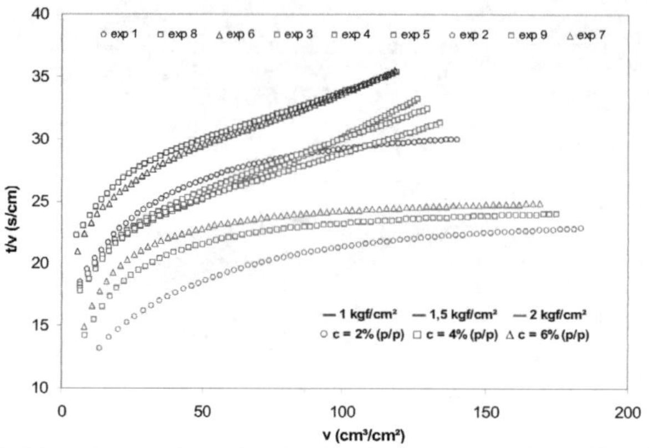

Figure 3. Experimental results obtained in the cross flow filtration loop

Table 1 - Parameter estimation on the cross flow filtration experiments (Scheid et al, 2010)

Experiment	c (p/p)	P (kgf/cm²)	α (cm/g)	R_m (1/cm)	ε
1	0.0202	1	2.30E+08	2.62E+09	0.56
2	0.0203	2	3.46E+08	3.86E+09	0.54
3	0.0402	1.5	7.56E+08	2.89E+09	0.55
4	0.0398	1.5	5.98E+08	3.21E+09	0.56
5	0.0411	1.5	5.32E+08	3.18E+09	0.55
6	0.063	1	2.58E+08	2.48E+09	0.55
7	0.0588	2	6.60E+07	4.56E+09	0.54
8	0.041	1	3.75E+08	2.54E+09	0.57
9	0.0422	2	1.36E+08	4.24E+09	0.54

Static Filtration Tests Through Non Isotropic Media

Drilling anisotropic reservoirs require special care to minimize invasion. Fluid design strategies should contemplate representative experiments to account for heterogeneities. A dedicated sintered steel filtration disk was specially developed for this purpose (Waldmann et al., 2011), as illustrated in **Figure 4.**

Normally, the ability of the fluid system to prevent invasion or the filtration properties are evaluated by standardized static filtration experiments. In these tests, the fluid is pressurized through a filter paper or into a consolidated inert porous medium. The volume which crosses the porous core is monitored along the time.

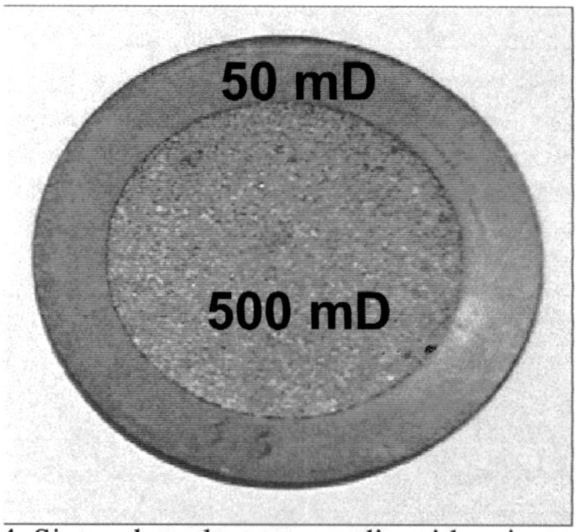

Figure 4. Sintered steel porous media with anisotropic permeability

Experimental tests were carried out considering 2 setups: setup 1 - static filtration tests with Newtonian fluid and setup 2 - static filtration tests with *drill-in* fluid. In step 1, the static filtration tests were carried out considering a 80% v/v glycerol in water solution flowing through sintered anisotropic porous media with air permeabilities of 50mD/500 mD. Tests performed with differential pressures of 100 psi and 200 psi showed good reproducibility and no problem was observed during the experiments (**Figure 5**).

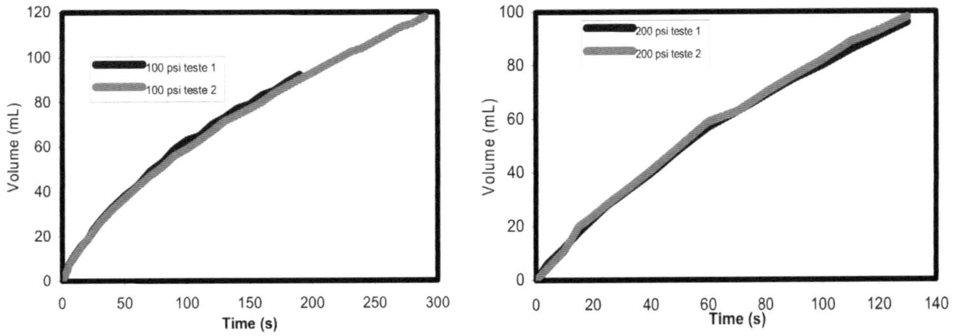

Figure 5. Filtration curve – Newtonian fluid

In step 2, the static filtration tests were carried out considering a *drill-in* fluid flowing through sintered anisotropic (50/500 mD) and isotropic porous media (50 mD and 500 mD). Tests performed with a differential pressure of 300 psi showed good reproducibility and no problem was observed during the experiments. The main idea in this approach is to observe and quantify the influence of porous media configuration on the filtercake permeability. Filtercake permeabilities were estimated considering a non isotropic filtration model proposed by Waldmann et al., 2011 and the results show a very good model adequacy (**Figure 6**).

Tests conducted through the porous medium with isotropic permeability equal to 50 mD presented a filtercake permeability equal to 2.69 x 10^{-4} mD and static filtration tests performed through the isotropic permeability equal to 500 mD presented a filtercake permeability equal to 9.37 x 10^{-3} mD. On the other hand, filtration tests performed considering non isotropic medium (50/500 mD) showed filtercake permeabilities values equal to 1.02 x 10^{-3} mD (50 mD section) and 1.17 x 10^{-3} mD (500 mD section). Results show the impact of anisotropy permeability on filtercake permeability. In this case, the filtercake permeability estimated by model presents same order of magnitude, the values are relatively similar and are directly influenced by parallel flow through porous media with different permeabilities. Figure 6 shows experimental filtrate volume data and model adequacy.

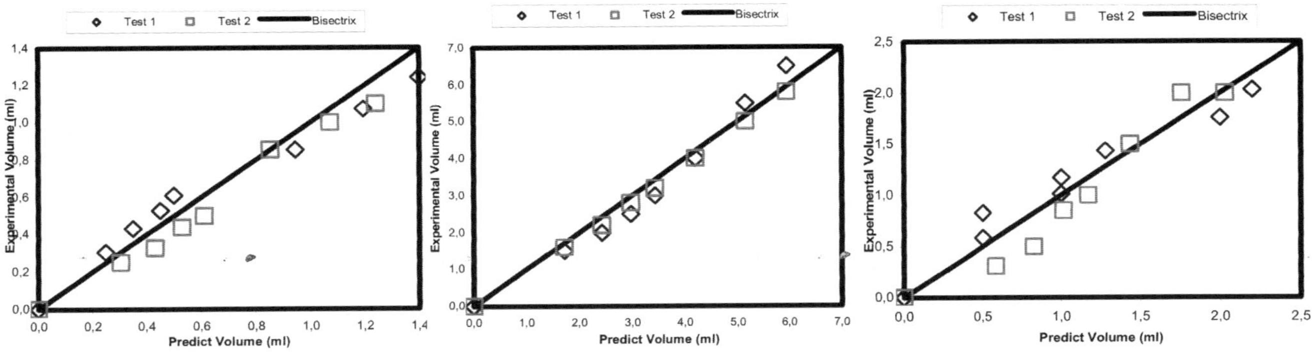

Figure 6. Filtrate volume data and model adequacy to 50 mD, 500 mD and 50/500 mD filter disks – *Drill in* fluid

Injection Experiments

This study intends to add the reservoir engineering's point of view to the design of non-invasive polymeric solids free drill in fluids. An innovative experimental study, including injection experiments and CT scanning evaluation (X-Ray/2D), was carried out to investigate the role of rheology from different polymeric solutions (one purely viscous polymer and one viscoelastic polymer) on the invasion of different rocks – fluid reservoir systems.

Laboratory investigation was planned and executed focusing on the analysis of drill-in fluid invasion in oil reservoirs. Detailed descriptions of the experimental setup and test protocol were presented in Moreno et al., 2007. Selected samples and test conditions were carefully planned with the purpose of isolating the effects of some phenomena. Redundant and independent information were acquired to allow comparative analysis on the influence of the chosen parameters. Mineral oil (~100 cp at 25°C) and lubricant oil (~300 cp at 25°C) were used to simulate formation oil and also aiming to access the influence of resident fluid viscosity level on invasion characteristics. Tests were performed under room temperature and pressure (25 °C and 1 atm) and polymer injection conditions were adjusted to simulate formation exposition during over-balanced drilling operations (100 psi and 200 psi). The injection fluids were polymeric solutions as follows: 3 lb/bbl xanthan gum or 4.5 lb/bbl PHPA (partially hydrolyzed polyacrylamide). The purpose of polymer selection, one natural (xanthan gum) and another synthetic (PHPA), was to observe the influence of the polymer's structure in the displacement process. Polymer concentrations were established in order to provide similar rheology behavior related to the shear rate range of interest.

Figures 7 shows final saturation profiles due to Xanthan Gum (XG) flow into light oil formation. Characteristics of a stable displacement can be recognized and the repeatability for the same test conditions, performed on independent samples, can also be observed. Saturation distribution behind advance front is quite constant and can lead to the conclusion that all mobile oil was contacted and displaced uniformly by the invasion fluid. Results show more retention of the Xanthan Gum (**Figure 7**). Oppositely to Xanthan invasion, PHPA invasion shows less polymer retention at the contact injection face (**Figure 8**). Although all the difficulties involved on performing repetitive experiments on similar, but independent, porous media samples, it is hard to neglect the fact that repetition can indicate real phenomena occurrence on displacement final profiles. Low oil removal close to the injection face can reflect polymer size entrapment and elongational phenomena. The continuum application of differential pressure can overcome elongational strength and push the polymer molecules throughout of the pore throat. This mechanism can produce extra friction and consequently extra reduction on flow rate, which inhibits the invasion.

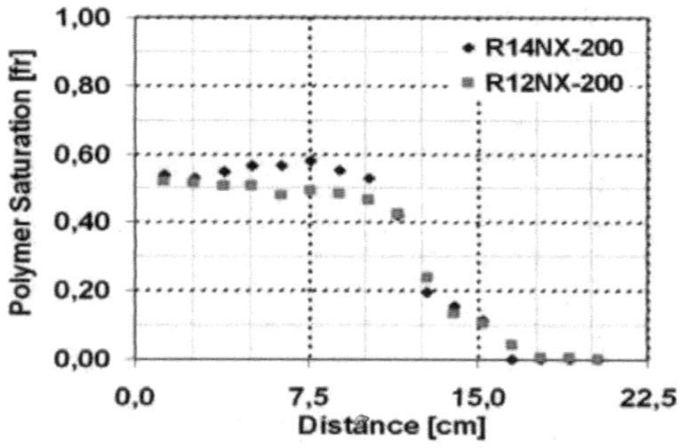

Figure 7. Repeatability of XG Flow into cores samples

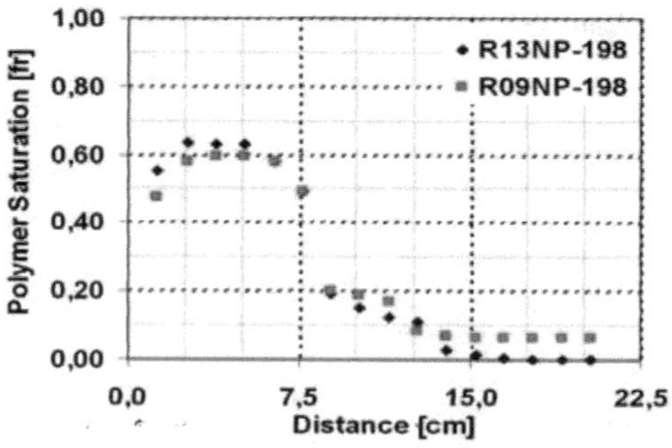

Figure 8. Repeatability of PHPA Flow into cores samples

Figure 9 shows the resulting relations of all the tests performed. Full colored symbols represent the tests under high difference pressure levels, while empty symbols are related with low differential pressure tests. Correspondent values are read as indicated by the same colored arrows. Since the invasion flow rate and contact time between drill-in fluid and the formation are interchangeable, results shown in **Figure 10** can be used to access the necessary time to have 30% of the pore volume invaded by the polymer fluid under the average conditions shown in **Figure 9**.

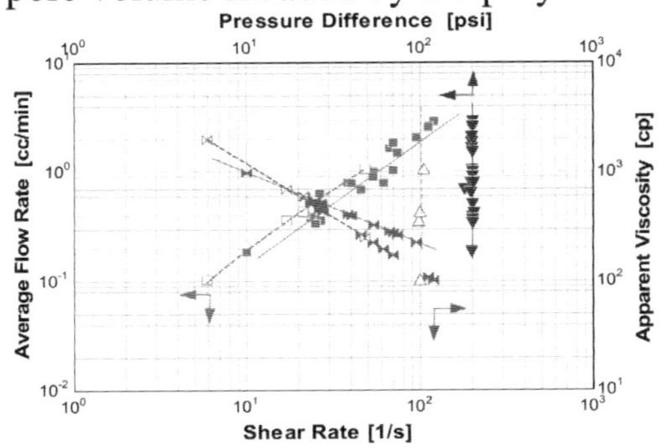

Figure 9. Cross Mapping Conditions

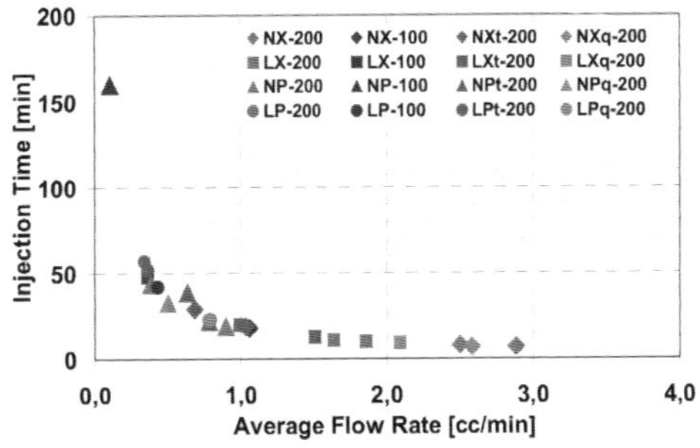

Figure 10. Invasion Time (30% VP) *versus* Invasion Flow Rate

Evaluation of imbibition mechanisms in low permeability porous media

In tight gas sands, the forces acting on the gas-liquid, gas-rock and rock-liquid interfaces become very relevant. Frequently, well construction operations may cause damage to this type of reservoir. The damage is believed to be due to drilling fluid filtrate invasion into very low permeability rock by the action of capillary forces. To decrease the interfacial tension, surfactants should be added to drilling and stimulation fluids formulations. This topic details an experimental evaluation of invasion control mechanisms in gas reservoirs, mainly related to capillary forces, through the use of tension reduction additives.

The methodology is based an experimental unit designed specifically to quantify the impact of capillary force on drilling fluid filtrate invasion (Andrade et al, 2008). The experimental setup consists in quantifying the capillary rise along the time through rock blades using two different non ionic surfactants in polymeric solutions. Surfactants types and concentration selection allowed isolating viscous effect in the capillary rise tests, once the rheological profile is very similar (Andrade et al, 2008). Rock blades with permeabilities ranging from 0.04 to 0.3 mD were positioned vertically and a scale conveniently positioned (**Figure 11**). The set is placed on Petri plate with volumetric capacity of 80 mL. When the volume of fluid on Petri plate touch the bottom of the blade, a chronometer is set in motion. So, it is possible to follow the fluid height evolution on blade along the time. **Figure 12** shows tests carried out with Surfactant A and B solutions at concentrations shown in **Table 2**. Results demonstrate a greater speed for solution containing surfactant A rise on porous media. This fact can be explained by the difference of surfactant structure. Surfactant A presents a linear structure while surfactant B presents a more complex structure. Large structures represent more molecule area and can cause mechanics blockage.

Table 2: Initial and final interfacial tension data

Copolymers	**CMC (% w/v)**	**Initial Interfacial Tension (mN/m)**	**Final Interfacial Tension (mN/m)**	**Molecule area (Å^2)**	**Mn**	**Ratio EO/PO**	**Mn/Mw**
Surfactant A	0.3	43.6	41.5	32.7	8000	6.2	1.6
Surfactant B	0.5	39.5	39.0	60.8	8253	3.4	1.6

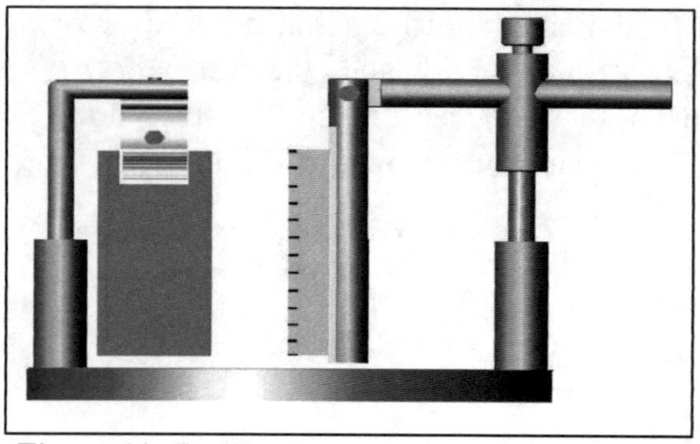

Figure 11. Capillar rise apparatus

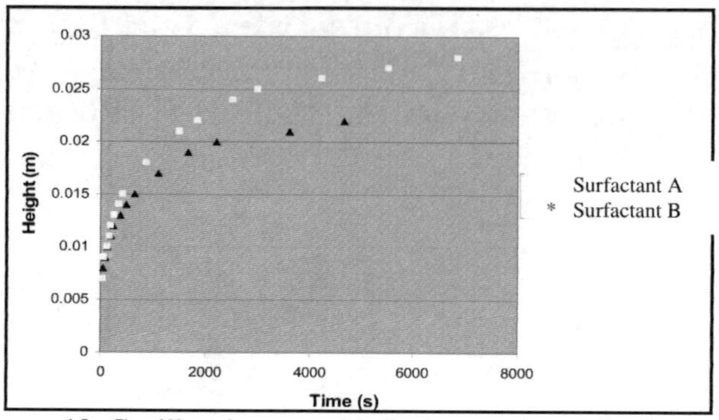

Figure 12. Capillar rise test with Surfactant A and Surfactant B

Numerical Simulation

Invasion profile evaluation

This methodology aims to provide a numerical tool to predict the drilling fluid filtrate invasion profile considering: a non isotropic porous medium, Newtonian and non Newtonian flows, compressible flow and filtercake erosion (dynamic filtration).

The mathematical model to represent the flow through porous media is based on the continuity equations, represented by mass conservation equation, Darcy's law (1856), and equations of state, which for liquids is the equation of compressibility (Ahmed, 2000). The method used for discretization of partial differential equations described in the mathematical modeling was the finite volume. In this method each calculated property is located in the center of the discretized cells and the flow located on the borders between cells. Details about radial flow modeling and numerical simulation validation for multi-layers methodology are described in Waldmann et al, 2005 and 2011.

Figure 13 illustrates the saturation and pressure profiles for 3 different porous media considering database illustrated in **Table 3** and 3D saturation profile for reservoir permeabilities equal to 950 mD and 2500 mD. Results indicate the higher filtrate invasion front is observed in the more permeable reservoir layer.

Figures 14 show the impact of filter cake permeability on the pressure and saturation profiles. The simulations were performed considering filtercake permeabilities of 0.2 mD and 0.02 mD, respectively. Results indicate that the pressure profile is significantly affected by additional friction loss generated by the filtercake. Lower permeability values result in higher friction losses and consequently, low invasion profile.

Table 3. Database setup used in the numerical simulations.

Variable	Values	Units
k_{mp}	2500	mD
k_{cake}	0.02	mD
μ_f	$2{,}5 \times 10^{-03}$	Pa.s
μ_o	$3{,}3 \times 10^{-02}$	Pa.s
r_{well}	$2{,}16 \times 10^{-01}$	m
ϕ_{mp}	$3{,}50 \times 10^{-01}$	-
ΔP	600	psi
time	12	hours

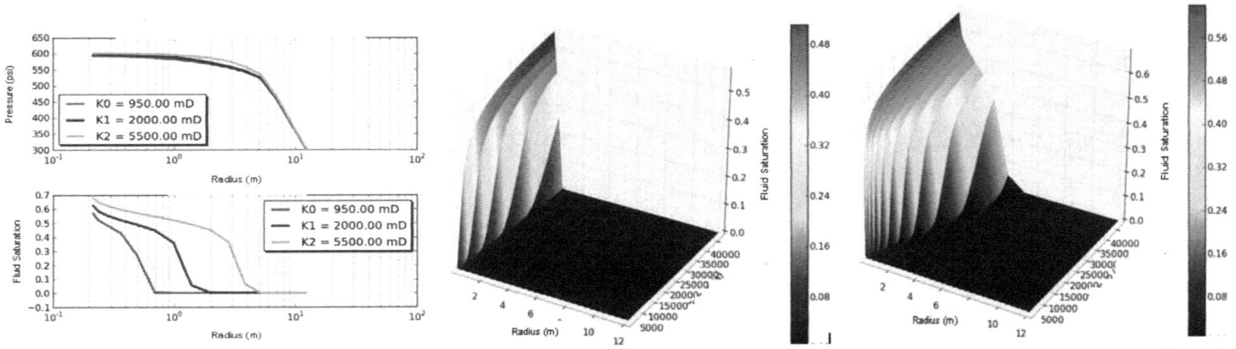

Figure 13. Saturation and pressure profiles for 3 different layers and 3D invasion profiles to 950 mD and 5500 mD

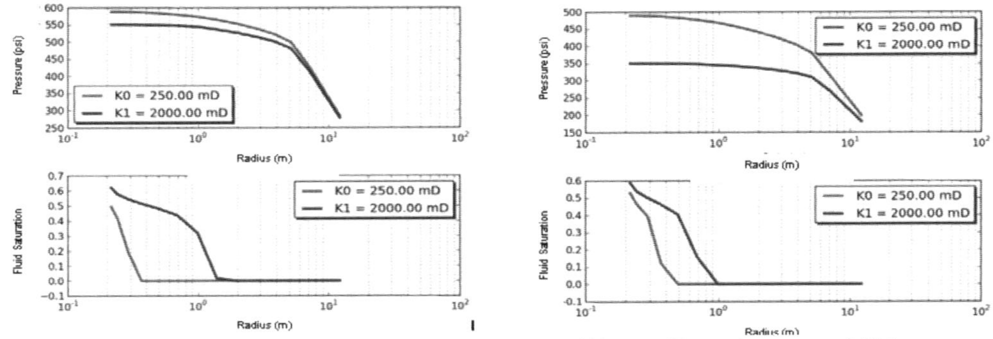

Figure 14. Saturation profile considering 2 different filtercake permeabilities

Field Study

Figure 15 shows a comparison between invasion profiles predicted by model and invasion profile measured by log resistivity tools during an exploratory drilling (12.25 in section). Results show a good agreement between predicted and measured values. Maximum invasion depth predicted is around 0.55 meters (22 inches) and invasion depth measured by log tools is around 25 inches (green curve).

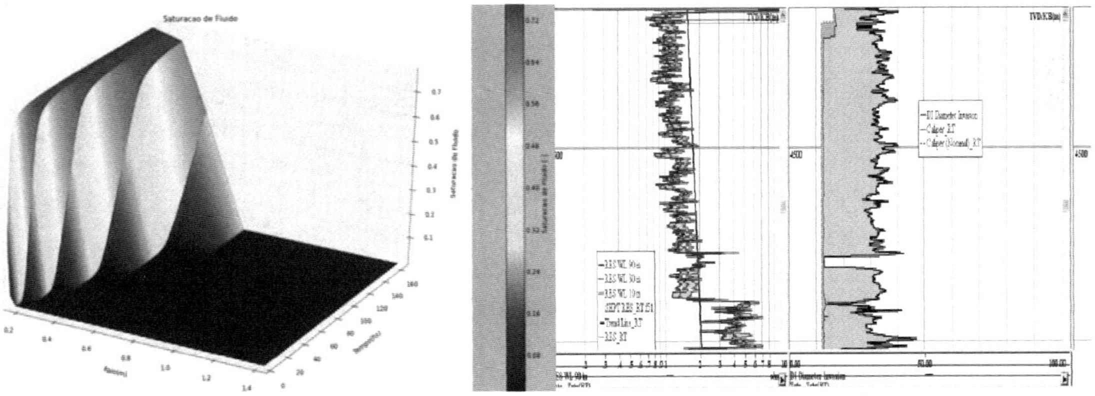

Figure 15. Field evaluation – Determination of invasion profile

Bridging agent optimization

The main idea in this methodology is to optimize different bridging agents size distribution used in drilling fluids design, creating the best particle size distribution curve that fits reservoir pore throat curve and minimizes drilling fluid invasion. In this methodology, the radial flow invasion modeling is coupled (Waldmann et al., 2005) allowing to predict the invasion profile into reservoir. This information is very important to optimize drilling fluid design to guarantee acceptable levels of invasion into the reservoir. The main input data required are: different types of bridging agent, reservoir properties (permeability and porosity), pore throat diameter models and mercury injection porosity data. The main outputs are: optimized bridging agent size distribution to minimize drilling fluid invasion and invasion profile into reservoir.

Field Study

The drilling job of Campos basin well A, located at an offshore Brazil development area represented a technological challenge to design an adequate drilling fluid. Due to the reservoir characteristics, a sandstone with high permeability and porosity (6500 mD and 32%, respectively), and high overbalances (775 – 900 psi), stuck pipe events and/or circulation losses could happen. The bridging agent adequate design allowed reservoir drilling without any operational problem, no static loss, maximum dynamic loss equal to 14 barrels/hour, no dynamic loss after drilling the reservoir. No circulation losses pills were required. **Figure 16** shows the optimum bridging agents size distribution designed to drill the reservoir from well A.

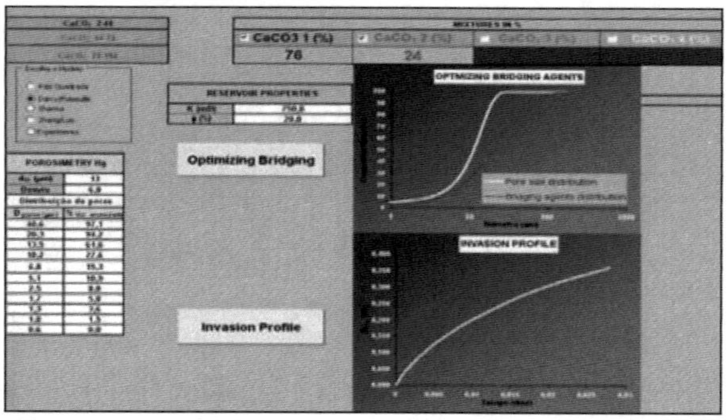

Figure 16. Bridging agent optimization methodology

Dynamic filtration modeling

This item describes a dynamic filtration model validated by data obtained from an experimental dynamic filtration loop, previously described at experimental work – dynamic filtration issues. The model predicts the filtercake buildup and filtrate flow rate in the drilling process using non-Newtonian fluids. It is also able to calculate filter cake properties such as permeability and thickness. The equations, derived in cylindrical coordinates, are based on the momentum and mass conservation principles described by Darcy's law (Scheid et al., 2010). **Figure 17** shows a comparison between dynamic filtration experimental data (test 3 shown at **Table 1**) and dynamic filtration model prediction. Results shows that model can be used to simulate the invasion of fluids in reservoirs, and even give information about the filter cake thickness buildup.

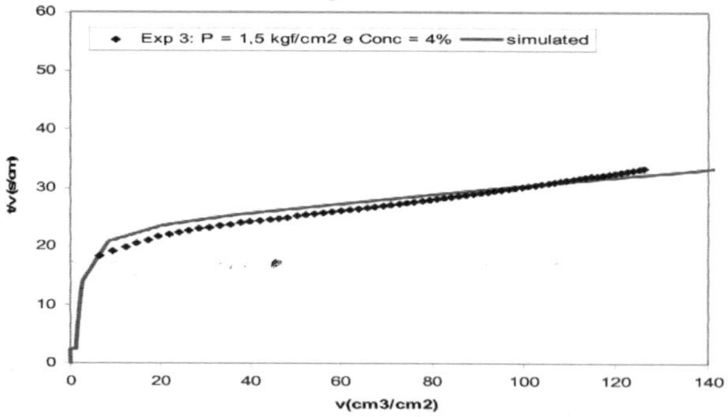

Figure 17. Experimental and simulated data from experiment 3

Final Remarks

The described simulation and experimental efforts represent important corporate initiatives to minimize the influence of the drilling activity in the evaluation and productivity of complex reservoirs. The continuity of a mid term project on invasion description and minimization starts to provide beneficial inputs for the activities. Joining expertise from the academy, software companies, service companies and operator was a key factor for the success.

References

1. Ahmed, T., Reservoir Engineering Handbook, 2th Edition, Gulf Professional Publishing, USA, 2008.
2. Andrade, A. R., Lomba, R.F.T., Leibsohn, A. L., Michel, R. C. And Giulherme, H. M., "Evaluation of Imbibition Mechanisms and Surfactants Performance in the Interaction af Aqueous Drilling Fluids with Low Permeability Gas Reservoirs, VII SEFLU CEMPO, Isla de Margarita, Venezuela, 2008.
3. Coelho, A.C.D., Camargo C., Kato, E.T. and Legrand, V.M.Q.F, 2005, "Utilizing Mini-DST for Formation Evaluation", SPE Latin American and Caribbean Petroleum Engineering Conference, SPE 94963, Rio de Janeiro, Brazil, 20-23 June 2005
4. Darcy, H., 1856, Les Fontaines Publiques de la Ville de Dijon, V. Dalmont, França
5. Dewan, J.T. And Chenevert M.E., 2001; A model for filtration of water-base mud during drilling: determination of filter cake parameters, Petrophysucs, vol 42, n. 3, p. 237-250.
6. Ding, Y. SPE, Herzhaf, B. And Renard, G. SPE, Institut Français du Pétrole, 2006, "Near-Wellbore Formation Damage Effects on Well Performance: A Comparison Between Underbalanced and Overbalanced Drilling", Journal SPE Production & Operations, Volume 21, Number 1, pp. 51-57, 2006, Society of Petroleum Engineers, SPE 86558 – PA.
7. Moreno, R.B.Z.L., Bonet, E.J., Waldmann, A.T.A. and Martins, A.L., 2009, "Tayloring Filtrate Rheology for Invasion Control in Heavy and Light Oil Reservoirs", 8th European Formation Damage Conference, SPE 122076-MS, Scheveningen, Netherlands, 27-29 May 2009.
8. Moreno, R. B. Z. L.; Bonet, E. J.; Waldmann, A. T.A. and Martins, A. L., " Methodology for Data Acquisition and Analysis of Polymer Solution Invasion Throught Saturated Porous Media", BJPG Brazilian Journal of Petroleum and Gas, v. 1, n. 2, p. 104-115, 2007.
9. Outmans, H.D., 1963, "Mechanics of static and dynamic filtration in the borehole": Society of Petroleum Engineering Journal, vol. 3, n. 3, p. 230.
10. Pereira, A.Z.I., Calderon, A., Chagas, C. M. and Pinto, E. A., "Recent Advances in Deepwater Horizontal Injector Wells Acidizing in Campos Basin", SPE European Formation Damage Conference, SPE 107787, Scheveningen, Holanda, 30 – 1 June, 2007.
11. Scheid, C. M., Calçada, L. A., Araújo, C. A. O., Waldmann, A. T. A. and Martins, A. L., "A Theoretical and Experimental Analysis of Dynamic Filtration in Drilling Operations", SPE International Symposium and Exhibition on Formation Damage, SPE 128036, Lafayette, Louisiana, USA, 10–12 February, 2010.
12. Waldmann, A.T.A., Martins, A.L., Aragão, A.F.L and Lomba, R.F.T., 2005, "Predicting and Monitoring Fluid Invasion in Exploratory Drilling", Journal SPE Drilling & Completion, Volume 20, Number 4, pp. 268-275, 2005, Society of Petroleum Engineers, SPE 86497- PA.
13. Waldmann, A.T.A., Martins, A.L. and Dannenhauer, C., 2011, "Understanding *Drill In* Fluid Invasion into Multi Layer Reservoirs Satured by Compressible Fluids", SPE European Formation Damage Conference, SPE 144156, Noordwijk, The Netherlands, 7–10 June 2011.
14. Waldmann, A.T.A., Rodrigues, D., Souza, E. A., Josino, F. S., and Martins, A.L., 2011, "Estudos Experimentais sobre a Avaliação das Propriedades de Fluidos de Perfuração em Meios Porosos Anisotrópicos", 35^0 ENEMP, Vassouras, RJ, Brasil, 23-26 october 2011.

Comparative Analysis of Gelled Fluids through LAOS

Jason Maxey, Halliburton Production Enhancement, Houston, TX

This invited paper was prepared for presentation in the 1st International Conference on Upstream Engineering and Flow Assurance, a part of the 2012 AIChE Spring Meeting, held in Houston, Texas, USA, 1-5 April 2012.

Abstract

Drilling muds have long been known to develop weak bonds between particles and emulsion droplets in suspension, commonly known as forming a gel structure. Drilling fluid design has long hinged on the balance between providing sufficient viscosity and gel structure to suspend solids (both cuttings and barite) and minimizing the rheological impacts on equivalent circulating density (ECD), pump initiation, and swab and surge pressures. A wide variety of products and solutions are offered to boost or thin viscosity, boost "low-shear" viscosity, and control gel strength; all of these aim to provide suspension and hole cleaning with minimal ECD and risk from pressure spikes associated with initiating fluid flow. The prima facie supposition that increasing viscosity or yield point naturally leads to better solids suspension has been disproved through hard experience, while the idea that drillings fluids that appear significantly less rheologically "robust" can still provide suspension is now being observed. The exact reasons for these apparent contradictions are as yet poorly understood.

Currently, the rheological impact of various products and treatments on microstructure formation are evaluated through a transient flow start-up test on a standard oilfield viscometer, with the gel strength recorded as the peak in this transient test. Both its ease of use and ubiquitous nature in the industry have led to a long reliance on this instrument as the best, most convenient source of rheological information on drilling fluids. Its shortcomings have been previously identified, and this paper endeavors to explore rheological differences through a variety of techniques on a lab-grade rheometer. Questions as to how and why various typical products affect viscosity and gel structure in invert emulsion fluids are explored, along with the implications to suspension and transient pressure spikes in drilling operations.

Introduction

The need for low viscosities for ease of flow in the annulus, the need for high viscosities to prevent fluid invasion and aid in hole cleaning, and the intermittent nature of the drilling process all compete for priority in the design of a successful drilling fluid. To meet these opposing criteria a mud must be a complex fluid, exhibiting viscoelastic properties to provide appropriate viscosities at the needed shear rates and also exhibiting viscoplastic properties through thixotropic and yielding properties to suspend solids in low-flow and stagnant conditions. For these reasons clays, which form associative networks or a microstructure in the fluid, are used as viscosifiers.

Traditionally, invert emulsion drilling fluids are prepared using various types of base oils emulsified with brine as the internal phase. Amine-treated bentonite (or other organophilic clay) is added, along with polymeric viscosifiers, to control rheological properties. Various lignitic and asphaltic materials may be added for filtration control, and barite or other weight material added to control density.[1-4] All of these components, including barite which is usually considered as inert in the system, interact to form the overall microstructure of the fluid. Any change in one of these components affects how everything interacts and thus affects the nature of the microstructure.

With modern options, the microstructure may be based around a traditional emulsion / organophilic clay interaction, or it may arise from emulsion only with no organophilic clays or solids added. These fundamentally different systems cannot be expected to behave the same way, and neither should the effect of various traditional oil-mud additives on microstructure (and thus rheology) be expected to be the same. It has long been observed that colloidal systems form associative microstructures based on several mechanisms, including:

- bridging,
- flocculation,
- steric stabilization,
- depletion attraction, and
- depletion stabilization.[5-7]

While these mechanisms are well known, they have been little applied to drilling muds beyond the thought that the bridging of emulsion droplets by organophilic clays as the main source of microstructure in traditional invert emulsion drilling fluids. This paper explores this concept through investigation of field muds with and without organophilic clays before and after treatment with various standard products.

Experimental Methods

Rheological testing was performed on an Anton-Paar MCR501 stress-controlled rheometer and compared to standard API measurements on a Model 35-type viscometer. In general, before testing, all fluids were brought to a test temperature of 120°F and then pre-sheared for two minutes immediately before beginning the test, significantly breaking the microstructure and providing a common starting point. When using a standard couette cell to produce a flow curve for oil-based muds, wall slip significantly distorts measured data below ~$1\text{-}s^{-1}$. In order to avoid this problem, parallel plates which had been sandblasted to a 300 grit finish were used for testing. The slight roughness of the plates serves to greatly reduce the effects of wall slip and allows for relatively easy testing of fluids to rates as low as $10^{-4}\ s^{-1}$. All tests were performed at a gap of 0.75-mm.

Many tests have been used over the years to evaluate the "strength" of the gel formed in muds and the yield stress[3]. The results of these tests often vary greatly, may or may not be correlated to one another for a given fluid, and ultimately may not lead to a significant understanding and differentiation of drillings fluids and their microstructure.

Treatment of a mud with various additives has long been recognized to affect, to a greater or lesser extent, the rheological properties of that fluid. A general understanding of the results of treatment has been made, but no real attempt to understand how these types of products influence microstructural growth (and thus broader rheological properties) has been undertaken. The focus of this paper is to begin to understand these microstructural influences through standard rheological techniques and through use of large amplitude oscillatory shear (LAOS) to examine yield stress and the energy required to break the microstructure in a single test.

LAOS

Several recent papers have examined drilling fluids through the LAOS technique.[4,8] The methodology followed here is similar to that previously reported by Maxey[9]. A good explanation of this method and the rheological parameters which can be obtained from it can be found in the recent papers by Ewoldt and McKinley[8,10] among other sources. In short it systematically connects, through a limited set of experiments, steady flow viscosity $\eta(\dot{\gamma})$, linear viscoelastic moduli G'(ω) and G"(ω), and nonlinear viscoelastic properties.[11] By examining the transient response of stress to an applied strain within the oscillation cycle, rather than by simply taking the absolute magnitudes of those oscillations, more information on the nature of the fluid can be obtained without disturbing the microstructure formed in the fluid.

For these tests, a triangular wave oscillation was applied at 0.1-rad/sec for strains of 10%, 50%, 100%, 500%, and 1000% (giving a maximum oscillation through ~180°). Triangular waves were used, rather than standard sinusoidal waves, for ease of setup and data collection. As a result, the traditional moduli (G' and G'') do not have the same physical meanings and thus will not be discussed here. However, this test can be used to determine at what strains the fluid becomes fully yielded (where elastic and viscous contributions are overwhelmed by the plasticity of the fluid) while remaining in a closed-loop test where the fluid is stressed but not completely disturbed. By testing under fully yielded conditions, the following parameters can be obtained:

1. Flow Stress (τ_F) which, at strains where flow is just becoming fully developed, is the yield stress (τ_Y) of the fluid;
2. Peak, or overshoot, Stress (τ_P); and
3. Energy Dissipation (E_D) in breaking microstructure

The typical response of a fluid to a triangular-wave LAOS at small and large strains is presented in Figure 1. At small applied strains, the stress response is roughly in the same form as the input – a triangular wave with some bending due to elasticity in the sample. For very large deformations the viscoelastic contributions are dwarfed by the yielding behavior and the stress response looks like a square wave.

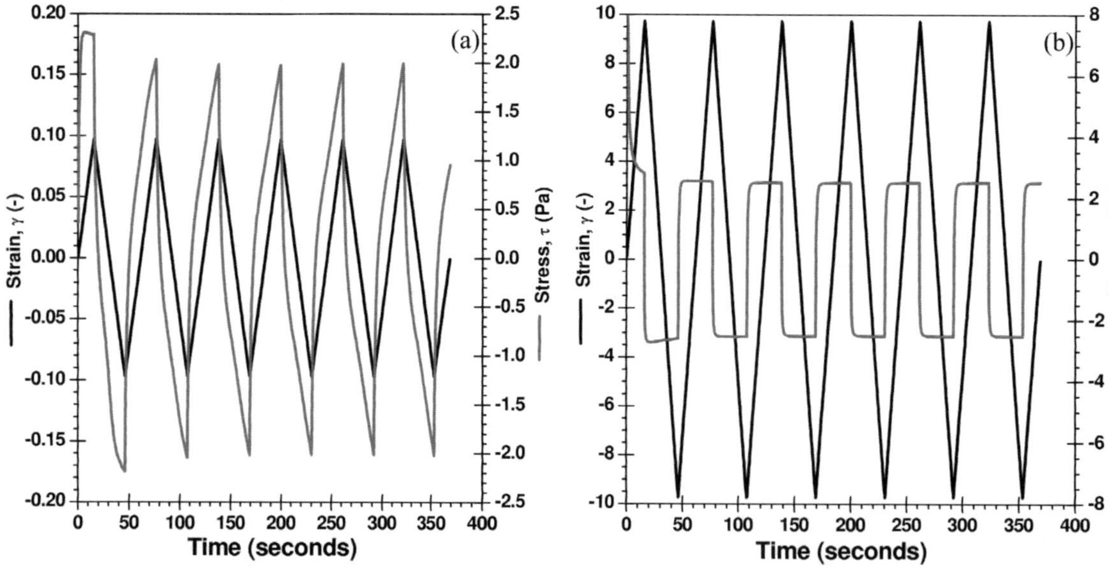

Figure 1. Triangular-wave LAOS raw data for a drilling fluid sample at (a) 10% and (b) 1000% strain amplitude.

A more convenient way to look at this data is using a Lissajous curve, plotting the stress as a function of strain. The data in Figure 1 is replotted in this way and presented in Figure 2. This allows one to view the test as an enclosed curve, which as an oscillatory test it is, and observe the transient response of stress as a function of strain. In this figure, each oscillatory cycle is presented as a different color curve. What is most obvious is that the first cycle is appreciably different from subsequent cycles; this is the result of the start-up response of the mud to flow initiation. For the small strain case (Figure 2a) stress rises quickly in the first cycle and then plateaus before the flow direction reverses. This is due to partial yielding of the fluid at this strain, which becomes quickly dominated by viscoelastic response as the gel structure is partially broken and finds a steady state. Subsequent cycles do not display this yielding behavior, only the viscoelastic response of the fluid.

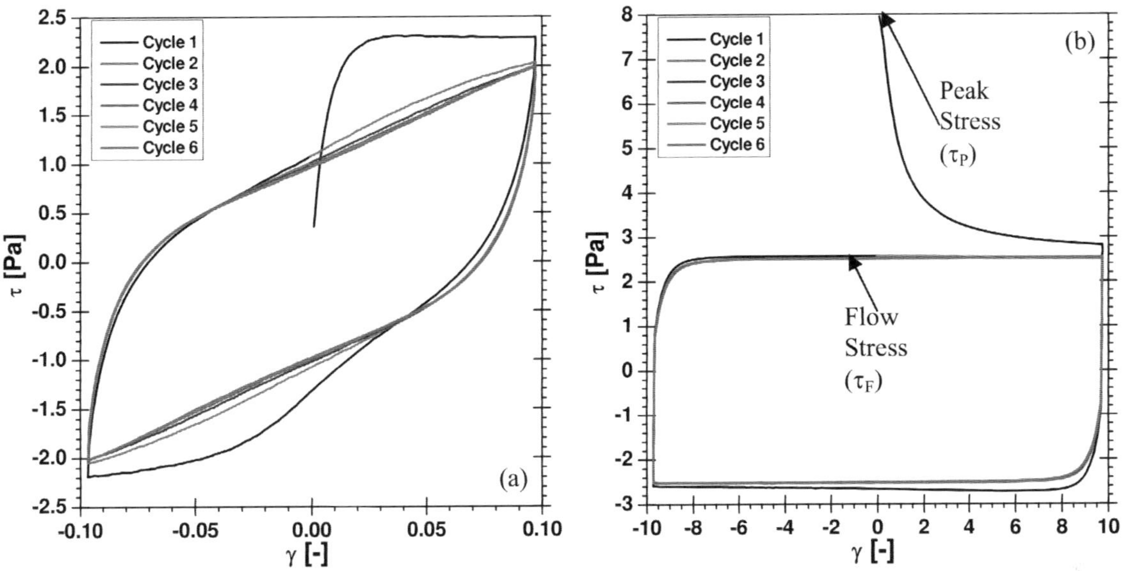

Figure 2. Raw data from Figure 1, for a drilling fluid sample at (a) 10% and (b) 1000% strain amplitude, replotted as Lissajous curves. A total of six oscillatory cycles are presented for each test.

The Lissajous curve for the 1000% strain test (Figure 2b) is quite different from the test at 10% strain. Rather than a slow rise to a plateau in the first cycle, a large spike is observed as a large stress (large amount of energy) is required to yield the fluid and initiate flow. Subsequent cycles are square, owing to the dominance of yielding, plastic behavior over viscoelastic responses at this strain, and approach a steady state with the microstructure broken. The magnitude of the peak stress, τ_P, varies from fluid to fluid, and is tracked in this paper as a parameter of interest. The stress plateau observed here does not change appreciably (at constant oscillatory frequency) once the fluid is fully yielded and is thus a reliable measure of the yield stress of the fluid. All data compared here is performed at 1000% strain amplitude, which was found to be fully yielded for all mud samples tested.

The stress required to initiate flow represents an energy burden for breaking microstructure in the fluid. The LAOS test is also ideal for examination of this energy. The energy for a single cycle can be found by simply integrating the stress as a function of strain

$$E_D = \int \tau d\gamma \tag{1}$$

where E_D is the energy dissipation per unit volume. By evaluating E_D for each cycle individually, the energy required to initiate flow (cycle 1) can be compared to the energy required for flow at steady state (usually by cycle 6). This energy is that which is required to break microstructure in the fluid, is greatest in the first cycle (where most bonds are broken) and less in subsequent cycles as all microstructural bonds are broken over time. By comparing E_D in each cycle verses that at minimum value (usually steady state), and excess energy dissipation, E_D^E, is found

$$E_D^E = \sum_i \left(E_{D,i} - E_{D,min} \right) \tag{2}$$

which shows the total energy per unit volume required to break microstructure and initiate flow at this strain. Here the methodology differs slightly from previous work, in that the excess energy dissipation is calculated at the minimum cycle energy dissipation rather than at long-time steady state. This is due to the observance that many fluids have a tendency to completely break gel microstructure after only a few cycles and then dynamically reform microstructure in subsequent cycles. The dynamic growth of structure in drilling fluids is well documented at low shear rates[3] and so only the point to where the structure is maximally broken (the dissipative energy minimum) is considered.

Results

Test Fluids

For testing of the comparative effects of various common field mud additives, two invert emulsion drilling fluids were used as base fluids. One of which was a traditional oil-based mud (OBM) containing organophilic clays and other "black powder" products – OBM #1 – and the other being a high performance organophilic clay free invert emulsion drilling fluid – OBM #2. These are the same fluids as Mud #6 and Mud #1, respectively, from previously published work[9]. These two fluids were selected for two reasons: their relatively similar basic properties including oil water ratio (OWR), density, and low gravity solids (LGS) content, and because they form microstructure in fundamentally different ways. To each mud a relatively common type of product was added to a 1 lab barrel (350-ml) sample and mixed on a multimixer for 30 minutes. The sample was then hot rolled at 150°F for 16 hours and re-sheared on the multimixer for 15 minutes before testing. Products added for testing included:

- lignite,
- polyamide polymer,
- carbon nanotubes,
- standard bentonite,
- dimer/trimer fatty acid (low-shear rheological modifier),
- organophilic clay (dry processed to amine treat),
- organophilic clay (wet processed to amine treat),
- fatty acid ether (an emulsifier),
- asphalt,
- fatty acid ester (low-shear rheological modifier),
- petroleum distillates (emulsifier),
- sepiolite, and
- 5-mm calcium carbonate.

Flow Curves

As a starting point to examine the effects of various common mud products on the rheology and microstructure of drilling fluids, flow curves on each fluid were measured at 120°F over shear rates of 10^{-4} s^{-1} to 1200 s^{-1}. The flow curves for the organophilic clay based muds (OBM #1) are presented in Figure 3 and those for the organophilic clay-free muds (OBM #2) are presented in Figure 4. From this data it is observed that while some products have a marginal effect on viscosity, others can greatly increase or reduce the viscosity. In addition, the magnitude of the change in viscosity is dependent on the differences in how the base fluid forms microstructure.

For example, the addition of carbon nanotubes has only a moderate effect on OBM #1, with the bulk of the effect being observed at very low shear rates. However, when added to OBM #2, the carbon nanotubes greatly increase the viscosity and increased the observed yielding plateau by ~50%, from ~2.5 Pa to ~5 Pa. Similarly, the addition of dimmer/trimer fatty acid to OBM #2 significantly boosted viscosity at high and mid shear rates, but was identical to the base mud at very low shear rates. However, addition to OBM #1 resulted in minimal changes at high rates but a very large viscosity increase at low shear rates. Additions of organophilic clays, as expected, increased the viscosity in both muds as did addition of the polyamide polymer.

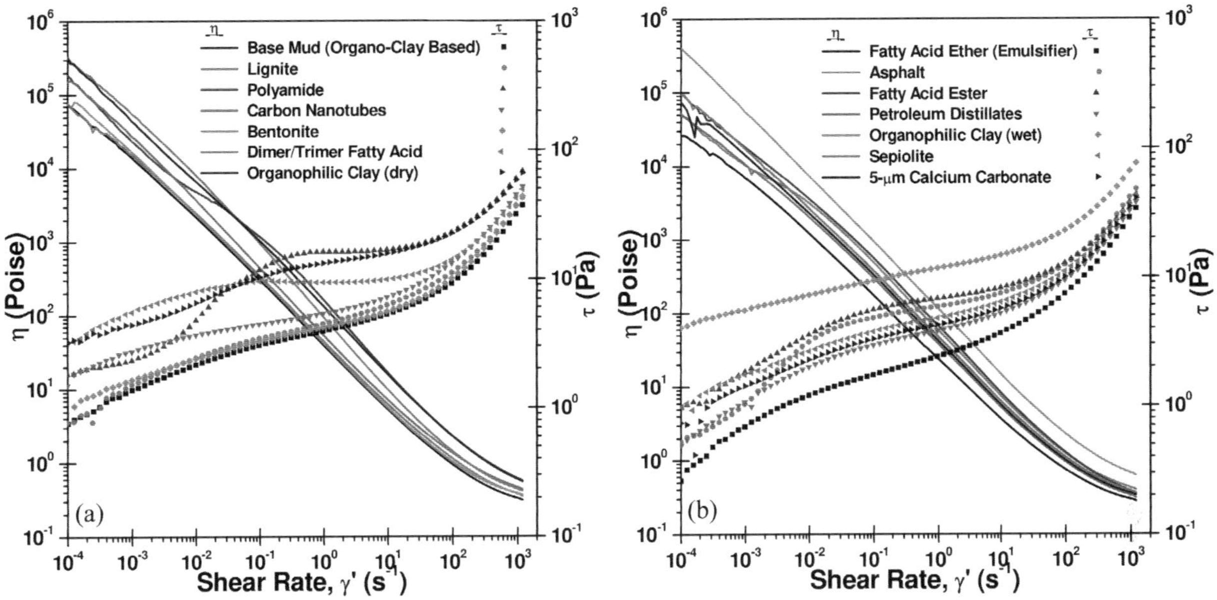

Figure 3. Flow curves at 120°F using sandblasted parallel plates for the organophilic clay-based invert emulsion mud, OBM #1, with various additives.

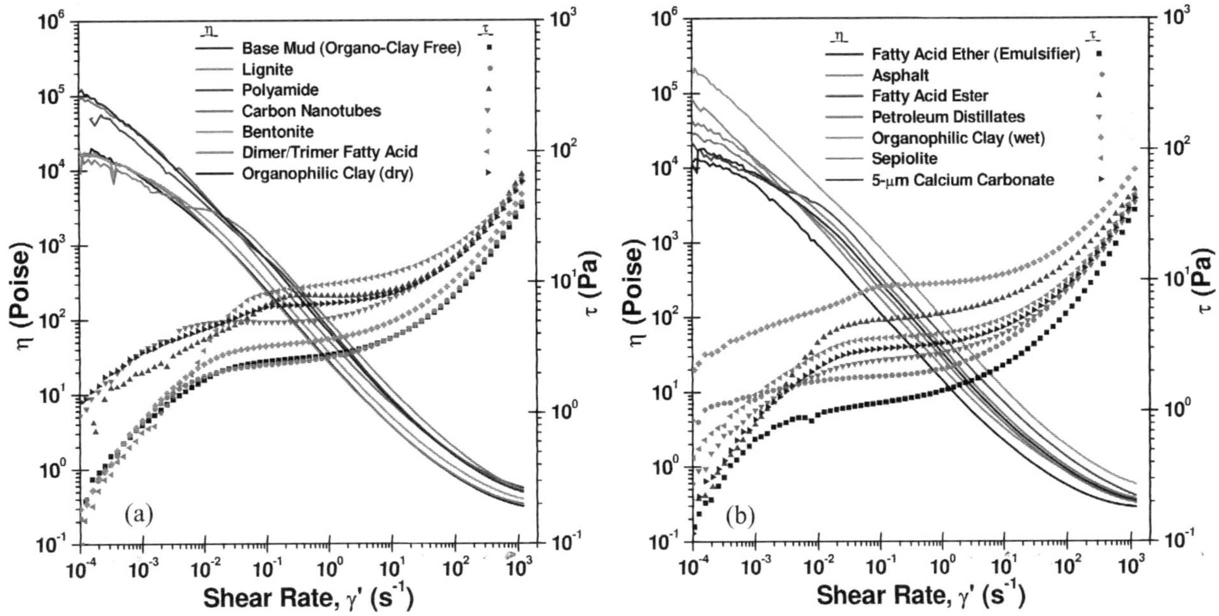

Figure 4. Flow curves at 120°F using sandblasted parallel plates for the organophilic clay-free invert emulsion mud, OBM #2, with various additives.

Aside from these general observations, and the fact that some additives have differing effects on the organophilic clay-based and clay-free fluids, no real information on how these products effect microstructural growth can be gleaned from just the flow curves. Therefore, further testing to examine the gel strengths of the fluids was performed.

Gel Tests

The gel strength of the various treated mud samples were tested using standard API procedures on a Model 35-type viscometer. In addition, a similar test was repeated on the MCR501 rheometer, where the sample was sheared, allowed to rest for 10-seconds, 10-minutes, or 30-minutes, and then a shear rate of 0.1 s^{-1} was applied (as opposed to the 5.11 s^{-1} rate applied in the API test) and the maximum overshoot recorded as the measured gel strength. When the 0.1 s^{-1} shear rate is applied, it is allowed to continue for a minimum of five minutes, usually long enough to achieve steady state. With this data, the energy required to break the gel structure and return to steady state can be calculated by integration of the stress as a function of strain (as done in the LAOS tests described above). The tests were performed on the rheometer in addition to the standard API test in order to address questions as to the validity of gel strengths measured by the API method. This is discussed in the following sections.

MCR501 Rheometer Gel Measurements

Gel strength tests as described above using the MCR501 rheometer are presented in Figure 5 for OBM #1 and in Figure 6 for OBM #2. A similar methodology for

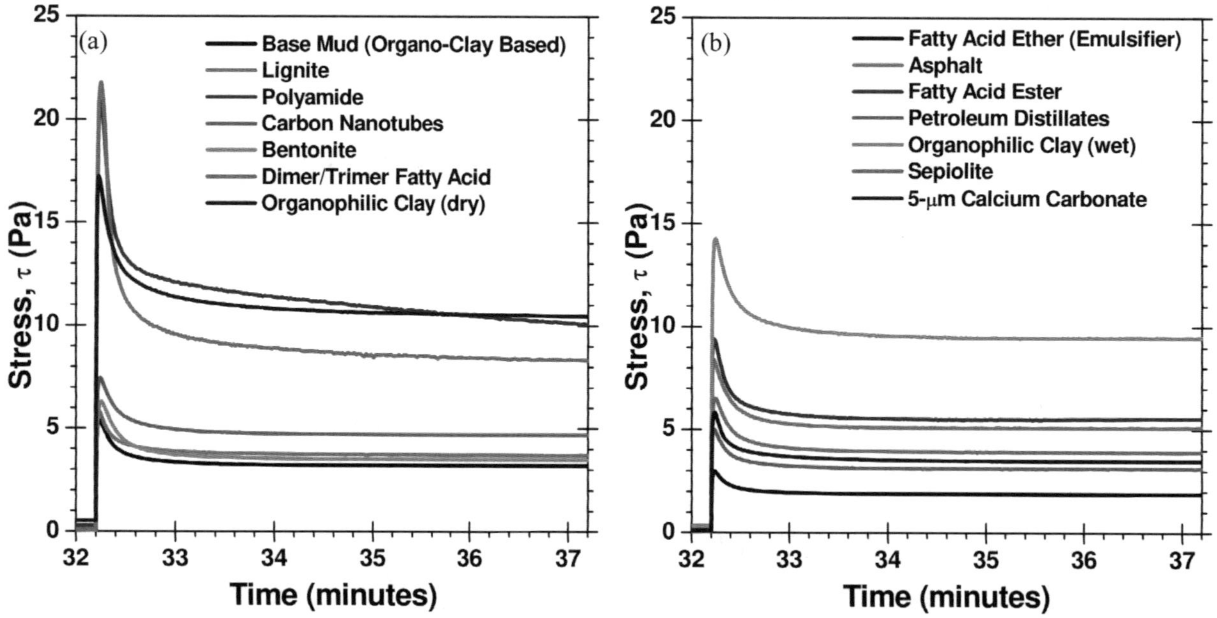

Figure 5. Gel strength measurements on the MCR501 rheometer for OBM #1 at 120°F after (a) a 10-minute gel period.

testing at low shear rates applied to an instrument such as a Brookfield viscometer should give similar results. With these tests, some idea as to the effects of additives on microstructure begins to be seen. From the flow curves the viscosification effects were observed, but in the gel tests the differences in how these additives influence microstructure formation in fluids based on organophilic clays and those without organophilic clays is observed.

In gel tests on the MCR501 rheometer, after the gel period the shear rate was raised from 0-s^1 to a stable 0.1-s^{-1} in ~0.01 seconds (fast transition, so minimal motor effects). The peak values were recorded and are presented, along with the API gel strengths, in Table 1. Contrasting effects of additives in the two mud systems can be observed in such additives as the dimmer/trimer fatty acid. In both muds, a large viscosification was observed when this was added; however, the increase in gel strength measured by the MCR501 rheometer was much greater in OBM #1 than in OBM #2, indicating a synergistic effect with the growth of microstructure in the organophilic clay-based mud. The addition of carbon nanotubes, while not demonstrating a significant viscosification of OBM #1, results in slightly higher gel strengths than in OBM #2, again interacting differently with the dominant microstructures in both mud systems.

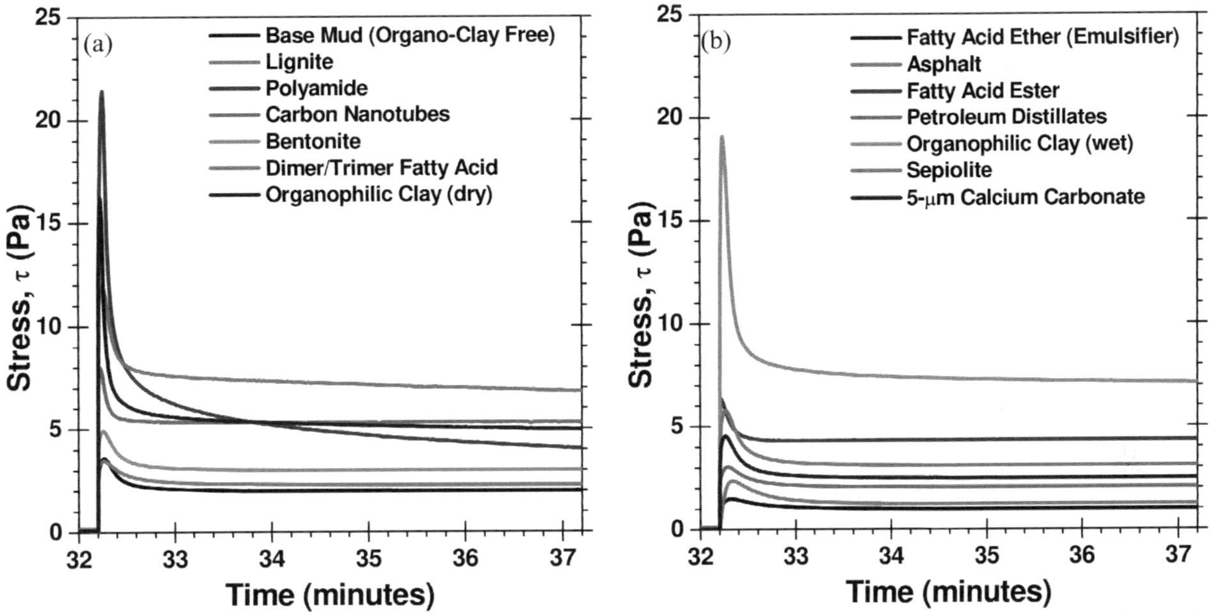

Figure 6. Gel strength measurements on the MCR501 rheometer for OBM #2 at 120°F after a 10-minute gel period.

Table 1. Gel strengths from API standard tests and gel tests on the MCR501 for OBM #1 at 120°F, after 10-minute and 30-minute gel periods.

	OBM #1				OBM #2			
	FANN® 35 Gels (Pa)		MCR501 Gel Tests (Pa)		FANN® 35 Gels (Pa)		MCR501 Gel Tests (Pa)	
	10-min	30-min	10-min	30-min	10-min	30-min	10-min	30-min
Base Mud	11	11	5.4	6.2	10.1	11	3.6	3.6
Lignite	11.5	11.5	5.9	6.6	10.1	11	3.5	3.2
Polyamide	25.9	26.8	20.9	17.0	25.9	26.8	21.4	13.1
Carbon Nanotubes	13.4	12	7.4	9.8	15.3	17.2	8.0	8.3
Bentonite	12.4	11.5	6.3	7.9	12.9	13.9	4.9	4.0
Dimer/Trimer Fatty Acid	16.8	13.9	21.8	28.6	21.1	21.5	12.0	7.9
Organophilic Clay (dry process)	18.7	15.3	17.2	20.6	20.1	21.5	16.2	18.2
Fatty Acid Ether	5.75	6.7	3.0	3.1	7.18	8.14	1.5	1.7

	OBM #1				OBM #2			
	FANN® 35 Gels (Pa)		MCR501 Gel Tests (Pa)		FANN® 35 Gels (Pa)		MCR501 Gel Tests (Pa)	
	10-min	30-min	10-min	30-min	10-min	30-min	10-min	30-min
(Emulsifier)								
Asphalt	9.58	13.9	8.4	9.3	8.62	11.5	2.4	4.2
Fatty Acid Ester	10.1	14.4	9.4	11.2	14.4	15.3	6.3	4.6
Petroleum Distillates	6.7	9.58	5.0	5.7	9.58	9.58	3.1	3.4
Organophilic Clay (wet process)	13.9	18.7	14.3	16.8	22.5	23	19.1	23.9
Sepiolite	8.14	12	6.5	7.9	14.4	15.3	5.8	6.3
5-µm $CaCO_3$	7.66	11.5	5.8	6.7	12.9	13.4	4.5	4.3

Addition of standard bentonite, while giving no significant viscosification of OBM #1 and showing modest viscosification to OBM #2, actually demonstrates a significantly larger increase in gel strengths in OBM #1. This result would, for example, tend to indicate that bentonite in an organophilic clay-free system produces a weak microstructure, such as would be expected from depletion attraction effects, while the more crowded system simultaneously results in increased viscosity. In an organophilic clay-based system, however, stronger bonds are formed, possibly due to partial incorporation of the bentonite into the bridging structure of the existing amine-treated organophilic clays in the mud. A comparison of the energy dissipated in breaking the microstructure in gel tests on the MCR501 are presented in Table 2.

While this test is somewhat more beneficial than the standard API gel strength test, there are still some problems with it. For example, the gel strengths measured on the polyamide polymer addition decrease significantly from the 10-minute gel period to the 30-minute gel period. This is not likely due to a breakdown in microstructure in the mud over time, but is rather a sign of physical problems with the test (such as fluid fracture resulting in multiple flow regimes in the gap and only partial breakage of the microstructure). This would be less of a problem if using a vane stirrer rather than parallel plates for testing, but does demonstrate that in any test care should be taken to obtain good data.

Triangular-Wave LAOS Results

LAOS experiments were performed on all mud samples using triangular-waves at a frequency of 0.1-rad/sec and multiple strains. Since all fluids exhibited fully developed yielding behavior at strain amplitudes of 1000%, data from these tests will be analyzed here. Some debate may be considered as to the proper strain at which to compare these fluids. One argument would suggest that using a uniform yielding strain for all comparative tests provides a simple basis to examine the energy dissipated in microstructure breakage with a minimal number of tests. This is the approach taken in this paper. Another argument would suggest the best comparison would be at the lowest strain at which each fluid fully yields, the yield strain. This has the advantage of potentially better portraying the actual energetic requirements for microstructure breakage when tripping pipe or during pump startup in drilling operations; however, this method requires many more tests and greater care to identify the yield strain. A third argument would promote the use of the gel point, the strain at which G' and G'' are equal, as a comparative point. This would not require an extensive test matrix, but would not reveal information on the actual yield stress or the overshoot experienced during flow initiation.

The energy dissipation, E_D, is plotted as a function of oscillatory cycle at 0.1-rad/sec, 1000% strain, and 120°F in Figure 7a (for OBM #1) and Figure 7b (for OBM #2). By observing the progress of E_D from peak in the first cycle to steady state we can draw conclusions as to the strength of microstructural bonds, and quantify this with the excess E_D value. This also further differentiates both the basic mud systems and how various additives affect their microstructures. While the base mud curves are relatively similar, with OBM #1 having a slightly higher E_D per cycle, the E_D^E for OBM #1 is significantly higher (Table 2). This indicates a stronger microstructural association in the organophilic clay-based mud – likely due to a bridging of emulsion droplets by the amine-treated clays – and a weaker microstructural association in the organophilic clay-free mud – likely indicating a predominantly depletion attraction mechanism. These differences in microstructural mechanisms allow for design of muds which have similar flow properties and gel strengths but have lower energetic requirements for breaking gel structure and returning to steady flow conditions.

When the polyamide polymer is added to each mud, significantly different results are observed from results of flow curve and gel strength comparisons. From the viscometric and gel strength tests, there was little differentiation how the polyamide affects the two mud systems. Both resulted in fairly similar flow curves (both in form and magnitude) and the gel strengths measured by MCR501 were also very similar. In addition, τ_P for both mud systems with polyamide is nearly identical. However, the magnitude of E_D in each oscillatory cycle is significantly higher in OBM #1 (~650 J/m^3) than in OBM #2 (150-400 J/m^3), demonstrating a

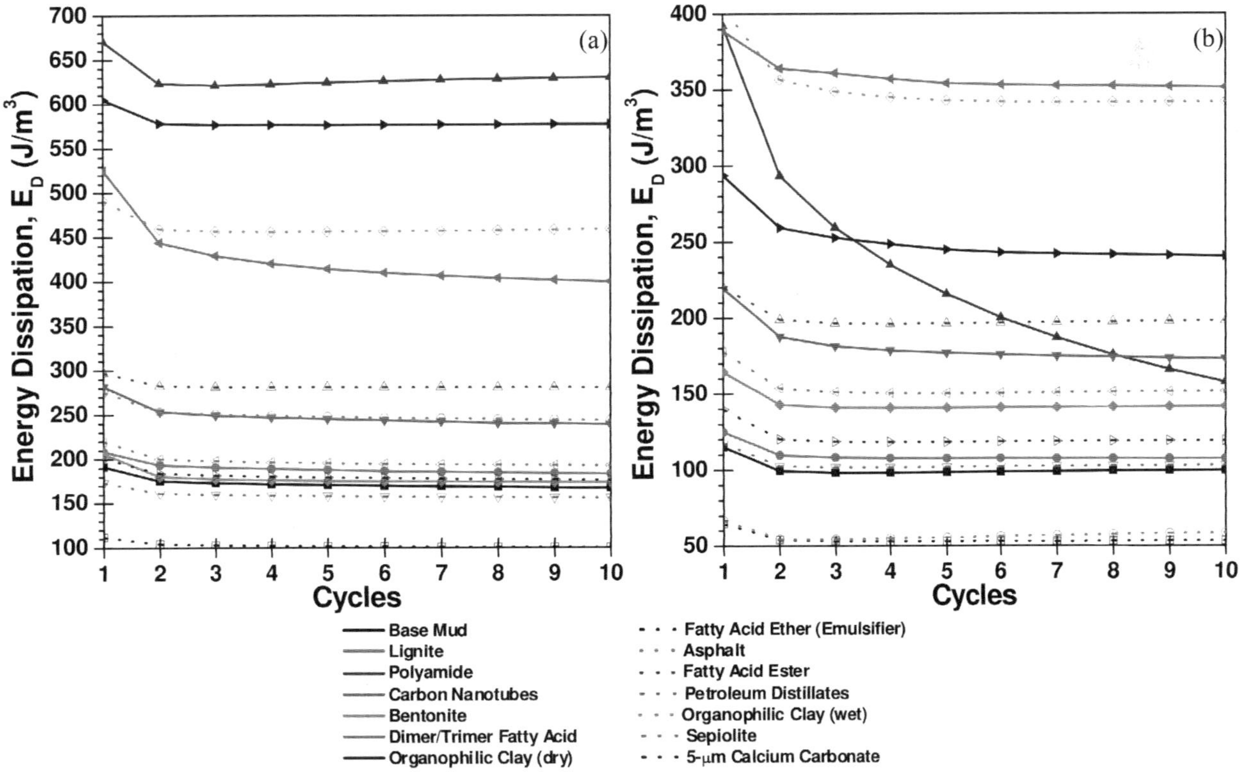

Figure 7. E_D from triangular-wave LAOS as a function of oscillation cycle for OBM #1 (a) and OBM #2 (b) at 120°F at 0.1-rad/sec and 1000% strain.

stiffer response than previously indicated. However, in OBM #1 E_D rapidly reaches a minimum and returns to steady state, while for OBM #2 it continuously decreases through the 10 monitored cycles.

As a result, the E_D^E in OMB #1 is ~7% of that for OBM #2. The energetic requirements to break the microstructure formed in OBM #2 with polyamide polymer strongly indicate that the polyamide bridges emulsion droplets, with a majority of the polymer involved in such connections. When added to OBM #1, however, the existing organophilic clays would appear to out-compete the polyamide polymer in creating emulsion droplet bridges, leaving the polymer to condense the microstructure through depletion attraction and only minimally change E_D^E.

Additions of organophilic clay (amine treated through either a dry or wet process) have similar effects on the two mud systems. When added to OBM #2, the organophilic clays form bridging networks between emulsion droplets and the excess E_D increases ~500%. However, when more organophilic clays are added to OBM #1, the excess E_D actually decreases, possibly as a result of depletion stabilization of the emulsion droplet resulting in a less "sticky" microstructure. The value of τ_P for both systems is again very similar; however, τ_Y is significantly higher in OBM #1, despite a lower excess E_D.

Table 2. Peak stress, yield stress, and excess E_D from triangular-wave LAOS OBM #1, compared with the energy dissipated in breaking microstructure in gel tests on the MCR501.

	OBM #1					OBM #2				
	LAOS			Gel Tests		LAOS			Gel Tests	
	τ_P (Pa)	τ_F (Pa)	E_D^E (J/m^3)	10-min (J/m^3)	30-min (J/m^3)	τ_P (Pa)	τ_F (Pa)	E_D^E (J/m^3)	10-min (J/m^3)	30-min (J/m^3)
Base Mud	9.4	4.4	52.9	29.3	41.5	7.6	2.6	17.9	21.4	26.5
Lignite	9.2	4.8	62.0	29.7	39.9	6.5	2.8	23.2	15.2	16.1
Polyamide	33.3	16.8	51.6	124.0	141.0	33.1	4.1	704.4	220.7	145.1
Carbon Nanotubes	14.1	6.3	92.2	36.5	61.4	12.5	4.6	85.9	38.6	51.5
Bentonite	11.7	4.6	52.2	36.9	57.0	9.9	3.7	26.4	25.9	24.7
Dimer/Trimer Fatty Acid	41.2	10.3	255.4	176.7	282.9	24.7	9.2	73.9	62.4	52.9
Organophilic Clay (dry process)	25.8	15.1	30.4	85.2	115.7	25.4	6.2	103.3	155.5	191.3
Fatty Acid Ether (Emulsifier)	5.2	2.6	26.2	14.8	16.9	3.4	1.4	13.0	6.2	7.7
Asphalt	13.7	6.4	64.1	45.2	55.1	3.8	1.6	12.5	14.2	36.0
Fatty Acid Ester	13.9	7.3	24.8	53.6	70.7	12.4	5.2	27.2	28.9	25.0
Petroleum Distillates	8.0	4.1	23.0	25.7	34.3	6.0	2.7	16.6	13.2	19.4
Organophilic Clay (wet process)	21.9	12.0	38.7	63.5	83.4	32.9	8.9	88.6	163.7	236.0

	OBM #1					OBM #2				
	LAOS			Gel Tests		LAOS			Gel Tests	
	τ_P (Pa)	τ_F (Pa)	E_D^E (J/m^3)	10-min (J/m^3)	30-min (J/m^3)	τ_P (Pa)	τ_F (Pa)	E_D^E (J/m^3)	10-min (J/m^3)	30-min (J/m^3)
Sepiolite	10.9	5.0	52.0	34.6	50.4	11.9	4.0	31.4	36.1	47.3
5-mm $CaCO_3$	9.8	4.6	50.8	31.4	41.9	8.8	3.1	23.9	27.9	29.0

The addition of two emulsifiers (the fatty acid ether and petroleum distillates) to OBM #2 demonstrates different results. For the addition of the petroleum distillates, minimum effect on the mud was observed, either in flow curves, gel strengths, or LAOS measurements. However, addition of the same amount of fatty acid ether emulsifier significantly decreased viscosity, decreased gel strengths by 50%, and decreased LAOS parameters. This could be due to the emulsifier causing a steric stabilization of the emulsion, reducing the connectivity of the microstructure and thus reducing rheological properties.

Finally, a comparison of the E_D^E from LAOS with the energy dissipated in gel strength tests on the MCR501 rheometer after 10-minute and 30-minute gel periods can be made in Table 2. In many cases, the excess energy dissipation calculated from LAOS is significantly different, usually higher, than that observed from gel strength testing. This could be a result of differences in oscillatory and rotational testing. In a rotational test, the fluid is constantly sheared in a single direction. When testing yielding fluids using parallel plates, some degree of shear banding or flow fracture is possible in the gap, the measured stress response will be different and thus the calculated energy to reach steady state will not include complete breakdown of the gel structure. In an oscillatory test, however, shear banding is less likely to occur and the full breakdown of the gel structure will be observed and measured in the test. The use of a vane stirrer would likely bring the two measurements into better agreement.

Conclusions

Differences in the mechanisms for microstructural growth allow for the design of drilling fluids with substantially similar flow properties and gel strengths, but which exhibit great differences in the energetic requirements for breaking that gel structure and returning the fluid to steady flow conditions. The addition of various common oil mud additives can affect the way in which the gel microstructure forms in the invert emulsion fluid. Standard rheological techniques as employed in both field testing and laboratory testing can give a superficial view of these effects; however, when combined with LAOS measurements insight as to microstructural growth mechanisms can be inferred and fluids further differentiated. Through better understanding of these mechanisms for microstructure formation in invert emulsion fluids, better planning of how the mud will behave downhole (including under pump initiation, tripping pipe, drill ahead and set-in situations) can be achieved.

In addition, this understanding may also lead to better understanding of the mechanisms of barite settling (sag). Fluids which are designed to build microstructure without the benefit of organophilic clays or other strong natural emulsifiers exhibit significant disruption in performance when such materials are added to the system. Standard oilfield rheological testing methods are insufficient to determine potential mechanisms for microstructural formation. The API standard gel strength measurement, while useful, is known and demonstrated to be highly inaccurate and cannot provide reliable data to monitor more subtle changes in gels.

Acknowledgments

The author would like to thank Halliburton for granting permission to publish the paper.

References

1. Herzhaft, B., et al., "Influence of Temperature and Clays/Emulsion Microstructure on Oil-Based Mud Low Shear Rate Rheology", *SPE Journal*, 211-217 (2003).
2. Gandelma, R.A., et al., "Study on Gelation and Freezing Phenomena of Synthetic Drilling Fluids in Ultradeepwater Environments", SPE/IADC 105881, 2007 SPE/IADC Drilling Conference.
3. Maxey, J., "Thixotropy and Yield Stress Behavior in Drilling Fluids", AADE-07-NTCE-37, AADE 2007 National Technical Conference.
4. Maxey, J., R. Ewoldt, P. Winter, and G. McKinley, "Yield Stress: What is the "True" Value?", AADE-08-DF-HO-27, AADE 2008 Fluids Conference.
5. Araki, T. and H. Tanakaasd, "Dynamic depletion attraction between colloids suspended in a phase-separating binary liquid mixture", *J. Phys.: Condens. Matter* **20**(7), (2008).
6. Otsubo, Y., "Rheology Control of Suspensions by Soluble Polymers", *Langmuir*, **11**(6), 1893-1898 (1995).
7. Pickrahn, K., B. Rajaram, and A. Mohraz, "Relationship between Microstructure, Dynamics, and Rheology in Polymer-Bridging Colloidal Gels", *Langmuir*, **26**(4), 2392-2400 (2010).
8. Ewoldt, R., P. Winter, J. Maxey, and G. McKinley, "Large amplitude oscillatory shear of pseudoplastic and elastoviscoplastic materials", *Rheologica Acta*, Online First preprint.
9. Maxey, J., "A Rheological Approach to Differentiating Muds by Gel Structure", AADE-10-DF-HO-27, AADE 2010 Fluids Conference.
10. Ewoldt, R. H., A. E. Hosoi and G. H. McKinley, "New measures for characterizing nonlinear viscoelasticity in large amplitude oscillatory shear," *Journal of Rheology* **52**(6), 1427-1458 (2008).
11. Dealy, J. M. and K. F. Wissbrun, *Melt rheology and its role in plastics processing: theory and applications* (Van Nostrand Reinhold, New York, 1990).

Using Polymer Elasticity to Scale Up the Lab Characteristics to Field Application of Friction Reducers

A.M. Gomaa, J. Zhou, H. Sun, and Q. Qu, Baker Hughes, Houston, Texas

This invited paper was prepared for presentation in the 1st International Conference on Upstream Engineering and Flow Assurance, a part of the 2012 AIChE Spring Meeting, held in Houston, Texas, USA, 1-5 April 2012.

Abstract

In order to improve our ability to predict the behavior of the drag-reducing polymers used in slickwater fracture treatments, a better understanding of the polymer elastic characteristics is needed. Elasticity of the polymer under turbulent conditions forces the near-wall vortical structures to become weaker, thicker, longer, and fewer. This significantly reduces the frictional drag of the solution. Therefore, elasticity of the polymer should be considered when scaling up lab results to field data.

Two types of polymers were used at concentration range from 0.5 to 1 gpt. A coiled tubing setup (ID: 0.18 to 0.43 in.) and straight tubes (ID: 0.67 to 1.16 in.) were used to provide the lab data. An oscillatory rheometer was used to measure the elastic and viscous properties.

This paper presents a new model to scale up the friction pressure drop based on polymer relaxation time, pipe diameter, and velocity. Field data was used to support the new model, and an average error of 0.045 psi/ft was observed. Also, a new parameter was introduced to eliminate the coiled tubes' curvature effect and allow it to predict the field data directly.

Experimental results show that elastic and viscous moduli increased by increasing the polymer concentration; however, relaxation time was independent of the polymer concentration. This led to no significant effect on the pressure drop reduction at different polymer concentrations. The elastic modulus of the polymer solution was increased by increasing the shear rate. Therefore, at the same diameter, a higher pressure drop reduction was obtained by increasing the injection rate. Combining lab small scale testing and this model provided a reasonable prediction for field friction pressure drop.

Introduction

Slickwater fracturing is commonly used in stimulation treatments of tight and shale gas reservoirs. Slickwater fracturing involves pumping a low-viscosity fluid (typically fresh water) down the wellbore; to maintain proppant transport, rates of 120 bbl/min or higher, may be required. This process typically requires injecting friction reducers to minimize pipe wall friction and reduce the surface horsepower that would otherwise be required to achieve these high rates, Kaufman *et al.* (2008). Toms (1949) discovered that by dissolving a minute amount of long-chain polymer molecules in water, the frictional drag of turbulent flow could be reduced dramatically. Friction reducers have a few properties in common: They are large polymers; the longer the polymer chain the more effective the material; they tend to build non-Newtonian gel structure; and they are shear-sensitive, White (1964). A copolymer of polyacrylamide emulsified in hydrotreated light petroleum distillates is the most widely used as friction-reducing agent in oilfield applications.

Emulsified polymers typically comprise water-in-oil emulsions of high-molecular-weight copolymers of acrylamide with different anionic monomers. In such emulsions, the active polymer is "packed" inside water droplets dispersed in the continuous oil phase. Upon contact with aqueous-based ambient fluid, emulsion polymers undergo what is commonly known as an "inversion process," during which the high-molecular-weight polymer is released from the emulsion droplet into the ambient fluid. Turbulent friction is reduced when the polymer has migrated to the near-wall boundary layer, where its interaction with eddies of turbulence promotes the reduction of friction, Zelenev *et al.* (2009).

Prediction of friction losses produced by fluid injection down the wellbore has always been an important aspect of stimulation treatment design. Initially, these predictions were used to estimate injection horsepower requirements. Introduction of realtime treating pressure analysis placed additional emphasis on the need for reliable friction loss estimates. Lord (1987) found that a nontrivial scale-up method does not exist. It is necessary to determine unique, method-dependent parameters using turbulent flow pressure loss measurements obtained with the specific fluid formulation.

Keck *et al.* (1992) presented two correlations for predicting friction pressures for HPG linear gel and delayed-crosslink fracturing fluid systems. The first correlation developed from field-sized yard tests with Prandtl's law of friction for turbulent Newtonian flow. The second correlation predicts the increased friction loss caused by proppant. Both equations avoided the problem of diameter scale-up. Webster and Humphrey (1993) proposed empirical friction factor formulas derived from experimental study in the turbulent flow regime for helical coil. Their correlation was limited to Newtonian fluids. Azouz *et al.* (1998) experimentally investigated the tubular frictional pressure loss in straight and coiled tubing. They found that tubing curvature has a significant effect on the pressure losses. Also, for borate-crosslinked HPG, the pressure gradient is dependent on both the fluid pH and the length of CT, whereas for borate-crosslinked guar gum, the pressure gradient is a function of pH and is not very sensitive to the length of tubing.

Zamora *et al.* (2005) developed a basic set of equations to calculate frictional pressure losses for pipes and annuli. This set of equations applies to laminar, transitional, and turbulent flow of Herschel-Bulkley fluids, with Bingham-plastic and power-law behavior as special cases. Friction reducer based on polymer is usually used at concentration range from 0.5 to 1.75 gpt, Sun *et al.* (2010), and Zhou *et al.* (2011). This low polymer concentration affects rheological behavior in a way that does not follow Herschel-Bulkley or Bingham-plastic models fluids. Therefore, the objective of this study is to create a mathematical model that can be used effectively to scale up the lab testing data to obtain the field pressure drop caused by typical oilfield concentrations of friction reducer polymers.

Role of Polymer Elasticity

Cowan *et al.* (2001), Wagner *et al.* (2003), Li *et al.* (2006), and Cunha and Andreotti (2007) linked the performance of friction-reducing polymers with their rheological behavior in extensional flow fields, and especially their ability to reach an extended conformation and resist degradation of molecular weight due to the action of shear force. White and Mungal (2008) stated that commonly used friction reducers are typically high-molecular-weight polymers that are believed to cause friction reduction by interacting with eddies of turbulent flow. This agrees with Kim *et al.* (2007) and Kim *et al.* (2008), who dynamically simulated the turbulent flow of water and stated that near-wall vortical structures are closely related with production of Reynolds shear stress as shown in **Fig. 1a**.

De Gennes (1990) confirmed that drag reduction does not come from a purely viscous effect of the dilute polymer solution because if the viscosity were a dominant parameter for drag reduction, the drag would decrease regardless of the polymer concentration. However, experimental studies have shown the existence of a threshold concentration for drag reduction. Joseph (1990) showed that elasticity plays a predominant role in drag reduction where the coiling and stretching of flexible polymers in turbulent flow produces an elastic component in the polymer stress, as shown in **Fig. 1b**. Polymer elastic stresses can balance Reynolds shear stresses in a bulk turbulent flow at some scale and above some polymer concentration. Therefore, drag is reduced by intrinsic elastic counter-torques that retard the rotation of turbulent vortices. Sreenivasan and White (2000) mentioned that polymer molecules absorb the small-scale turbulence energy into elastic energy and prohibit turbulence cascade, reducing drag. This makes the vortical structures in polymer solutions weaker, thicker, longer, and fewer, as shown in **Fig. 1c**.

This demonstrates that polymer elasticity plays a predominant role in the mechanism for drag reduction. Therefore, including the elastic characterization is a key to developing a model to scale up from lab data to field application.

Experimental Studies

Materials

Two friction reducers were used in this study, identified as Polymer A and B. Both are based on polyacrylamide and have been applied successfully as friction reducers in the oilfield. Polymer A's backbone is more susceptible to oilfield oxidizer breakers, so its use reduces or eliminates formation damage as compared to Polymer B.

More than 100 field data points were collected to verify and support the required mathematical model to predict the field application friction reduction. Field data covers a wide range of pipe diameter (3.826 to 6.276 in. ID) and wide range of injection rates (23 to 120 bbl/min). These wide ranges of field data increase credibility when validating the applicability of the developed mathematical model.

Measurements

The rheology of the friction reducers was measured using a Thermo-Haake Rheostress 300 at room temperature. A Haake rheometer, which had a shear stress range from 10^{-5} to 900 Pa, and a frequency range from 10^{-5} to 100 Hz, was used with a parallel plate assembly and a gap of 1 mm for all measurements. Two types of experiments were conducted to determine the viscoelastic properties of the friction reducer: amplitude sweep test to define the linear viscoelastic region, and frequency sweep test to determine the shear-rate-dependent properties.

A coiled tubing loop was constructed as shown in **Fig. 2a** to evaluate the drag reduction of the friction reducers inside coiled tubes. Pressure transducers were connected to a computer to monitor and record the pressure drop across a core during the experiments every one second. A centrifuge pump with maximum allowable flow of 12 gal/min was used to circulate the polymer solution through the loop. **Fig. 2b** shows the straight-tube setup, where 10-ft-long tubes were used to evaluate drag reduction inside straight tubes.

Table 1 shows the specifications of tubing used in the lab-scale flow loop testing setup. The coils were made by spooling different diameters of straight stainless steel tubing into drums with 2-ft diameter to yield a change in curvature ratio. The four coils can be interchanged through quick connections.

Procedure

The friction reduction testing process involved the following steps:

1. Prepare the specified amount of tap water in the mixing tank and circulate water in the flow loop for 1 minute to fill the tube at the experiment desired injection rate. Check flow meter and pressure sensors to ensure proper operation and stability.
2. Start recording the differential pressure, temperature, and flow rate every second, and allow circulating the tap water for 5 minutes.

3. Inject friction reducer emulsions in the form of a slug from a syringe directly into the bottom orifice of the tank to simulate on-the-fly addition. Circulate the tap water with the friction reducer for 5 minutes at the same flow rate used for the tap water alone.
4. Drain all fluids into disposal tank.
5. Clean the whole system with tap water.
6. Evaluate friction-reducing properties by plotting pressure drop as a function of time. Equations used to calculate the friction reducer properties are presented in Appendix A.

Results and Discussion

Rheology Measurements

A stress sweep test confirmed that the applied stress value of 0.1 Pa was within the linear viscoelastic limit. Therefore, a sinusoidal shear stress of 0.1 Pa was applied at an oscillation frequency range of 10^{-2} Hz to 10 Hz to determine frequency limits of elastic or viscous regimes. **Fig. 3** shows the elastic modulus (G') and the viscous modulus (G") as a function of frequency for Polymer A at a concentration of 0.5 gpt and room temperature. As the frequency increased from 10^{-2} to 10 Hz, the G' increased from 10^{-5} to 0.2 Pa and G" increased from 1.5×10^{-3} Pa to 0.1 Pa. The G' and G" crossover point was achieved at a frequency of 2.05 Hz, where both (G' and G") had the same value. Polymer A at concentration of 0.5 gpt and at a frequency less than 2.05 Hz behaved as a viscous fluid (viscous regime), where the viscous modulus was dominant over the elastic modulus. However, at frequency higher than 2.05 Hz, Polymer A behaved as an elastic material (elastic regime), where the elastic modulus was dominant over the viscous modulus, **Fig. 3**. It can be concluded that elasticity increased with increasing frequency and/or shear rate.

Relaxation time is defined as the inverse of the frequency value at the G' and G" crossover point. **Table 2** shows the values of relaxation time for Polymers A and B at all tested concentrations. Based on this table, relaxation times for these polymers were independent of polymer concentrations. The chemistry of both polymers was nearly identical; therefore, there was no difference between the relaxation time of Polymer A and B.

Figs. 4 and 5 present the effect of increasing polymer concentration on the performance of G' and G" as a function of shear rate, respectively. Increasing the polymer concentration increases the values of G' and G". However, G" increased slightly more than G' with increasing polymer concentration.

A literature survey observed that friction reducers based on polymers were only effective under turbulent flow conditions. There is no drag reduction under laminar flow conditions. This is because in laminar flow, the shear rate environment is typically low, and the elastic component of the polymer is significantly low, as shown in **Figs. 3-5**. However, in turbulent flow, the polymer elastic regime will dominate. Therefore, a drag reduction is always observed in the turbulent regime.

Friction Reduction

The phenomenon of friction reduction of Polymers A and B under turbulent flow conditions was evaluated at different injection rates, pipe diameters, and polymer concentrations using lab-scale flow loop shown in **Fig. 2**. **Fig. 6** shows the reduction in the pressure drop when 1 gpt of Polymer A was introduced to the system at different injection rates. At an injection rate of 10 gal/min, the pressure drop of water flow inside 0.245-in. ID tube was nearly 567 psi. After Polymer A was injected, pressure drop decreased to 243 psi within 10 second. This demonstrates the time needed to invert the water-in-oil emulsion to oil-in-water and allow the polymer to completely hydrate.

Fig. 7 shows the effect of the injection rate on the friction reduction using Polymer A at 1 gpt concentration. It was clear that increasing the injection rate improved the friction reduction of Polymer A. This observation agrees with the rheology performance of Polymer A, where increasing the shear rate by increasing the injection rate enhanced the elastic characterization of Polymer A, which in turn enhanced the ability of Polymer A to overcome the turbulent flow vortices. Therefore, at higher injection rate, a higher friction reduction was observed.

Fig. 8 shows the effect of polymer concentration on the pressure drop performance of Polymer A. It was clear that at test conditions, polymer concentration has no effect on the friction reduction — the entire pressure drop was nearly the same. Based on **Table 2**, relaxation times for Polymers A and B were independent of polymer concentration. This agrees with Min *et al.* (2003), who observed that the relaxation time of the polymer solution is essentially associated with the transport of the elastic energy. The polymer stores the elastic energy from the flow very near the wall and then releases it there when the relaxation time is short, showing no drag reduction. However, when the relaxation time is long enough, the elastic energy stored in the very near-wall region is transported and released in the buffer and log layers, significantly reducing drag.

Drag-Reduction Envelope

Prandtl-Karman (PK) plots are generally used to depict drag reduction studies where the ordinate and abscissa axes are ($1/f^{0.5}$) and $\log(N_{Re} f^{0.5})$, respectively. Plotting drag reduction data as PK coordinates is useful in investigating the onset of drag reduction in pipes. **Fig. 9** shows the drag reduction envelope as PK coordinates for 0.5 gpt of Polymer A in straight and coiled tubing. At N_{Re} less than 3200, laminar flow regime exists, and all fluids follow the laminar trend (i.e., the blue line in **Fig. 9**). When turbulent flow is fully developed, different behavior is obtained depending upon whether the fluid is Newtonian or polymeric. When wall shear stress is below a critical value, drag is not reduced and the curve of PK coordinates follows the "Lower Turb" curve (shown as a red line in **Fig. 9**). After critical shear stress is exceeded for a polymer solution, the friction factor will decrease at a given Reynolds number relative to a Newtonian fluid. The "Upper Turb" curve (shown as a green line in **Fig. 9**) represents the maximum possible drag reduction, in which the friction factor is sensitive to the polymer solution employed, Kamel (2008).

The drag reduction envelope provides two key pieces of information. First, the intersection of the data line or its extrapolation with the bottom baseline ("Lower Turb") indicates the onset of drag reduction. Second, a data line with a slope higher than the bottom baseline indicates that higher drag reduction can be achieved by increasing flow. In addition, a drag reduction envelope for coiled tubing depends on its curvature ratio. Thus, a different envelope results for different curvature ratios, Kamel (2008).

Fig. 9 shows that the onset of drag reduction for Polymer A inside straight tubes (ST) was lower than coiled tubes at any curvature, and the onset of drag reduction of coiled tube is reduced by reducing the tube curvature. **Fig. 9** also verifies that injection rate is the main parameter affecting drag reduction: Both the straight tube data (ST) and the coiled tube data have a higher slope than the "Lower Turb" line. Finally, the data for the coiled tube that has a curvature ratio of 0.0179 is outside the drag reduction envelope.

Friction Velocity Model

The friction velocity procedure is one of the models often used to scale up lab results to field applications. This model typically correlates a function describing drag reduction level with polymer velocity. Definition of these parameters is based on the assumption that turbulent flow friction loss of the polymer solution can be reduced only to the maximum level described by the difference between the solvent turbulent flow friction loss and the polymer solution laminar flow friction loss, Lord (1987).

Fig. 10 shows a log-log plot of Polymer A friction velocity and average velocity for coiled (CT) and straight (ST) tubes. Data obtained from straight tubes tend to behave in linear trend (power law model), whereas for coiled tubes another linear trend was observed for all tubes except the one that has curvature ratio of 0.0179, which was outside the drag reduction envelope shown in **Fig. 9**. The straight tube trend and coiled tube trend were not parallel to each other. Therefore, if the friction velocity model is used, the effect of tube curvature cannot be corrected.

Applying the friction velocity model to fit the field data for Polymer A and B was not successful. As shown in **Fig. 11**, the field data was too scattered. The friction velocity scale-up procedure is believed to be superior to empirical procedures because its origin is a semi-logarithmic resistance law based on: an assumed constitutive equation for viscoelastic flow that requires specification of one material time scale; a particular analytical model for mean velocity profile extending from the laminar sublayer to the outer turbulent layer; and functions developed from the most reliable turbulent velocity profile data available, including the limiting drag reduction ,Lord (1987).

New Model Based on Energy Dissipation of Viscoelastic Fluid

The ultimate objective of conducting flow tests is to develop a mathematical procedure that can scale up the lab test data to obtain the friction pressure of polymer friction reducers under different field conditions. Lab-scale data are customarily used to develop such a correlation. The theory of the developed model is based on energy dissipation of viscoelastic fluids that follow Maxwell model. If a Maxwell fluid is subjected to an oscillatory simple shear, the rate of energy dissipation per unit volume ($\dot{W}_v$) can be expressed as follows (Darby 1976):

$$\dot{W}_v = \frac{1}{4}\frac{\rho f_r v_o^2}{\sqrt{1+f_r^2\lambda^2}} \qquad (1)$$

Where, f_r is oscillation frequency, ρ is the fluid density, λ is the fluid relaxation time, and v_o is the amplitude of the oscillatory velocity.

Energy dissipation is synonymous with friction pressure loss, so:

$$\dot{W}_v = \frac{\Delta P\, q}{l} = \frac{\pi}{4}\frac{\Delta P\, v\, d^2}{l} \qquad (2)$$

where, l is the length between pressure ports (ft); Δp is the pressure drop (psi); d is the internal diameter of straight or coiled tubing (in.); and v is the fluid velocity (ft/sec).

Based on equations 1 and 2, a strong relation between a log of Y axis representing the pressure drop gradient × diameter^4 × $(1 + De^2)^{1.5}$" (where De is the Deborah number) and X axis representing relaxation time × velocity was obtained. This relation can be used to predict field performance from lab-scale straight tube experiments, as shown in **Fig. 12**. Data obtained from the lab-scale straight tube (ID: 0.677 to 1.162 in.) and field applications (ID: 3.826 to 6.276 in.) follow the same curvature trend, **Fig.12**. Therefore, the lab-scale straight tubes will be used to create this trend, which will be used for predict the friction pressure drop for field applications.

Another benefit from this trend is that it corrects for the curvature effect that exists in the data obtained from lab-scale coiled tube experiments, and uses the correction to predict the field data. The coiled tube curvature was corrected by multiplying the Y axis by $R^{0.22}$, where R is the coiled tubing curvature ratio. **Fig. 13** shows the corrected lab-scale coiled tube data can fit the obtained trend for use in predicting field friction reduction performance.

Procedures to Apply the New Model

1. A series of lab tests should be conducted to provide information for the mathematical model to predict the field data. Mainly two tests should be conducted:
    a. Small-diameter flow loop test, either straight or coiled tubes.
    b. G' and G" measurements.
2. For the small-diameter flow loop test, the tested polymer solution will be injected through coiled or straight tubes at different injection rates (at least three points are needed). The pressure drop across a certain length will be recorded as a function of injection rate. Based on the tube diameter and injection rate, the polymer solution velocity can be calculated (see Appendix A). For G' and G" measurements, the polymer solution will be tested using an oscillatory rheometer. A frequency amplitude test will be applied to the polymer solution, and polymer relaxation time will be measured.
3. The lab testing will provide data about the pressure drop, velocity, tube diameter, and relaxation time. These four parameters will be used to predict the field data by plotting on log-linear scale where:
    - The X axis will be (linear scale): relaxation time × velocity.
    - For straight tubes, the Y axis (log scale) will be pressure drop gradient × diameter^4 × $(1 + De^2)^{1.5}$, where De is the Deborah number.
    - For coiled tubes, the Y axis (log scale) will be pressure drop gradient × diameter^4 × $(1 + De^2)^{1.5} \times R^{0.22}$, where R is the coiled tubing curvature ratio.
4. The lab test point will fit the form of Eq. 3, and the constants "a" and "b" will be defined:

$$Log\,(y) = ae^{\frac{b}{\log(x)}} \tag{3}$$

5. After determining the constants "a" and "b" from the lab data, Eq. 3 can be used directly to calculate the pressure drop for any field or other tube data.

Table 3 shows the error analysis in predicting the field data from coiled and straight tubes, and the results were in the reasonable range. When using all data obtained from straight tubes, an average error of 0.045 psi/ft and standard deviation of 0.043 psi/ft were obtained. When using all data obtained from coiled tubes, an average error of 0.049 psi/ft and standard deviation of 0.042 psi/ft were obtained. There was no significant impact on the error analysis when using the data obtained from coiled tubes (after curvature correction) or straight tubes. Increasing the number of lab data point will reduce the error observed from the developed model.

Conclusions

Drag reduction is not a purely viscous effect of a dilute polymer solution. Rather, polymer elasticity plays a predominant role in drag reduction. Therefore, polymer elasticity should be considered when evaluating friction reducers. Based on the results obtained, the following conclusions can be drawn:

1. At turbulent flow, the polymer elastic regime is dominant, while in laminar flow, the viscous regime is dominant. Therefore, drag reduction is always observed in the turbulent regime.
2. Relaxation time is a critical property in defining drag reduction.
3. The elastic modulus of the polymer solution increases with shear rate. Therefore, at the same diameter, a higher pressure drop reduction was obtained by increasing the injection rate.
4. The onset drag of reduction for Polymer A inside straight tubes (ST) was lower than the onset in coiled tubes.
5. For Polymers A and B, the drag reduction envelope confirmed that increasing injection rate is the main parameter affecting drag reduction.
6. By combining lab small-scale testing and new mathematical procedures, a reasonable prediction for field friction pressure drop was achieved.

References

Azouz, I., Shah, S.N., Vinod, P.S., and Lord, D.L. 1998. Experimental Investigation of Frictional Pressure Losses in Coiled Tubing. Journal SPE Production & Facilities, 13(2): 91-96.

Cowan, M.E., Garner, C., Hiester, R.D., McCormick, C.L. 2001. Water Soluble Polymers. Correlation of Experimentally Determined Drag Reduction Efficiency and Extensional Viscosity of High Molecular Weight Polymers in Dilute Aqueous Solution. J. Applied Polymer Sci., 82:1222.

Cunha, F.R., Andreotti, M. 2007. A Study of The Effect Of Polymer Solution In Promoting Friction Reduction In Turbulent Channel Flow. J. fluid Engineering.129:491.

Darby, R. 1976. Viscoelastic Fluids: An Introduction to their Properties and Behavior. Marcel Dekker, New York.

De Gennes, P. G. 1990 Introduction to Polymer Dynamics. Cambridge University Press.

Joseph, D. D. 1990. Fluid Dynamics of Viscoelastic Liquids.Springer.

Kamel, A.H.A. 2008. Flow Characteristics of Surfactant-Based Fluids in Straight and Coiled Tubing. Ph.D. Dissertation, University Of Oklahoma.

Kaufman, P., Penny, G.S., and Paktinat, J. 2008. Critical Evaluation of Additives Used in Shale SlickwaterFracs. SPE Paper 119900 presented at Shale Gas Production Conference, Fort Worth, Texas, November 16-18.

Keck, R.G., Nehmer, W.L., and Strumolo, G.S. 1992. A New Method for Predicting Friction Pressures and Rheology of Proppant-Laden Fracturing Fluids.Journal SPE Production Engineering.7(1):21-28.

Kim, K., Adrian, R., Balachandar, S., and Sureshkumar, R. 2008. Dynamics of Hairpin Vortices and Polymer- Induced Turbulent Drag Reduction. Phys. Rev. Lett., 100.

Kim, K., Li, C-F., Sureshkumar, R., Balachandar, S., and Adrian, R.J. 2007. Effects of Polymer Stresses On Eddy Structures In Drag-Reduced Turbulent Channel Flow. Journal of Fluid Mechanics. 584:281-299. DOI: 10.1017/S0022112007006611.

Li, C-F., Sureshkumar, R., Khomami, B. 2006. Influence of Rheological Parameters on polymer Induced Turbulent Drag Reduction. J. Non-Newtonian Fluid Mech. 140:23.

Lord, D.L. 1987. Turbulent Flow of Stimulation Fluids: An Evaluation of Friction Loss Scale-Up Methods. Paper SPE 16889 presented at Annual Technical Conference and Exhibition, Dallas, Texas, September 27-30.

Min, T., Yoo, J.Y., Choi, H., and Joseph, D.D. 2003. Drag Reduction by Polymer Additives in a Turbulent Channel Flow. J. Fluid Mech., 486:213–238.

Sreenivasan, K.R., and White, C.M. 2000. The Onset of Drag Reduction by Dilute Polymer Additives, and The Maximum Drag Reduction Asymptote. J. Fluid Mech. 409, 149–164.

Sun, H., Stevens, R.F., Cutler, J.L., Wood, B., Wheeler, R.S., and Qu, Q. 2010. A Novel Nondamaging Friction Reducer: Development and Successful Slickwater Frac Applications. Paper SPE 136807 presented at Tight Gas Completions Conference, San Antonio, Texas, November 2-3.

Toms, B. A. 1949. Observation on the Flow of Linear Polymer Solutions Through Straight Tubes at Large Reynolds Numbers. Proc. 1st International Congress on Rheology,**2**, 135–141.

Wagner, C. Amarouchene, Y., Doyle, P., and Bonn, D. 2003. Turbulent Drag Reduction of Polyelectrolyte Solutions: Relation with TheElongational Viscosity. Europhyslett. 64:823.

Webster, D.R. and Humphrey, J.A.C. 1993. Experimental Observations of Flow Instability in a Helical Coil. Trans., ASME, 115-436.

White, C.M., and Mungal, M.G. 2008.Mechanics and prediction of Turbulent Drag reduction with Polymers.Annu. Rev. Fluid Mech., 40:235.

White, G.L. 1964. Friction Pressure Reducers in Well Stimulation.Journal of petroleum technology, **16**(8), 865-868.

Zamora, M., Roy, S., and Slater, K. 2005. Comparing a Basic Set of Drilling Fluid Pressure-Loss Relationships to Flow-Loop and Field Data. AADE Paper 05-NTCE-27 presented at National Technical Conference and Exhibition, Houston, Texas, April 5-7.

Zelenev, A.S., Gilzow, G.A., and Kaufman, P.B. 2009. Fast Inverting, Brine and Additive Tolerant Friction Reducer for Well Stimulation. Paper SPE 121719 presented at International Symposium on Oilfield Chemistry, The Woodlands, Texas, April 20-22.

Zhou, J., Sun, H., Stevens, R., Qu, Q., and Bai, B. 2011.Bridging the Gap between Laboratory Characterization and Field Applications of Friction Reducers. SPE paper 140942 presented at Production and Operations Symposium, Oklahoma City, Oklahoma, March 27-29.

Appendix A: Equation used to define the friction reducers properties

Data gathered include flow rate, pressure drop across straight and coiled tubing, and fluid temperature. Data readings in the transition region due to change in flow rate or due to purging operation and any unstable points were eliminated. Only steady state data points in the stabilized region were averaged and used for analysis. Flow rate and pressure drop values were first converted to Fanning friction factor and generalized Reynolds number. These two dimensionless groups were used in characterizing fluid flow through straight and coiled tubing. The following equations (in field units) were used in data analysis. Fanning friction factor, f is a dimensionless variable used to determine friction pressure gradient. It is defined by the following expression:

$$f = 25.8\left[\frac{d\,\Delta p}{l\,\rho\, v^2}\right] \qquad \text{(A-1)}$$

where, l is the length between pressure ports (ft); Δp is the pressure drop (psi); d is the internal diameter of straight or coiled tubing (in.); v is the fluid velocity (ft/sec); and ρ is the fluid density (lbm/gal).

Average velocity, v in ft/sec is calculated from the following equation.

$$v = \frac{q}{2.448\, d^2} \qquad \text{(A-2)}$$

where, q is the flow rate (gal/min).

Fanning friction factor is plotted against Reynolds number, NRe.

$$N_{Re} = 928\left[\frac{\rho\, v\, d}{\mu}\right] \qquad \text{(A-3)}$$

where μ is the fluid dynamic viscosity (cP).

Shear rate inside lab scale tubes and filed pipes was calculated using the following equation:

$$\bar{\gamma} = \frac{8v}{d} \qquad \text{(A-4)}$$

The shear rate was calculated from the applied frequency, f_r, using Eq. A-5.

$$\bar{\gamma} = \frac{3n}{4}\,\frac{8\pi r f_r}{h} \qquad \text{(A-5)}$$

Where $\bar{\gamma}$ is the shear rate, s^{-1}; r is the radius of plate, mm; fr is the frequency, Hz; and h is the gap between the parallel plates, mm.

The Deborah number is a conventional dimensionless term used to quantify the viscoelastic behavior of fluids. A definition of Deborah number (D_e) is adopted by Kamel (2008) and was used in this study:

$$D_e = f_r \lambda = \bar{\gamma}\,\lambda \qquad \text{(A-6)}$$

Where $\bar{\gamma}$ is the shear rate, s^{-1}; fr is the frequency, Hz; and λ is the relaxation time, s.

Table 1. Specifications of Tubing Used in Lab-Scale Flow Loop Testing Setup

NO.	Tube Geometry	Inner Diameter, inch	Total Length, ft	Curvature
1	Coiled Tube	0.18	20	0.0075
2	Coiled Tube	0.245	20	0.0102
3	Coiled Tube	0.3198	50	0.013
4	Coiled Tube	0.43	20	0.0179
5	Straight Tube	0.677	10	-
6	Straight Tube	0.9	10	-
7	Straight Tube	1.162	10	-

Table 2. Relaxation Time of Polymers A and B at Different Loading Concentrations

Polymer Type	Polymer Concentration, gpt	Relaxation time, s
A	0.5	0.155
	0.75	0.154
	1	0.152
B	0.5	0.155
	0.75	0.154
	1	0.154

Table 3. Error Analysis Based on the Prediction Tube Size

	ST: (All Tubes)	CT: (All Tubes)	unit
Min	0.000	0.001	psi/ft
Average	0.045	0.049	psi/ft
ST Dev.	0.043	0.042	psi/ft

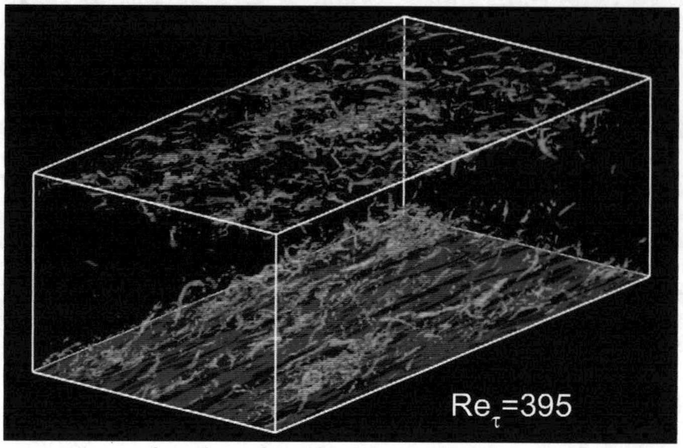

(a) Near-wall vortical structures are closely related with production of Reynolds shear stress. (Quasi-stream wise vortices, low-speed streaks, hairpin vortices, vortex packets, etc).

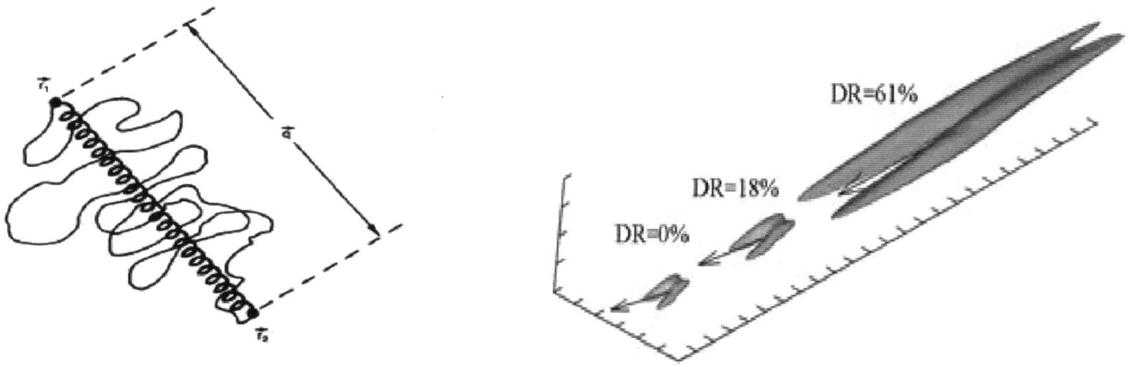

(b) Polymer structure in turbulent flow act like elastic springs, giving large, positive spanwise countertorque and acting against rotation at the heads of downstream and secondary hairpin vortices.

(c) Near-wall vortical wtructures in polymer solutions became weaker, thicker, longer, and fewer.

Fig.1. Role of polymer elasticity in reducing the drag after Kim *et al*. (2008)

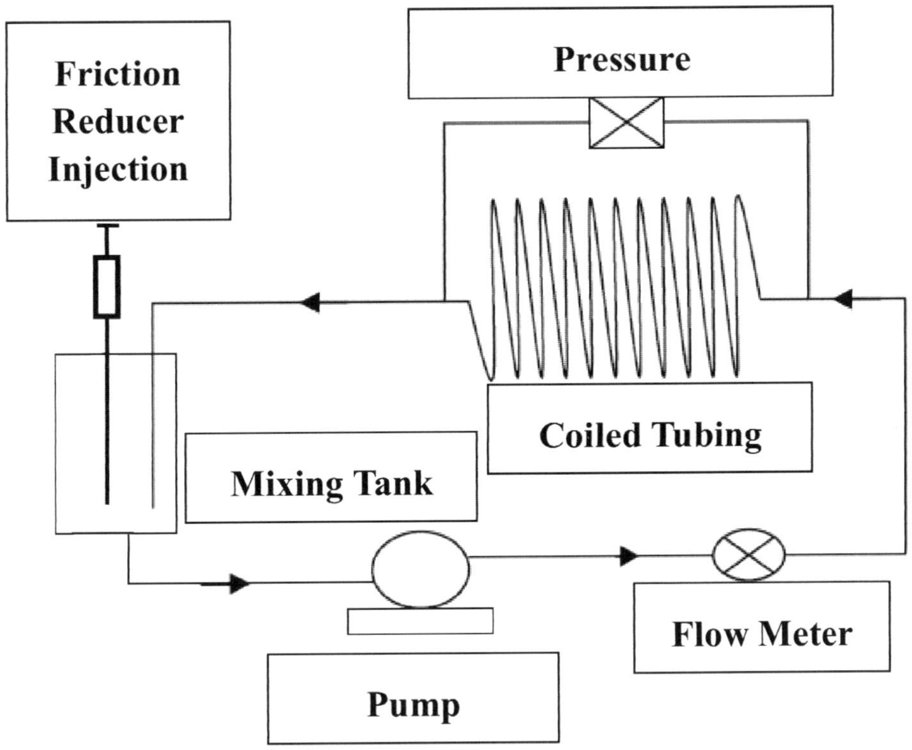

(a)

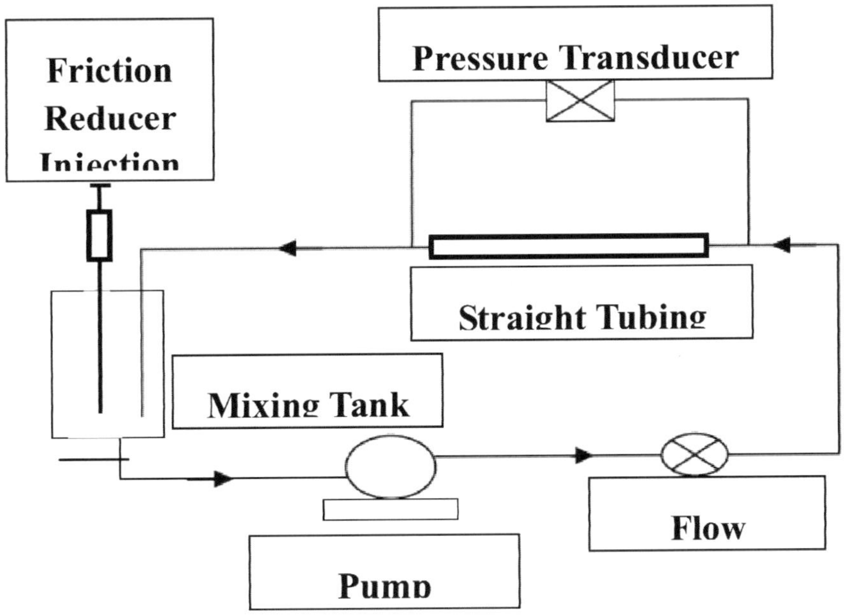

(b)

Fig.2. Schematic of lab-scale flow loop: (a) the coiled tube flow testing setup; and (b) the straight tube flow testing loop setup.

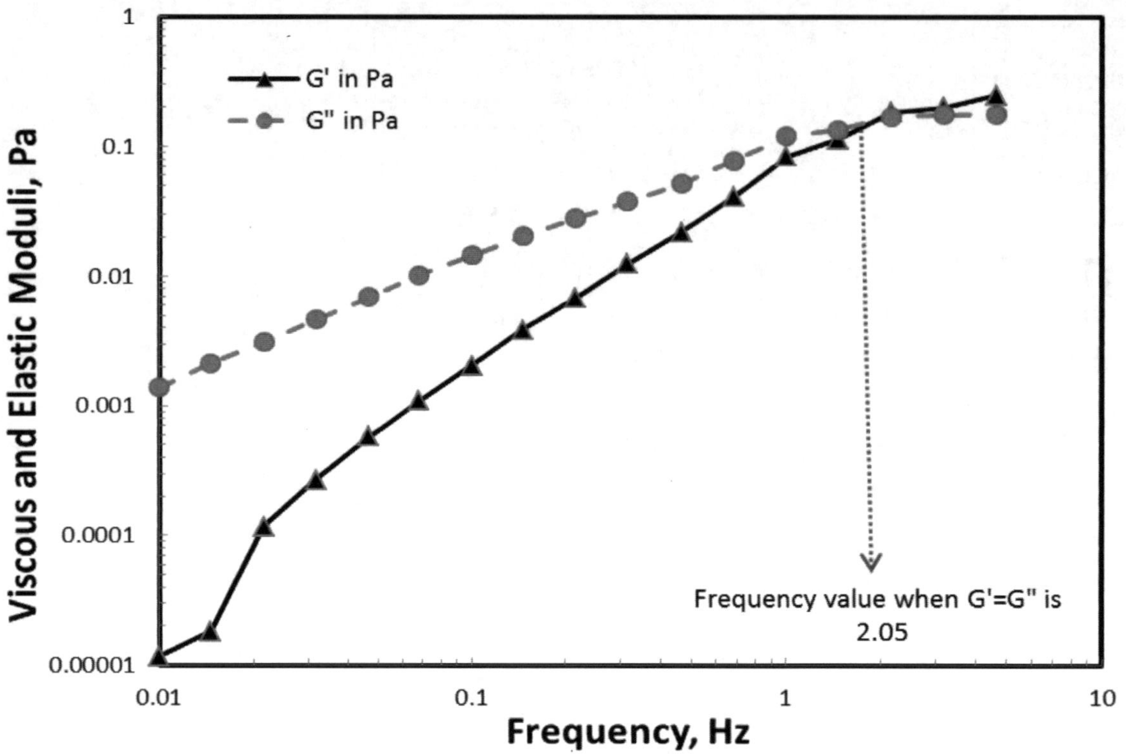

Fig. 3. Viscous and elastic modulus of Polymer A at a concentration of 0.5 gpt and room temperature.

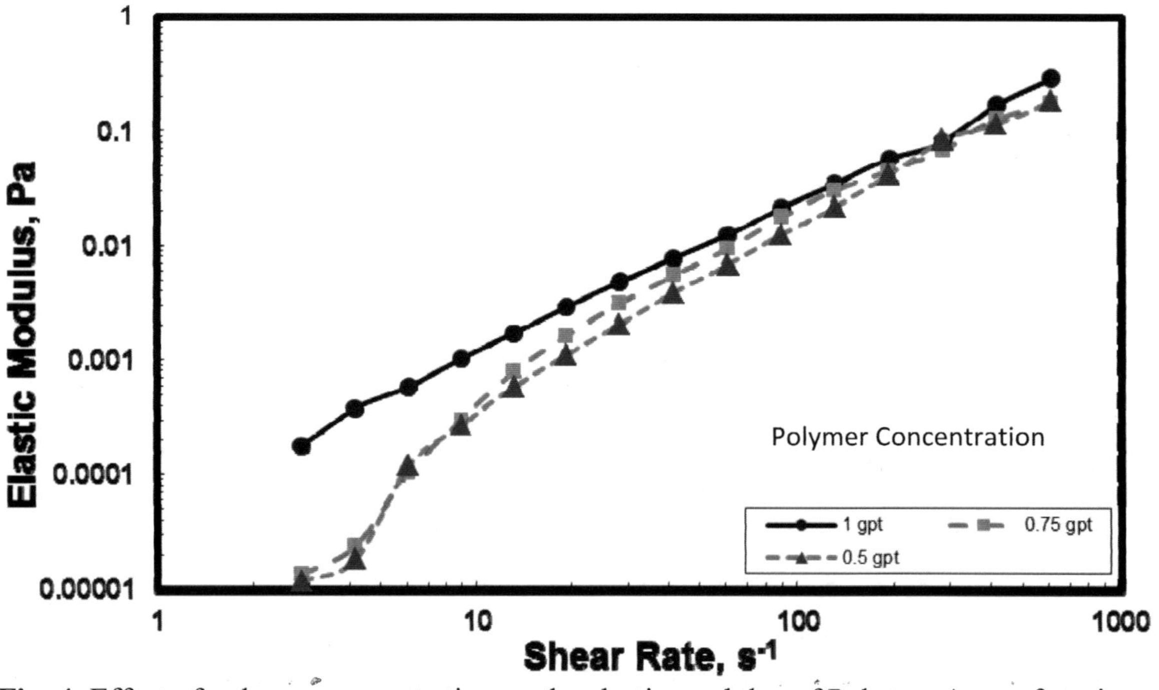

Fig. 4. Effect of polymer concentration on the elastic modulus of Polymer A as a function of shear rate at room temperature.

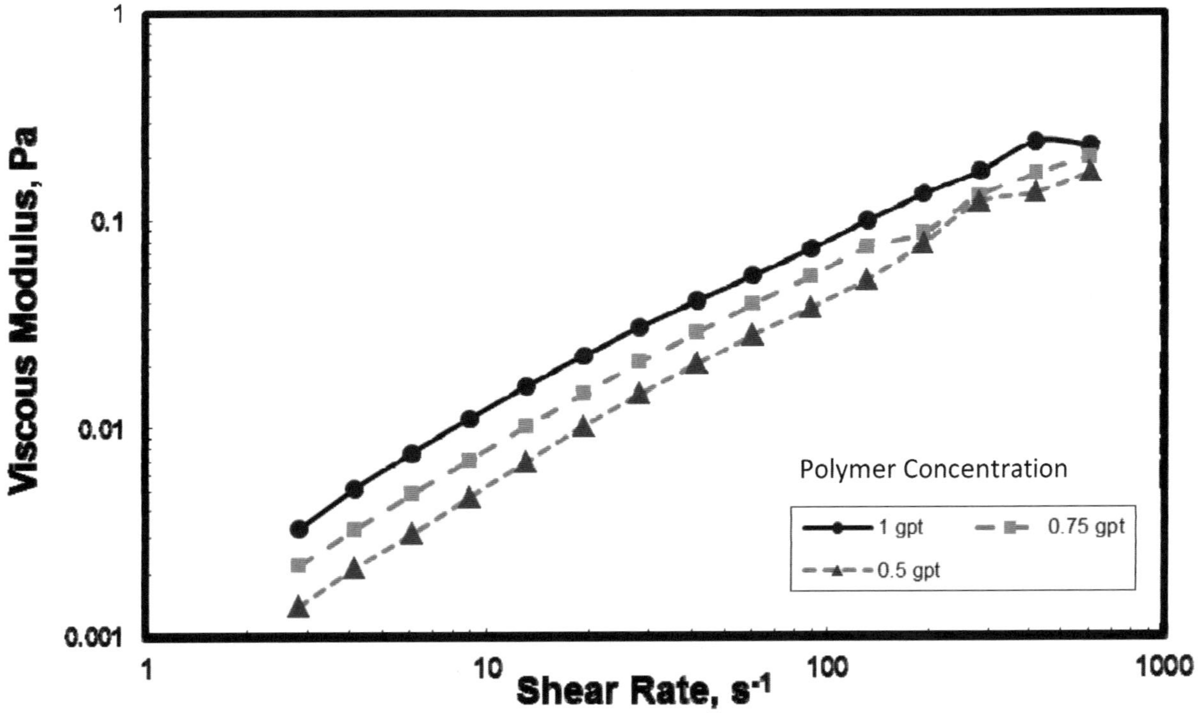

Fig. 5. Effect of polymer concentration on the viscous modulus of Polymer A as a function of shear rate at room temperature.

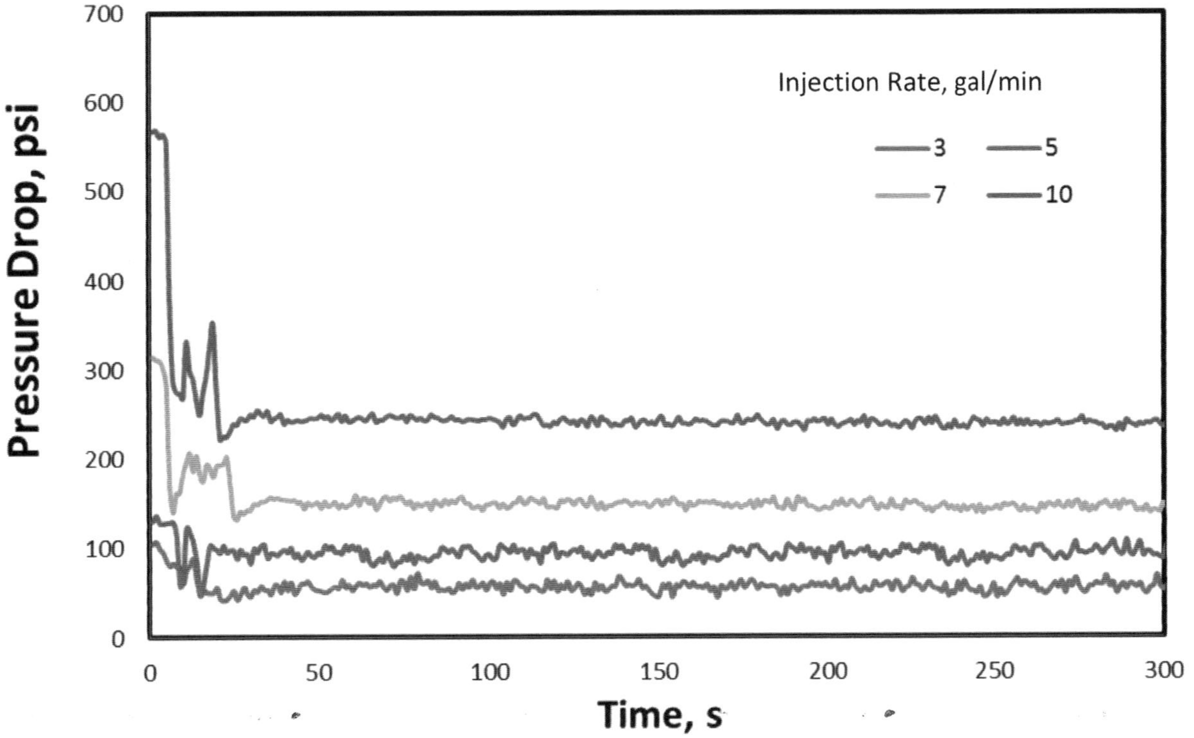

Fig.6. Effect of injection rate on the pressure drop reduction using Polymer A at 1 gpt through a 0.245-in. ID coiled tube at room temperature.

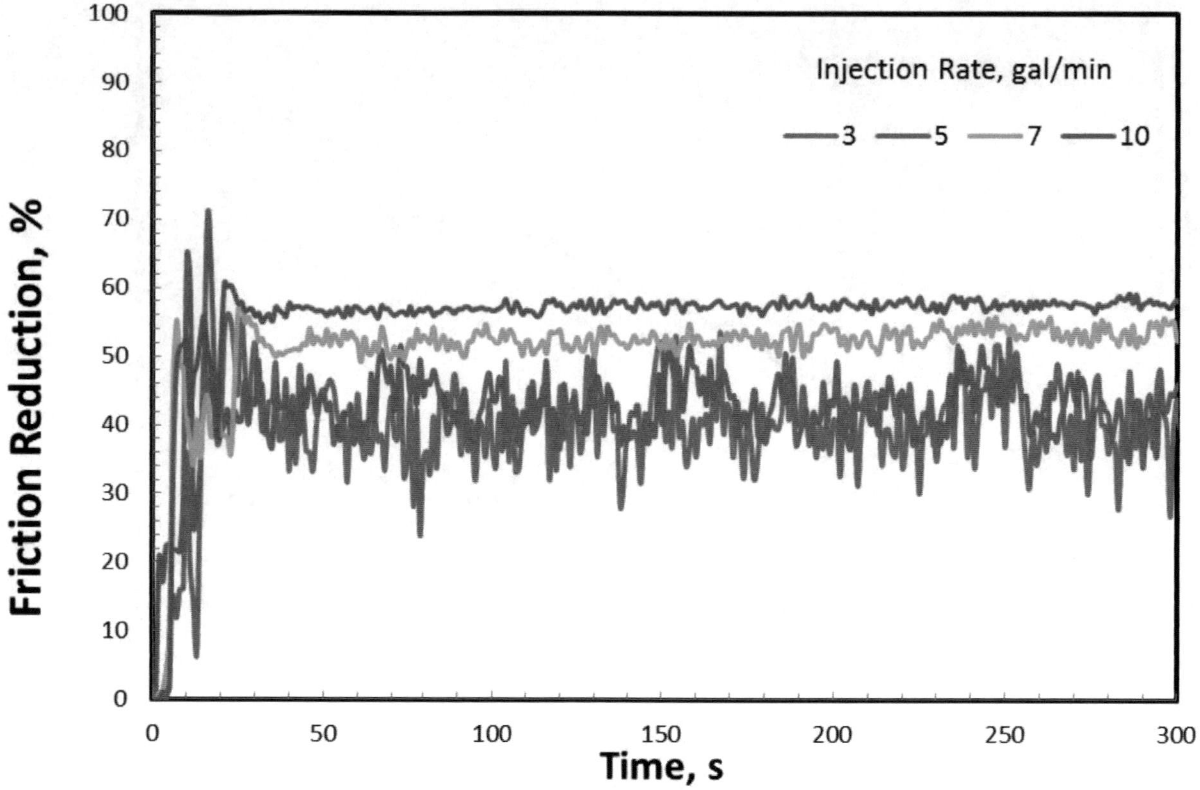

Fig. 7. Effect of injection rate on friction reduction using Polymer A at 1 gpt in 0.245-in. ID coiled tube at room temperature.

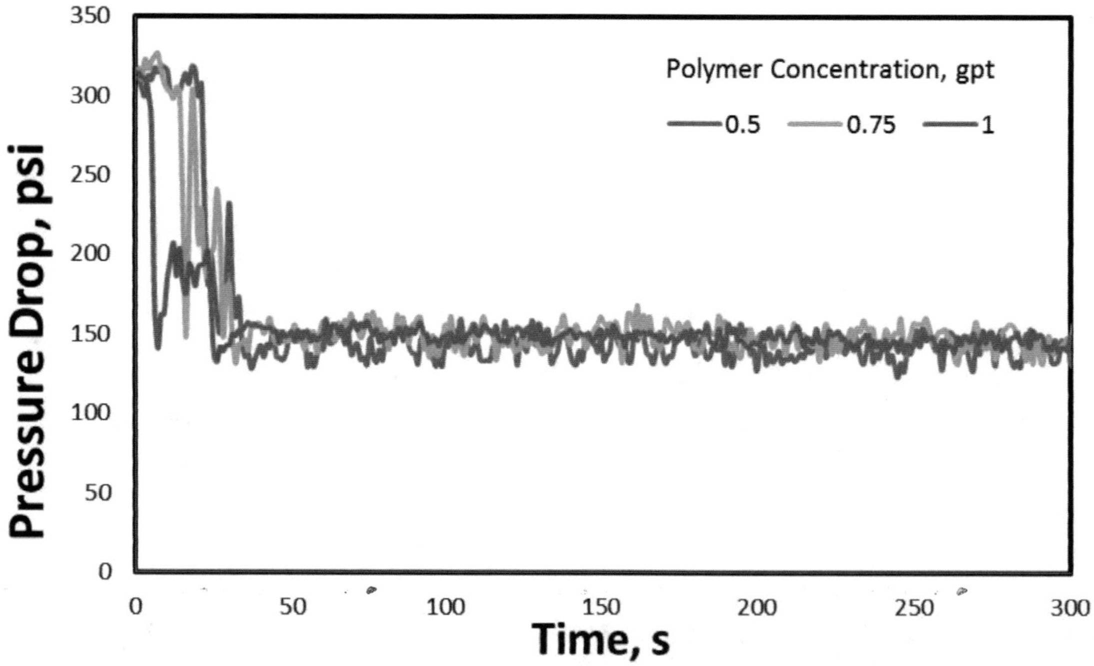

Fig. 8. Effect of polymer concentration on the pressure drop reduction using Polymer A at an injection rate of 7 gal/min in 0.245-in. ID coiled tube at room temperature.

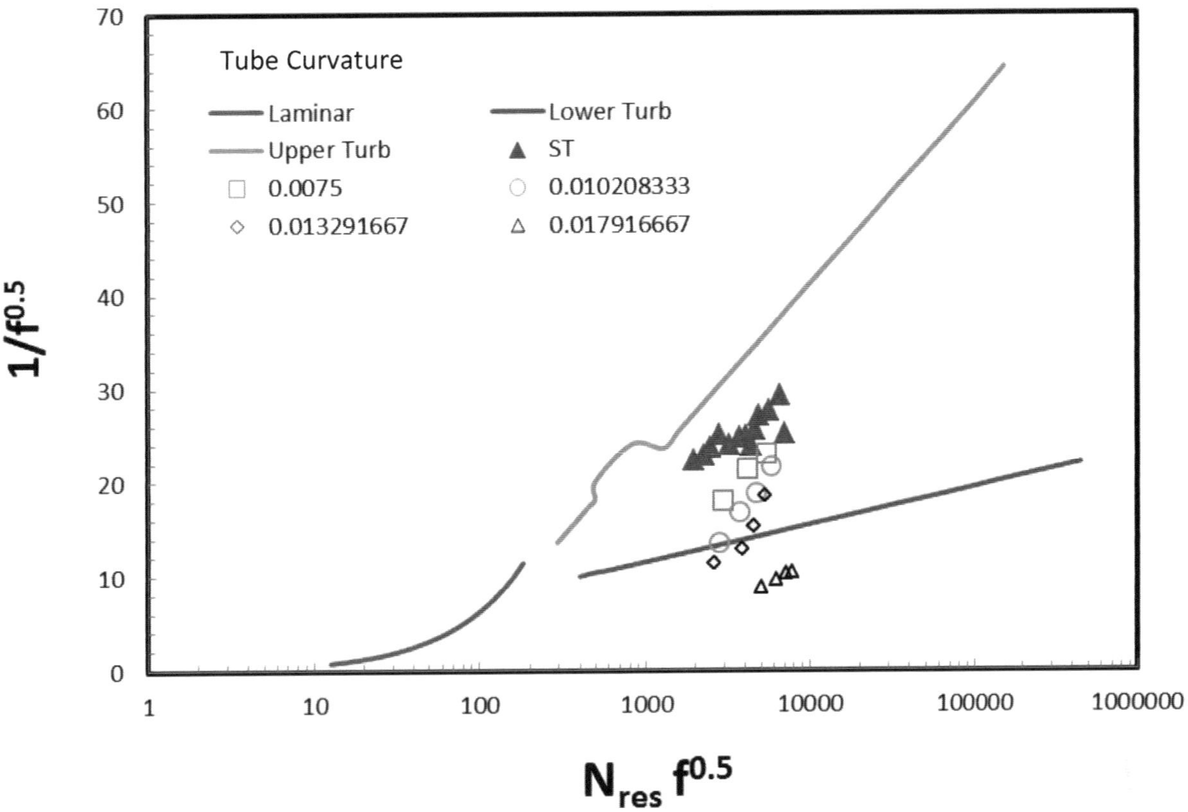

Fig. 9. The drag reduction envelope in Parndtl-Karman coordinates for straight and coiled tubing for Polymer A at 0.5 gpt concentration.

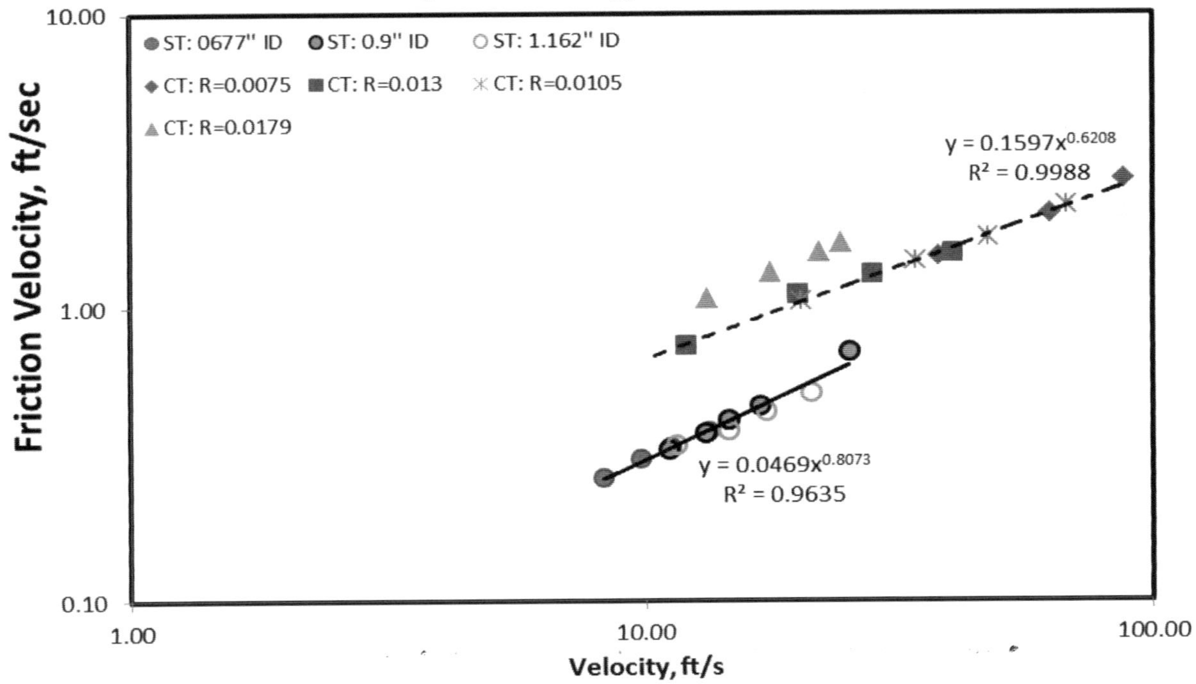

Fig. 10. Friction velocity as a function of velocity for coiled and straight tubes using\ Polymer A at 0.5 gpt.

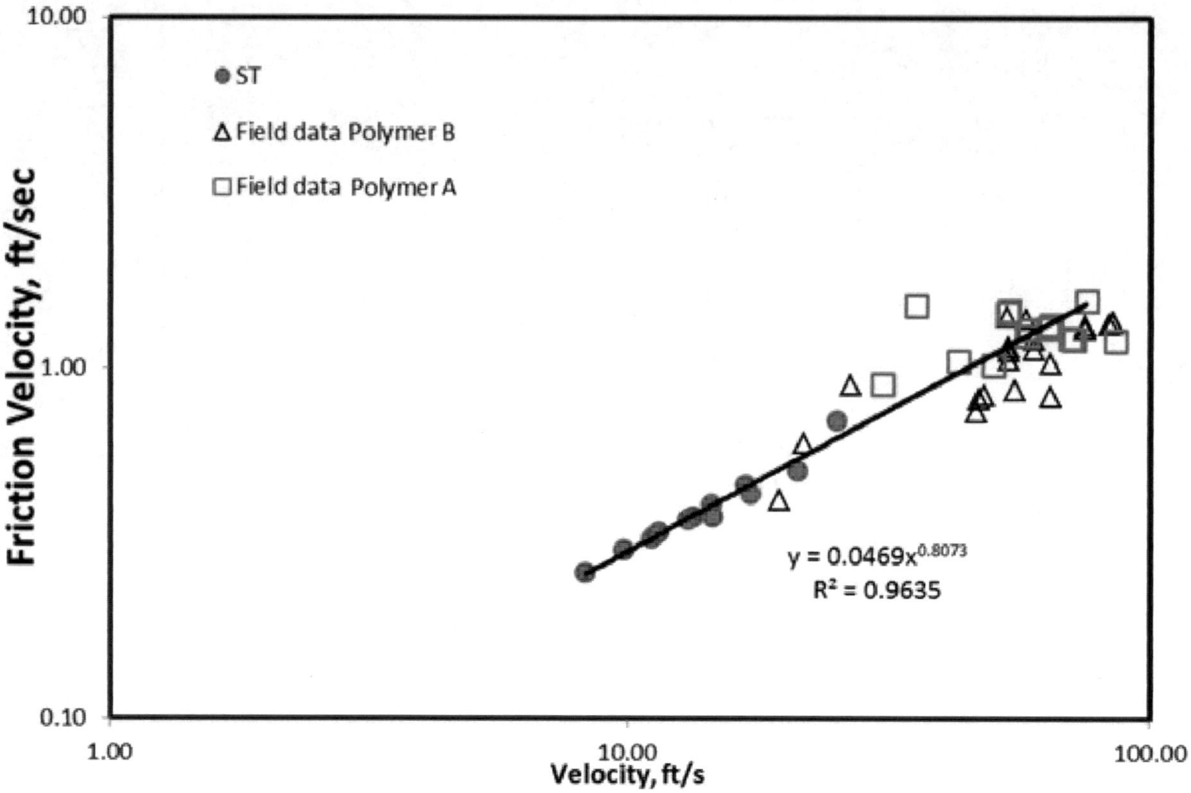

Fig. 11. Friction velocity model fitting the field data for Polymers A and B.

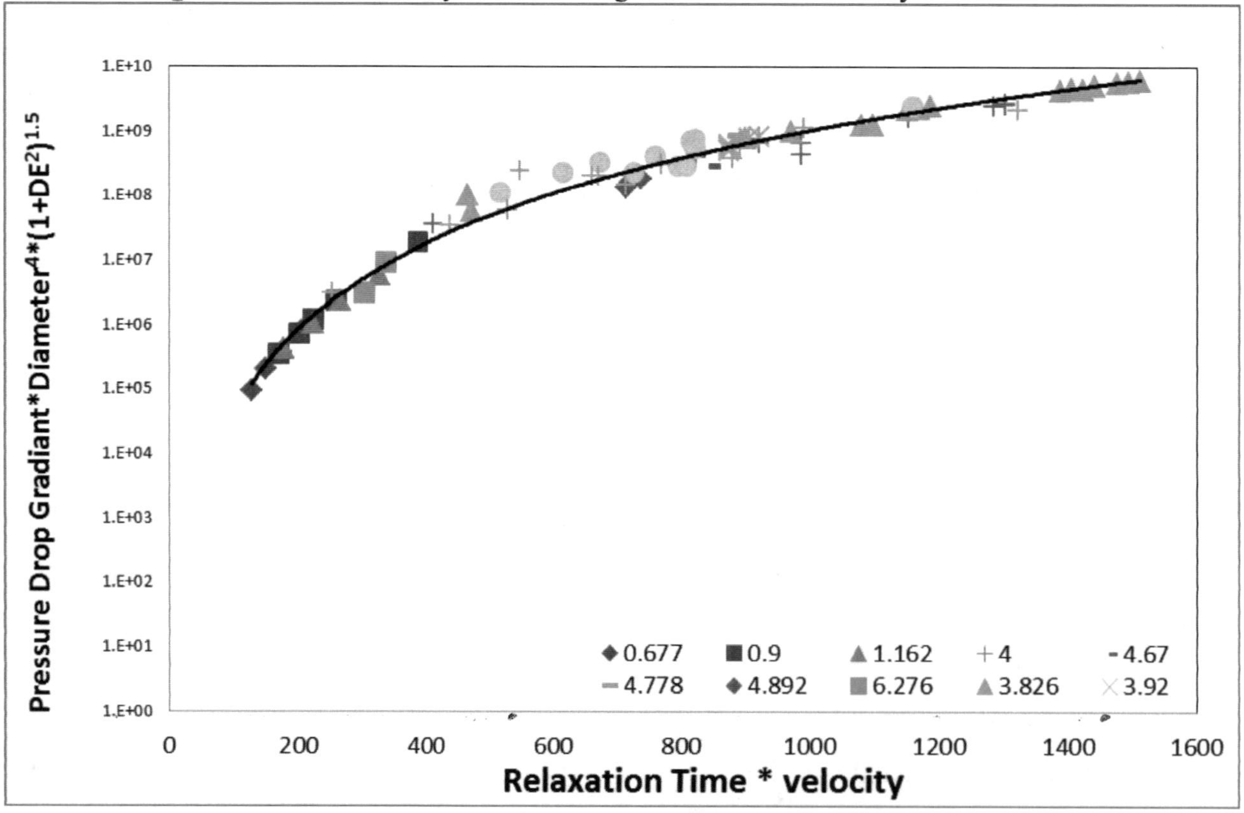

Fig. 12. New model fitting the field data for straight tubes.

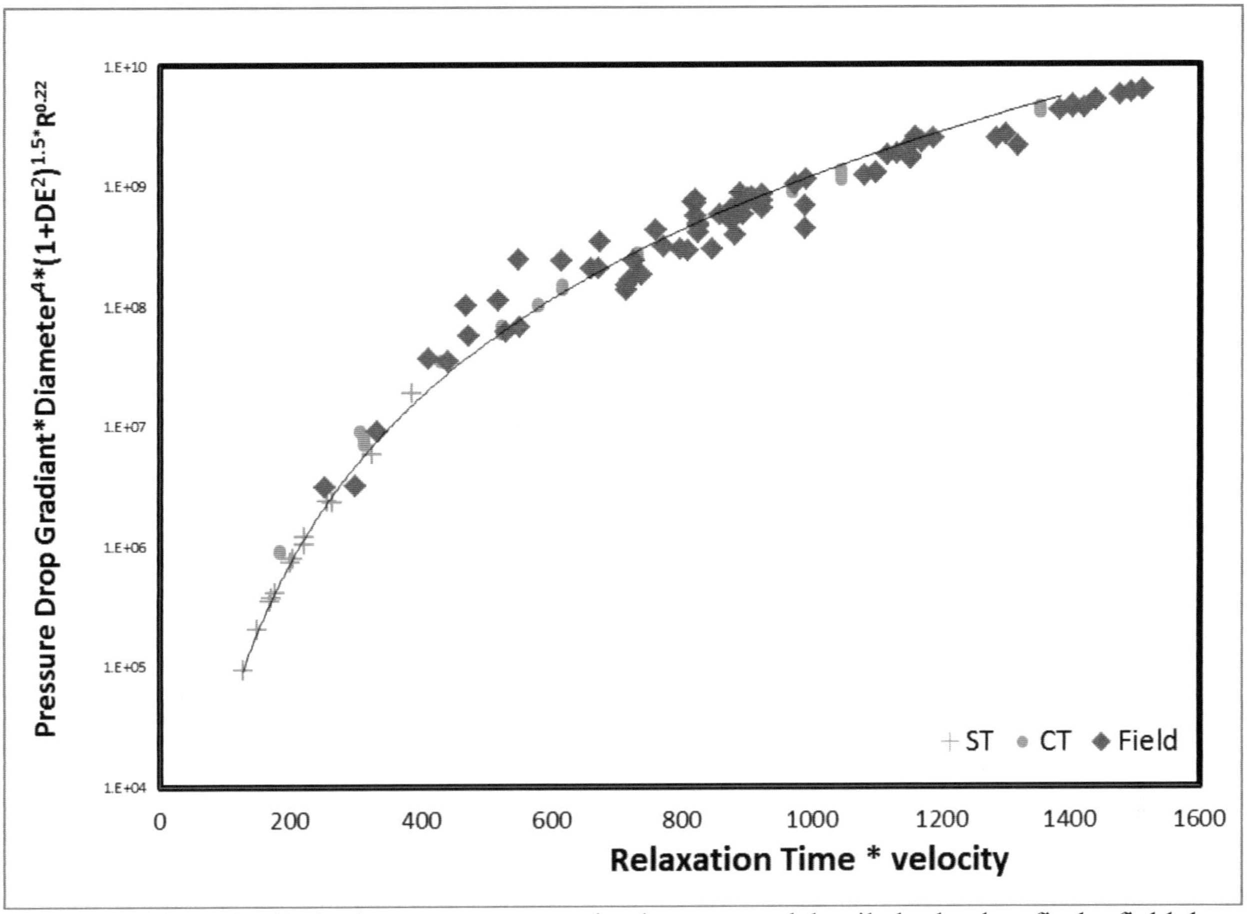

Fig. 13. Correction of coiled tube curvature to make the new model coiled tube data fit the field data and straight tubes.

Efficient Displacement of Synthetic or Oil-Based Mud and Transitional Phase Inversion

Lirio Quintero[1], Jean-Louis Salager[2], Ana Forgiarini[2], Gianna Pietrangeli[1], and Jonathan Brege[1];
[1]Baker Hughes, Houston, TX; [2]Universidad De Los Andes, Merida, Venezuela

This invited paper was prepared for presentation in the 1st International Conference on Upstream Engineering and Flow Assurance, a part of the 2012 AIChE Spring Meeting, held in Houston, Texas, USA, 1-5 April 2012.

Abstract

Effective displacement of invert emulsions, typically known as synthetic or oil-based mud (S/OBM) to completion brine is extremely important in order to minimize non-productive time (NPT), to lower brine filtration costs, reduce waste volume, prevent cement failures and reduce the risk of completion tool complications during the completion of a well. A good displacement spacer system should remove the S/OBM and debris, prevent viscous emulsions or sludge, be compatible with the drilling fluid in order to prevent channeling, and completely water-wet all metal surfaces in no more than a single circulation of completion brine.

The key spacer properties to obtain an effective displacement of the S/OBM are (1) high detergency or oil removal, (2) good mud/spacer compatibility without noticeable increase of viscosity at the fluids interface and (3) ability to change wettability of the rock and metal from oil-wet to water-wet.

Spacer fluids with such required properties need to have good microemulsification of the non polar components of the S/OBM and produce a transitional phase inversion from oil-continuous phase to water-continuous emulsion.

This paper discusses the phase behavior of microemulsion spacer systems, the transitional phase inversion, and the interfacial properties that result in successful displacements of synthetic-based drilling fluids in deepwater applications.

Introduction

Historically, oil and gas operators have directly displaced the drilling fluid from a wellbore with cement after running a casing string. Fluids incompatibility is a major issue when the displacement involves synthetic or oil-based mud due to the vast chemical dissimilarity between oil- and water-based fluid systems. If these incompatible fluids come into contact with each other, a highly viscous fluid-fluid mixture is usually formed at the interface between the S/OBM and cement slurry, decreasing the displacement efficiency. These inefficient displacements often lead to poor cement bonding, squeeze jobs, communication between zones and in some cases, blowouts.

Currently, operators commonly use one or more displacement spacers to physically separate the drilling fluid from the cement slurry, in order to prevent the commingling of potentially incompatible fluids. An additional fundamental difficulty in conventional oilfield cementing operations is the inability to effectively remove oil and to render the surfaces water-wet.[1] The failure to properly complete the displacement of a synthetic or oil-based mud from the wellbore and to water-wet the casing and formation before to cementing lead to problems in well completion and an increase of the operating costs.[2, 3]

The spacers most frequently used spacers today for displacing of S/OBM from the wellbore are water-continuous spacers pumped in turbulent flow mode. These spacers may contain viscosifiers, weighting material, surfactants, solvents, mutual solvents, and/or mixtures thereof. Drilling fluid properties, such as internal phase salinity, hardness, pH, solids content, oil-water ratio, and oil type add to the complexity of fluid-fluid incompatibilities. Wellbore conditions such as bottomhole temperature, wellbore angle and casing eccentricity complicate the displacement. Each of these properties can independently influence the performance and the overall success of the displacement operation.

During the displacement, the drilling fluid, spacers, and cement slurry commingle, creating a mixed interface between the spacer and S/OBM and between the spacer and cement slurry. This can result in settling of solids, poor cleaning, insufficient water-wetting, spikes in viscosity, etc. Significant fluctuations in viscosity can occur at the interface of a S/OBM and an aqueous spacer, which can lead to fluids channeling, which will be further exacerbated in a non-concentric wellbore.[4-7]

Selecting the proper spacer chemistry that meets the requirements of S/OBM displacement efficiency and changes of the surface casing wettability from oil-wet to water-wet is typically dependent upon the drilling fluid type, composition, and wellbore conditions.[8] Efficient displacement is defined as the complete removal of S/OBM from the wellbore and from the metal surfaces of the casing as illustrated in **Figure 1**. This Figure shows a metal sleeve surface exposed to an OBM, and the same sleeve after exposure for a few minutes to an efficient spacer.

Figure 1 Metal sleeve after being exposed to OBM and after a good cleaning with spacer

Fundamentals on Phase Inversion of Surfactant-Water-Oil Systems

Phase inversion of emulsions is well documented. In practice, the phase inversion of a dispersed system or emulsion could be triggered by a change in formulation or composition, such as an increase of water or oil concentration. This kind of phase inversion is generally called dynamic inversion.[9, 10]

A S/OBM formulation, such as an invert emulsion drilling fluid could be altered by the continuous addition of a water-based fluid containing surfactants under dynamic conditions. The magnitude of those changes can be observed using techniques such as microscopy and electrical conductivity measurements. The water-in-oil (W/O) emulsion has near-zero conductivity, whereas the oil-in-water (O/W) emulsion has high conductivity similar to the aqueous phase.

The phase inversion may occur suddenly or gradually, with change of composition, depending on the surfactant used in the water-base fluid. At the inversion point, the fluid system experiences its lowest interfacial tension. Under these conditions, spontaneous microemulsification occurs between the oil and water phase of the system.

Figure 2 shows a schematic of the changes experienced by a water-in-oil emulsion (OBM) resulting from the addition of a water-based fluid containing surfactants. When the system is approaching the phase inversion point, multiple emulsions are being formed. Multiple emulsions are the simultaneous occurrence of both emulsion types (O/W and W/O).[11, 12] In this case, the formulation transition is associated with the transition of phase behavior from Winsor type II to Winsor type I, passing through Winsor III. The phase behavior of a water-oil-surfactant system was described in the early work of Winsor.[13] Winsor I is formed by oil-swollen micelles in equilibrium with excess oil (i.e. two phases, with the surfactant in the lower phase). Winsor II is formed by water-swollen reverse micelles in equilibrium with excess water. In Winsor III phase behavior, a bicontinuous microemulsion or middle phase is in equilibrium with excess oil and water.[14-16] A single-phase microemulsion (Winsor IV) is formed when a sufficient amount of surfactant is added to solubilize the excess oil and water.[17, 18] The Winsor III and Winsor IV behavior is associated with ultra-low interfacial tension and maximum solubilization of oil and water per unit mass of surfactant. The Winsor I behavior is associated with an oil-in-water emulsion at the non-equilibrium condition. The Winsor II behavior is associated with an water-in-oil emulsion, also at the non-equilibrium condition.

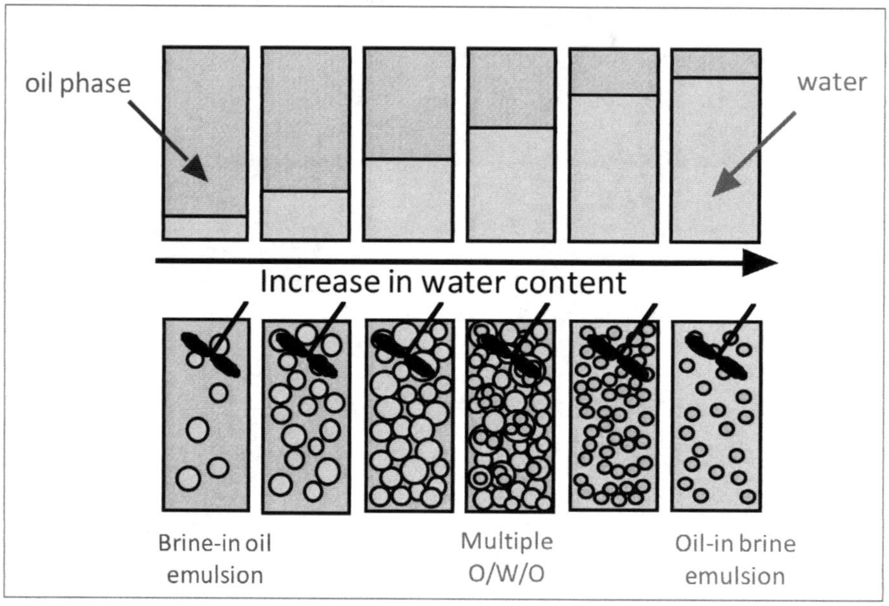

Figure 2 Schematic of the changes experienced by a water-in-oil emulsion (OBM) with addition of water-surfactant system

Materials and Test Procedures

The principal components of the spacer formulations used in this study are: (1) water; (2) the surfactant blend that produce an optimal formulation for oil microemulsification and removal; and (3) a biopolymer to adjust the rheological properties of the fluid and avoid fluid channeling during the displacement. Optional additives used are co-surfactants, potassium chloride, and/or barium sulfate (typically used to increase the density of the fluid). The composition of the spacer formulation is shown in **Table 1**.

Table 1 Composition of spacer formulation

Product	Composition (% by weight)
Water	88 – 92.25
Surfactant Blend	7 – 10
Biopolymer	0.75 – 2

The process of selecting a spacer formulation based on microemulsion technology needs to be done systematically, and normally includes phase behavior studies as functions of the composition and at various temperatures. Other important techniques that were used in this study include measurements of interfacial tension, contact angle, fluorescence microscopy, conductivity, and rheology.

Dynamic interfacial tension (IFT) is measured as a function of time when two immiscible fluids are in contact and one of the fluids contains surface-active material. The interfacial tension measurements were made using a SVT20 model Dataphysics Spinning Droplet Tensiometer. In these tests, the interfacial tensions were measured until the fluids reached equilibrium or until the drop of oil was completely solubilized by the microemulsion.

Conductivity measurements were used to evaluate the phase inversion from oil-continuous to water-continuous by monitoring relative changes in conductivity. A Tacussell-Prolabo Type CD 6N is used to measure the conductivity. This test is relatively straightforward in what is essentially a very basic titration. As the spacer inverts from oil-continuous emulsion (W/O) to water-continuous emulsion (O/W), this device measures the electrical conductivity of the fluid. The S/OBMs exhibit conductivity near zero, which increases as spacer fluid is added and the mixture shifts from oil-continuous to water-continuous. To determine the proportion of drilling fluid/spacer at which the inversion occurred, the 50 ml of S/OBM is placed in an empty mixing cup and mixed. As the S/OBM is mixed in the cup, spacer fluid is titrated in 1-mL increments, and the conductivity is recorded. The addition of spacer continues until the mixture reaches the phase inversion and a stable relative conductivity is achieved.

Fluorescence microscopy was used to study the mechanism of the phase inversion from oil-continuous (S/OBM) to water-continuous (O/W emulsion). The equipment used is a Light Microscope Leica DM R. Unweighted spacer fluid and solid-free drilling fluids were prepared to avoid interference from the solids during microscopy imaging studies. For fluorescence imaging, a green oil-soluble dye was added to the external oil phase. The dye is highly soluble in S/OBM and completely insoluble in an aqueous medium. The green fluorescence signal observed under the microscope indicates the presence of oil.

Contact angle measurements were used to evaluate the ability of the spacer fluid to change the surface wettability from oil-wet to water-wet during the displacement of NAF. The equipment used for these measurements is a Pocket Goniometer, Model PG-3. The measurements were made using glass slides treated with synthetic or oil-based drilling fluids, which turns the surface of the glass slides oil-wet. The process involved immersing the glass slide in the S/OBM for 10 minutes at selected temperatures. The glass slide was then slightly rinsed with oil that is similar to the S/OBM and then dried before the contact angle between the treated glass slide and a water droplet was measured

Fluid compatibility of S/OBM and spacers is important to accurate modeling of the displacement and aids in determining surge pressures, pump pressures, and equivalent circulation density (ECD). The difficulty in displacing S/OBM to any water-based fluid is due to the increased pressures observed because of the thick emulsion formed at the interface of the two fluids. Several drilling fluid/spacer mixtures were prepared and their rheology was measured at 75°F and 150°F using a TA Instruments Rheometer – Model AR-G2. From the expected rheological profile, one can model the displacement to determine appropriate flow rates, ECD, and pump pressures before to commencing the field application.

Vial tests were used to evaluate the cleaning efficiency of the spacer to remove the NAF from the glass surfaces. A series of vials with different ratios of spacer fluid to drilling fluid (from 90/10 to 10/90) were prepared and tested to determine the cleaning efficiency of the spacer. After dynamic mixing, the vials were allowed to statically age at temperatures of 40°F, 150°F, and 250°F for 16 hours. Afterwards, the vials were re-mixed by shaking and the contents poured out. A light rinsing with water to remove water-wet solids revealed the cleaning and water-wetting effectiveness of the designed pre-flush cement spacer with S/OBM contamination.

Results and Discussion

Phase Behavior

The phase behavior of systems that include base oil used in OBM, water, or low-salinity brine, together with a surfactant blend were evaluated over a temperature range of 77°F – 200°F. Short-chain alcohols and acids are optional additives in the formulation and were evaluated in the phase diagram studies.

The data obtained from the phase behavior studies was used to build the phase diagrams, allowing better understanding of capabilities and possible performance of the fluid formulated with a particular brine-surfactant-oil system. The Winsor I, Winsor II, Winsor III, and Winsor IV phases were identified in the phase diagrams built.

The selection of the type of system depends on the application of the formulation. For example, fluid formulations developed for wellbore displacement are Winsor I systems close to the Winsor III transition, that form an in-situ microemulsion (Winsor III or Winsor IV). In that case, it is important to have the phase diagrams, in order to determine how close the Winsor I systems are to the three-phase Winsor III systems where high-detergency and low interfacial tensions are present.

Figure 3 shows the phase diagrams built for brine-surfactant-oil system used to prepare one of the spacers for wellbore displacement. The ternary phase diagram was built with data obtained at 77°F and 180°F. These show that the boundaries of the transitions of Winsor for this particular system change with temperature. For this reason, the phase diagrams were developed as a function of temperature and salt concentration, which are the common variables in the application discussed in the paper.

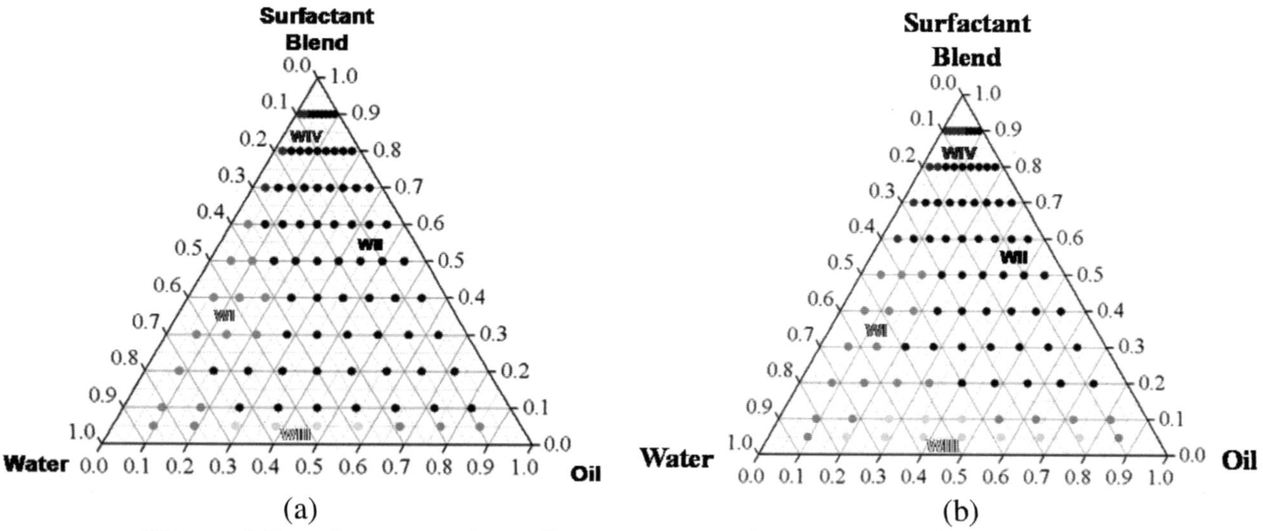

Figure 3 Pseudoternary phase diagram at (a) 77°F and (b) 180°F of a spacer system

Interfacial Tension (IFT)

The reduction of interfacial tension between the aqueous spacer and the oil phase is important in the selection of an aqueous surfactant solution to maximize cleaning and water-wetting efficiency. Ultimately, any reduction in the interfacial tension between the two phases will be dictated by the addition of surfactant(s) over a range of compositions and temperatures corresponding to expected field conditions. Selection of the proper surface active agent(s) or surfactants that can reduce the IFT between the aqueous and S/OBM phases will play a significant role in the overall effectiveness and ability of any displacement spacer system to water-wet, clean, and perform in a prescribed wellbore environment. **Figure 4** presents the dynamic interfacial tension between a spacer formulation selected from the phase diagram of Figure 3 and the oil phase of the OBM. Figure 4 shows the IFT measured when the spacer formulation (without solids) and the oil phase of the OBM (formulated with mineral oil) are in contact at 150°F. The spacer fluid exhibited two orders-of-magnitude lower interfacial tension with the OBM compared to the interfacial tension between water and the SBM (17 mN/m). Very similar results were obtained in IFT measurements using other oils, such as diesel oil and isomerized olefins, that are used to formulate S/OBM.

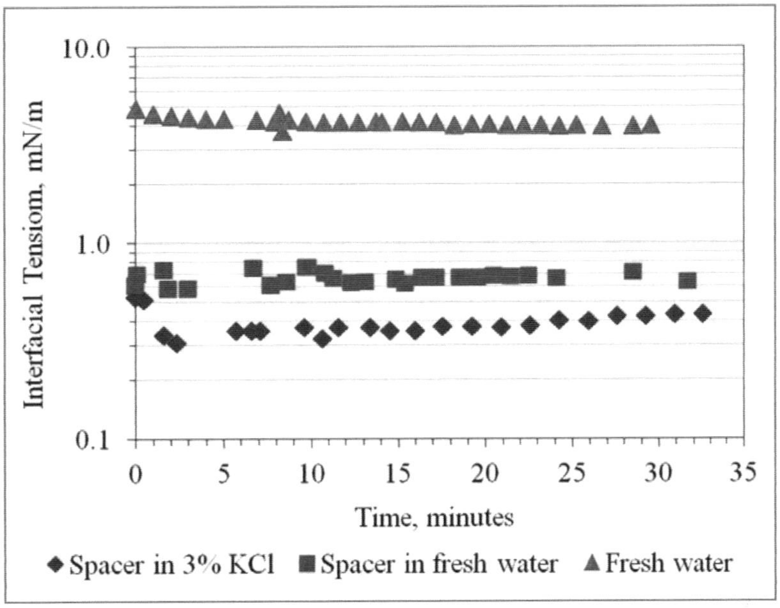

Figure 4 Interfacial tension between spacer and olefin-based oil

Conductivity Test

The conductivity of OBM titrated with spacers and the conductivity of the neat fluids are shown in **Figure 5** and **Table 2**. If the spacer is formulated with a surfactant-water system such that it goes through an optimum formulation when the OBM is added, a very acute phase inversion will be produced. This phase inversion is observed in the conductivity measurement of the OBM titrated with spacers, creating a steep slope, which leveled off when the inversion was completed and the mixture turned into a water-continuous fluid (**Figures 5a and 5b**). The results obtained with the two spacer formulations, one in a freshwater-based fluid and the second one in 3% potassium chloride (KCl) brine, were very similar. As can be seen in the data plotted in Figure 5a and 5b, once the concentration of the spacer reached approximately 45%, the mixture of fluids began to shift phase from the W/O emulsion to the O/W emulsion. This is the key for the spacer's ability to clean the S/OBM from downhole surfaces. Upon reaching approximately 50% of spacer, a very acute change in conductivity is produced, indicating that the phase inversion is completed and the mixture is water-continuous. Based upon the percentage of spacer fluid required to begin the phase inversion, one can draw conclusions about the volume of spacer required to change the system from oil-continuous to water-continuous. Both spacer fluids show a small increase in the percentage of fluid required to trigger the phase inversion when the temperature was increased from 77°F to 150°F. In addition to the cleaning of the OBM, once the phase inversion is completed, the wettability of the solid surfaces in the well should turn from oil-wet to water-wet, due to the very hydrophilic characteristic of the surfactants used in the spacer fluid.

Table 2 Conductivity values of fluid used in the mixtures

Fluid	Conductivity, mS
Water	0.00115
3% KCl in Water	34
Spacer formulated in fresh water	2.5
Spacer formulated in 3% KCl brine	3.3

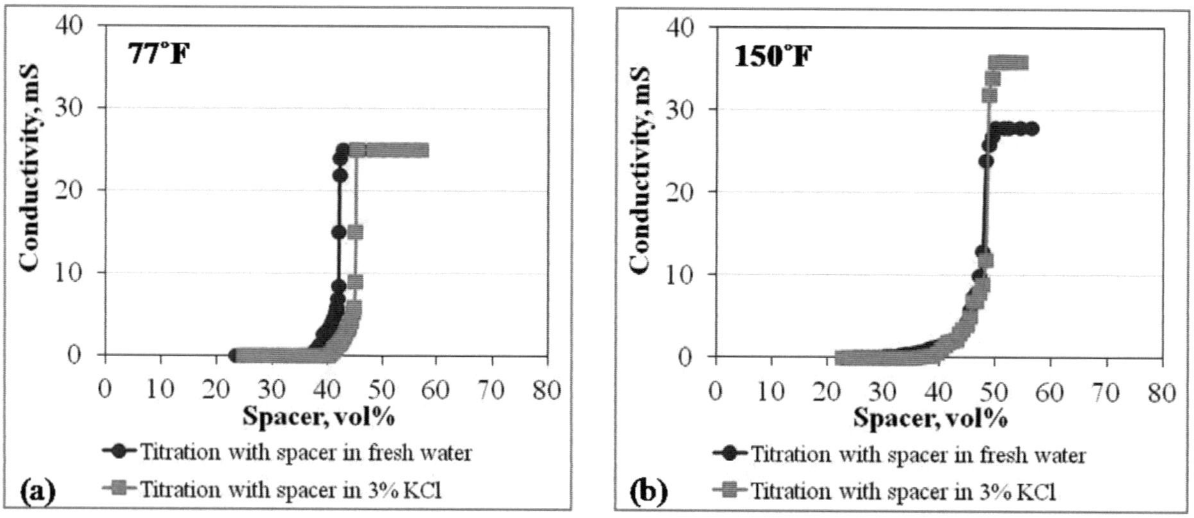

Figure 5 Conductivity of the spacers with OBM at (a) 77°F and (b) 150°F

Fluorescence Microscopy

Different ratios of spacer fluid to OBM mixtures were investigated using fluorescence microscopy to observe the steps from W/O emulsion to O/W emulsion passing from a micro-emulsification process that occurred at the phase inversion. **Figure 6a** show a microscopy photograph of the invert emulsion drilling fluid titrated with 10% of the spacer fluid. This clearly shows the presence of water droplets inside the oil-continuous fluid (W/O). As the titration continued the size of the water droplets in the internal phase increased (**Figure 6a and 6b**). At a spacer/OBM ratio around 42/62, spontaneous microemulsification began and an O/W emulsion coexisting with a W/O emulsion were observed (see **Figure 6d and 6e**) at the point of phase inversion. **Figure 6f** shows the O/W emulsions that resulted from the phase inversion. These microscopy images confirm the phase inversion of the fluids from oil-continuous to water-continuous.

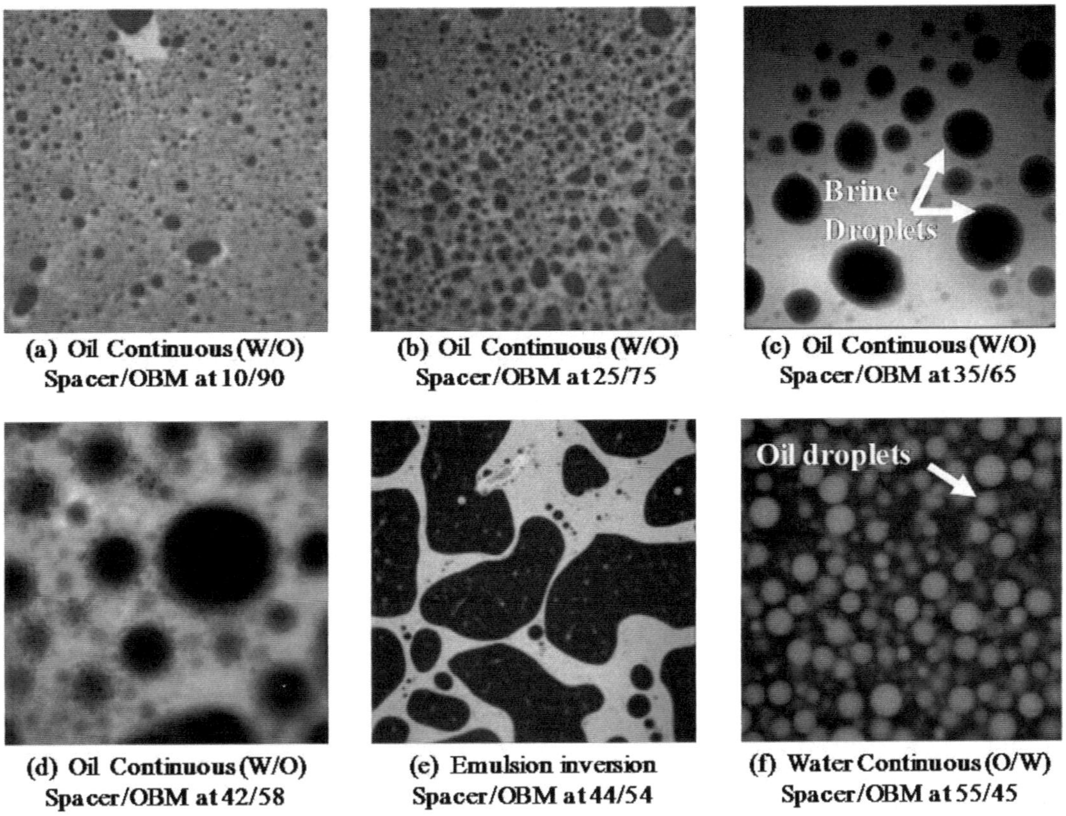

Figure 6 Fluorescence microscopy images of OBM/spacer samples

Contact Angle

Contact angle measurements were used to evaluate the ability of the spacer fluid to change the surface wettability from oil-wet to water-wet during the displacement of S/OBM.

Figure 7 shows the results of contact angle measurement on glass slide coupons. The coupon exposed to OBM exhibited a contact angle of 74° (

Figure 7a) which is much higher than the contact angle of the clean coupon (25°). This result proved that the originally water-wet surface becomes oil-wet after being exposed to the OBM.

Figure 7b shows that the oil-wet surface turned highly water-wet after treatment with the spacer fluid, as the contact angle is reduced to near 0°.

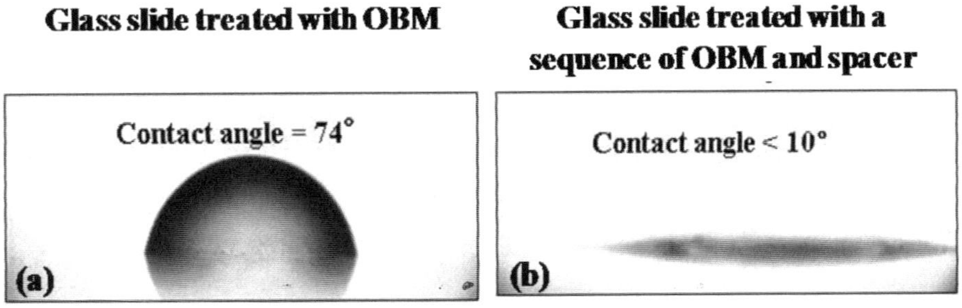

Figure 7 Contact angle between water drop and glass slide: (a) exposed to the OBM and (b) exposed to sequence of OBM and spacer

Fluid Compatibility: Rheology of Spacer/Drilling Fluid Blends

The major objective of these tests is to determine if the spacer/OBM blends produce viscosities higher than the viscosity of the neat OBM and spacer fluids. If spikes in viscosity of the blend are observed, the spacer formulation is not suitable for field application because high viscosities generate hydraulic problems during the pumping of the spacer for fluid displacement.

Figure 8 shows the viscosity as a function of the shear rate of the OBM, the spacer formulations and the blends of spacer/OBM. The spacers were formulated with the appropriate concentration of biopolymer in order to obtain a viscosity a little higher than the viscosity of the OBM

The spacer/OBM blends show viscosities lower than the neat fluids which results from using a spacer formulation selected from the area of the phase diagram corresponding to microemulsions Winsor III or Winsor IV.

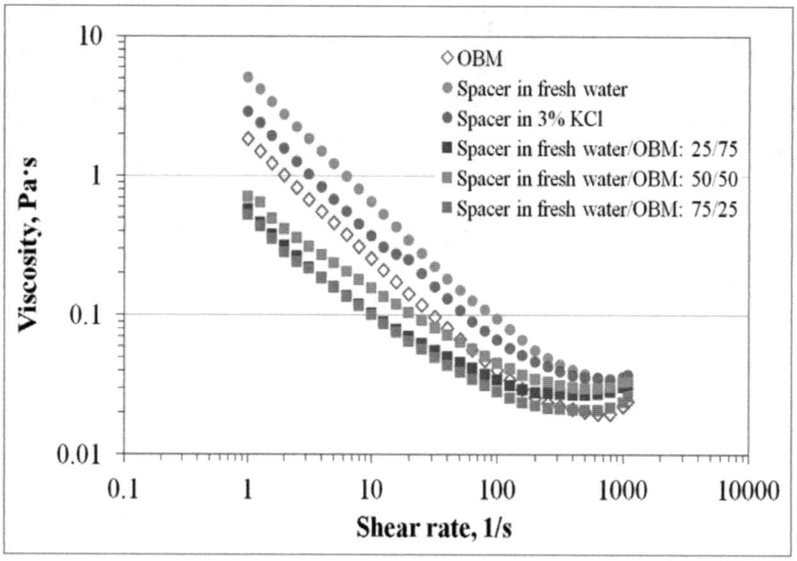

Figure 8 Rheology of various spacer/OBM ratio and the net fluids

The results of the measurements of the viscoelastic moduli G'and G" in the three spacer/OBM blends (70/30, 50/50, and 30/70) are presented in **Figure 9**. Results for all three fluid mixtures show the presence of viscoelasticity. The G' values, which are the elastic modulus, are greater than G" values, which are the viscous modulus. Typically, for an OBM, the ratio G'/G" is close to 3.5. When the spacer is mixed with the OBM the ratio decreases from about 3.5 to 2.3 (spacer/OBM:30/70) until the transition is reached (spacer/OBM:50/50). After the inversion, the ratio G'/G" increases and the fluid becomes more elastic, which decreases the necessary force required to cause irreversible deformation in the fluid's internal structures.

Another important bit of information from Figure 9 is the crossover point variation. When the spacer concentration in the mixture increases from 30% to 50%, the crossover point is moved to the left, indicating that the fluid is becoming more viscous than elastic (Figure 9b). However, after the inversion (Figure 9c), when the external phase is water, the crossover point increases (Figure 9a) and the internal structures recovered, indicating that the fluid is becoming more elastic and closer to an OBM behavior.

The lower viscoelasticity is expected around the phase inversion. The results are consistent with the conductivity and microscopy observations around the fluids ratio of 50/50.

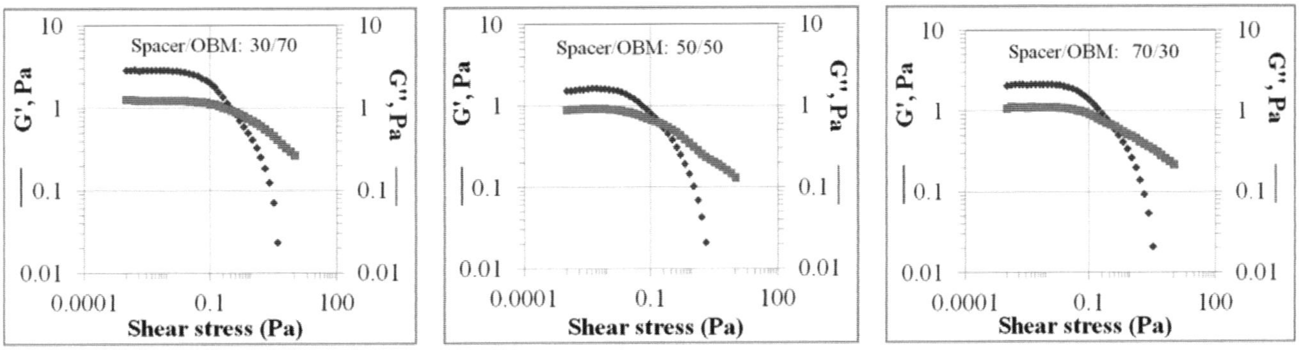

(a) **(b)** **(c)**

Figure 9 Rheological properties of the spacer/OBM blends: 30/70, 50/50 and 70/30

Cleaning Evaluation of Oil-Based Mud

A series of vial tests that simulated the drilling fluid/spacer interface were carefully prepared with S/OBM and the spacer fluid at ratios between 50/50 and 90/10. **Figure 10** shows the appearance of the vials after following the sequence described in the test procedure (mixing, ageing, pouring out and a slight rinsing). After the dynamic mixing, the vials were allowed to statically age at temperatures of 150°F, and 250°F for 16 hours. A light rinsing with water to remove water-wet solids revealed the cleaning and water-wetting effectiveness of the selected spacer with OBM contamination

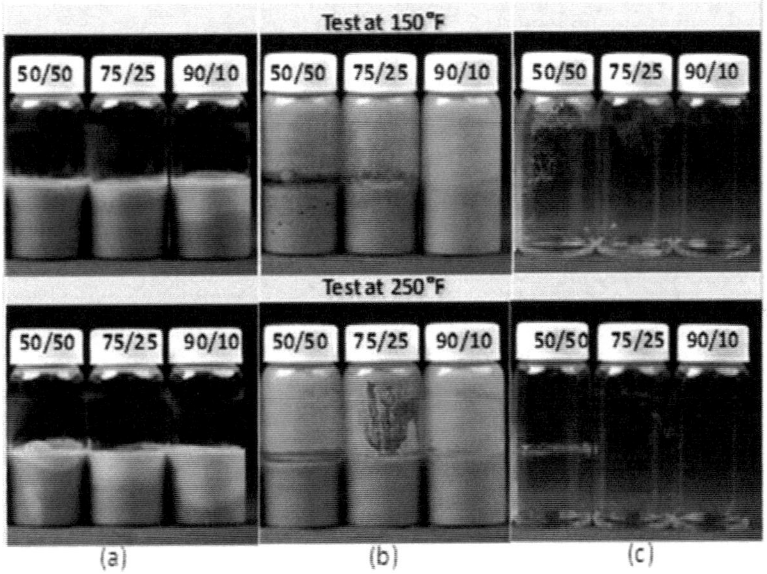

Figure 10 Cleaning test at various ratios of spacer/OBM in glass vials (a) Initial, (b) after 16 hour aging, (c) after slight rinse

Figure 11 illustrates the effect of changes in base oil type and composition using vial tests. In the case of the tests with oil-based fluid formulated with an olefin-based oil, the clean appearance of the vials at all the fluid ratios evaluated is an indication that the spacer is capable of inverting the oil-continuous fluid to a water-continuous fluid even at very high levels of SBM. There was no observable oil residue in the glass surface and the glass surface seemed to be water-wet after covering the steps of the test procedure. In the case of oil-based mud formulated with diesel, the same spacer formulation exhibits good performance with spacer/OBM proportions higher than 75/25. This is not surprising, because the equivalent alkane carbon number (EACN) is one of the variables that affect the phase inversion in a water-surfactant-oil system. The EACN of the olefin used is 15.3 and the EACN of the diesel is 11.9.

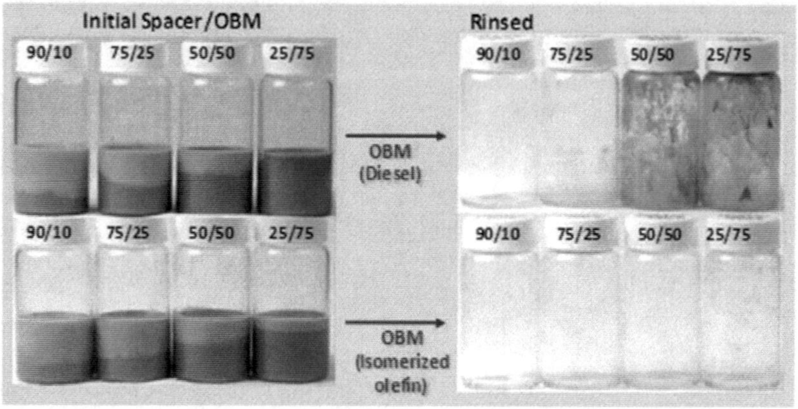

Figure 11 Cleaning effectiveness of the pre-flush spacer with two types of base oils at 250°F

Conclusions

- Relative conductivity, contact angle, and microscopy test results prove that the spacer fluid formulated based on phase behavior around the optimal formulation or Winsor III system produces a phase inversion of the S/OBM from oil-continuous to water-continuous.
- Vial cleaning tests prove that the spacer formulation evaluated in this study is not sensitive to changes in temperature up to 250°F, and that the spacer water-wets surfaces even with high S/OBM contamination, especially with SBM.
- In summary, the data discussed in this paper gives evidence that the spacer system formulated with surfactants that produce a rapid transitional phase inversion (1) removes the S/OBM, (2) is compatible with the drilling fluid, and (3) completely water-wets all solid surfaces, which achieves successful displacements of synthetic or oil-based drilling fluids.

Acknowledgments

We thank the management of Baker Hughes for allowing us to publish this paper. Special thanks to Daniel Ellis, and Qusai Darugar, from Baker Hughes; and Maria Antonieta Arandia, from Universidad de Los Andes; for their contribution in the lab tests.

References

1. Kelessidis, V.C., Guillot, D.J., Raggerty, R., at al. (1996) "Field Data Demonstrate Improved Mud Removal Techniques Lead to Successful Cement Jobs". SPE 26982, *SPE Advanced Technology Series,* Vol. 4: 53-58.
2. Newhall, C. (2006) "Improving Cement Bond in the Appalachian Basin with Adjustments to Preflush and Spacer Design". *Paper SPE 104576 presented at SPE Eastern Regional Meeting*, Canton, Ohio, October 11-13.

3. Javora, P.H., Baccigalopi, G., Sanford, J. et al. (2006) "Effective High-Density Wellbore Cleaning Fluids: Brine-Based and Solids-free". *Paper IADC/SPE 99158 presented at the IADC/SPE drilling Conference,* Miami, FL, 21-23 February.
4. Curtis, J. (2003) "Environmentally Favorable Terpene Solvents Find Diverse Applications in Stimulation, Sand Control and Cementing Operations". *Paper SPE 84124 presented at the SPE Annual Technical Conference and Exhibition,* Denver, Colorado, October 5-8.
5. Robles, J., Criado, M.A., Jensen, E., et al. (2005) "Dynamic Mud-Cake Removal Evaluation Under Annulus Hydrodynamic Conditions". *Paper SPE 95058 presented at the SPE Latin American and Caribbean Petroleum Engineering Conference*, Rio de Janeiro, Brazil, June 20-23.
6. Frigaarrd, I.A., Allouche, M., and Gabard-Cuoq, C. (2001) "Setting Rheological Targets for Chemical Solutions in Mud Removal and Cement Slurry Design". *Paper SPE 64998 presented at the SPE International Symposium on Oilfield Chemistry*, Houston, Texas, February 13-16.
7. Biezen, E., Van Der Werff, N., and Ravi, K. (2000) "Experimental and Numerical Study of Drilling Fluid Removal from a Horizontal Wellbore". *Paper SPE 62887 presented at the SPE Annual Technical Conference and Exhibition*, Dallas, Texas, October 1-4.
8. Quintero, L., Christian, C., Halliday, et al. (2008) "New Spacer Technology for Cleaning and Water Wetting of Casing and Riser". *Paper AADE-08-DF-HO-01 presente at the AADE Fluids Conference and Exhibition*, Houston, Texas, April 8-9.
9. Salager, J.L., Marquez, L., Peña, A., and al. (2000) "Current Phenomenological Know-How and Modeling of Emulsion Inversion". Ind Eng. Chem. Res. **39**: 2665-2678.
10. Salager, J.L., Forgiarini, A., Lopez, J.C., and al. (2004) "Dynamimic of Near-Zero Energy emulsification". *Paper # 203 presented at 6th World Surfactant Congress CESIO.* Berlin, Germany June 21-23.
11. Salager, J.L., Forgiarini, A., Marquez, L., et al, (2004) Using Emulsion in Industrial Process. Advances in Colloid and Interface Science, 108-109: 259-272.
12. Salager, L (2006) "Emulsion Phase Inversion Phenomena", Emulsion and Emulsion Stability. Chapter 4, Ed by Johan Sjoblom, CRC press by Taylor & Francis Group LLC,185-226.
13. Winsor, P.A., (1954) "Solvent Properties of Amphiphilic Compounds", Butterworth, London.
14. Scriven S. (1976) "Equilibrium Bicontinuous Structure". Nature, 263, 123-124.
15. Clausse, M. Peyrelasse, J. Heil, J., et al. (1981) "Bicontinuous Structures in Microemulsions". Nature, 293, 636-638.
16. Holmberg, K., (1998) "Quarter Century of Progress and New Horizons in Microemulsions" in Micelles, Microemulsion, and Monolayers, Science and Technology, ed. by D.O. Shah, Marcel Dekker, Inc, New York, 161.
17. Ezrahi, S., Aserin A., and Garti N. (1999) "Aggregation Behaviour in One-Phase (Winsor IV) Microemulsion Systems", in Handbook of Microemulsion Science and Technology, ed by P. Kumar and K.L. Mittal, Marcel Dekker, Inc, New York 185.
18. Salager, J.L and Anton R. (1999) "Ionic Microemulsions", in Handbook of Microemulsion Science and Technology, ed by P. Kumar and K. L. Mittal, Marcel Dekker, Inc, New York 247.

(28e) Mass Transfer Effects in Bore-Hole Filter-Cake Cleanup

Jeffrey Harwell, Chemical Engineering, University of Oklahoma, Norman, OK and David Sabatini, Civil Engineering and Environmental Science, University of Oklahoma, Norman, OK

A microemulsion-based fluid designed for removal of bore-hole filter-cake removal was studied to determine the dominant mechanism of cake dissolution. Two distinct regimes were identified as a function of temperature and pressure. At low pressure calcium carbonate dissolution to release carbon dioxide created interfacial turbulence that dominated the interphase mass transfer to such an extent that carbonate dissolution rates predicted the cake break-up time. At increased pressure, however, carbon dioxide generation no longer correlated with the cake breakup time, and oil-solubilization, dominated by surfactant diffusion rates, became the rate determining step.

(35a) Advances in Produced Water Treatment

Timothy P. Daigle, Savanna Hantz, Brian Phillips and Rafique Janjua, Fluor Offshore Solutions, Sugarland, TX

Today's Oil and Gas Industry is being confronted with having to manage production with increasing amounts of produced water from production wells. In fact, many engineers throughout the industry believe that good produced water management is the single most important aspect to extending the life of a producing reservoir.

With the increasing regulations and changing requirements found throughout the industry, the need for new techniques and better technologies is even more important. This manuscript will help identify some of the latest techniques and most advanced technologies that are changing the produced water treatment industry throughout the world. The presentation will highlight the significant use of hydrocyclones, compact flotation units and polymers that are making coalescence easier and more reliable.

This manuscript will review the latest produced water treatment technologies used throughout the topside offshore production industry. It will highlight the latest technologies used in the produced water treatment industry today and discuss their potential use in the subsea environment. The paper will illustrate those discussed concepts to perform subsea produced water treatment and will look at the many challenges and gaps towards making this technology viable and effective for subsea.

The paper will provide the newest advances in produced water treatment and discuss the gaps and challenges to applying their use at the seabed in ultra deepwater environments.

Research and compiled information will be shown and presented to highlight the latest technologies for produced water treatment and their potential use at the seabed in the ultra deepwater production systems.

As the produced water from oil and gas production continues to increase, new advances in produced water treatment systems will help the industries water handling abilities and seek to extend the life of a producing reservoir. This paper will focus on the three main benefits:

1.) Significantly reducing the cost of the fluid transportation system from the well site to the host platform facility.

2.) New produced water treatments systems are getting more reliable and compact, which is improving the water handling capabilities of many production systems and providing a smaller and lighter footprint on production facilities.

3.) How can the new and improved produced water treatment systems be utilized at the seabed to help reduce the amount of treatment required on the surface and to improve subsea production.

(35b) Advanced Pipeline Monitoring for Flow Assurance with Fiber Optics

John D. Hedengren, Chemical Engineering, Brigham Young University, Provo, UT and David Brower, Astro Technology, Inc., Houston, TX

Bass Lite deepwater field in the Gulf of Mexico, at water depths of approximately 6,750 feet, commenced operation in February 2008. Natural gas is produced from Bass Lite via a 56-mile subsea tieback to the Devils Tower Spar. This project involved several innovations, one of which was the incorporation of a sensing system that measures real-time temperature, pressure and strain along the pipeline length. This is a first of its kind innovation that is in actual operation.

This report details the installation of the advanced instrumentation and resulting operational data from startup. To obtain the data, fiber optic sensors were used in this application for the long-distance signal transmission capability. In addition, fiber optic sensors are attractive in deepwater because of their multiplexing capability, immunity to electromagnetic interference, and ruggedness. Topside software provides analysis of the optical signals and displays data at platform locations and on-shore sites in real-time. As part of this project, a new web-based signal protocol was developed to increase the reliability of transmission over an intermittent connection with high latency. As the data is streamed to an on-shore server location, remote viewing capability has allowed operators, technicians, and management to retrieve current and historical data. In addition to viewing streaming data, analysis is performed to detect pig location, hydrate build-up, plug locations, and sensor validation. The Advanced Process Monitoring (APMonitor) software has enabled effective trouble-shooting and boundary value management.

The Upstream industry presents many opportunities for utilizing measurement technology to monitor the long term reliability of production systems. In particular, deep-sea pipeline monitoring poses a challenge due to the remote environment, intermittent weather incidents, and gradual fatigue factors. There is a desire for improved monitoring of existing and new projects to give insight into the conditions that lead to failure. Analytical models utilize the data to monitor the operational integrity for flow assurance and riser integrity. By using an optimization framework, a mathematical model simulates the corresponding measured values to present detailed information about the production system dynamics. This optimization framework uses a receding horizon of process measurements to capture the changing process conditions. The Advanced Process Monitoring (APM) approach has been utilized in the Downstream and Chemicals industry for a number of years and finds new application in monitoring of Upstream production systems. APM attempts to optimally estimate the true state of the dynamic system, given a real-time stream of measurements and a model of the physical process.

This monitoring is accomplished with a variety of measurement devices that give axial, hoop, and reference pipeline strains at multiple sensor locations. When viewed individually, they offer some insight into the dynamics of the production platform. When intregrated together with an analytical model, the combined set provides a holistic view for flow assurance. Pressure, temperature, and flex strains are updated at 10 Hz frequency. This is important to accomplish objectives related to flow assurance, hydrate buildup monitoring, plug detection, and well-head valve monitoring.

As an extension to this project, Advanced Process Control (APC) is currently under investigation for platform gas processing, pipeline pressure, and pipeline temperature control. Preliminary results from this feasibility study are presented as a simulation of actual operational improvements. In particular, with a 56 mile sub-sea tieback line, there is significant flow, pressure, and heat transfer dead-time. Using a model-based approach, the delay is explicitly considered, leading to tighter control during start-up, shut-downs, and throughput changes.

As deepwater fields in the Gulf of Mexico, West Africa, offshore Brazil, and elsewhere are developed it becomes increasingly important to detect crucial engineering properties of subsea equipment and environmentally induced conditions. Such technology as described in this paper provides actual operational information and data for improved design. Advanced detection and mitigation of structural failure can greatly reduce the risk of unplanned hydrocarbon release.

FMC Technologies

Real-time Monitoring of a Subsea Boosting System

Phaneendra Kondapi and Anita Sengebusch
FMC Technologies, Houston, TX 77067

Abstract

Real time processing of the data from sensors along the flow path is vital in making decisions on flow optimization and operating the flow boosting equipment. The real-time flow metering, advisory and optimization software reads data from the sensors, performs flow calculations and provides the field operators and engineers with data on flowrates, densities, viscosities, pressures and temperatures of oil, water and gas along the flow path at several points of interest to optimize performance of the boosting system.

The application of real-time virtual metering software provides operation flexibility in managing a subsea boosting system having multiphase subsea electrical submersible pumps and recycle lines. The virtual metering software contains models of the pumps and the recycle lines and based on the sensor data, the pump models calculate variables like pump head, flowrates, shaft power etc. The virtual metering software also displays the pump operating point against pump operating curves (Speed and Power) on web pages which are enabled by the inline web browser links on the Human Machine Interface (HMI) screens. The operator is fed with the information needed to operate the field such that the pump operating point stays within the boundaries of the customer specified preferred operating envelope.

Objectives

The objective of the paper is to estimate the boosting system operating points and send the information to the operators to monitor the values to stay within boundaries of the operating envelope in order to maintain the safe operation of the pump.

Introduction

The virtual metering software is used to calculate the flow rates and other process information at the inlets, inter-stage and outlets of the subsea boosting systems using the sensor data supplied by the subsea control system. The production tree sensors that could be utilized for calculating production flow rates include:

- Pressure and temperature transducer located upstream of the choke valve
- Pressure and temperature transducer located downstream of the choke valve

- Pressure and temperature transducer located down hole
- Choke valve position transmitter

Based on these sensor measurements, accurate flow rates for all phases could be calculated real-time using the virtual metering thermo-hydraulic model, which also takes into account the production fluid properties, the properties and geometry of the production tubing (from perforations to wellhead), choke valve characteristics, and heat transfer with the ambient environment. Wellbore and subsea system models of the pumps are used to calculate the flow rates and fluid properties at the pump station.

Model Inputs

The input data required to configure and build the model is:

- Geometry of Well (Tubing and Casing), Jumpers and Flowlines
- Wall material properties (Well and other piping)
- Sea water velocity and temperature profile
- Geothermal gradient
- Recycle loop geometry and wall properties
- Production choke curve data
- Recycle loop choke curve data
- Pump performance data
- PVT fluid properties
- Pressure / temperature from the down hole PT/TT sensors
- Pressure/temperature from the PT/TT sensors upstream and downstream of the production choke and choke position
- Pressure/temperature from the PT/TT sensors at the inlet and outlet of the pump station, pump speed and recycle choke position
- Flow information (GOR, WC etc.,) from the subsea meters

Real-time virtual metering software output data

The calculated data from model that is being sent as output to the display are:

- Actual volumetric flow rates at pump inlets, inter-stage and outlets (oil, gas, water, mixture)
- Standard volumetric flow rates at pump inlets (oil, gas, water, mixture)
- Viscosity at pump inlets and inter-stage (oil, gas, water, mixture)
- Density at pump inlets and inter-stage (oil, gas, water, mixture)

- Pump head (calculated from pump performance correlations)
- Fluid pressure and temperature at pump inlets, inter-stage and outlets

In addition to the above, the software HMI web pages also display the pump operating point on the pump curves both on Head curves as well the as the Power curves. The HMI templates for Operating Parameters and Operating Points on Head and Power Curves are shown in the following figures.

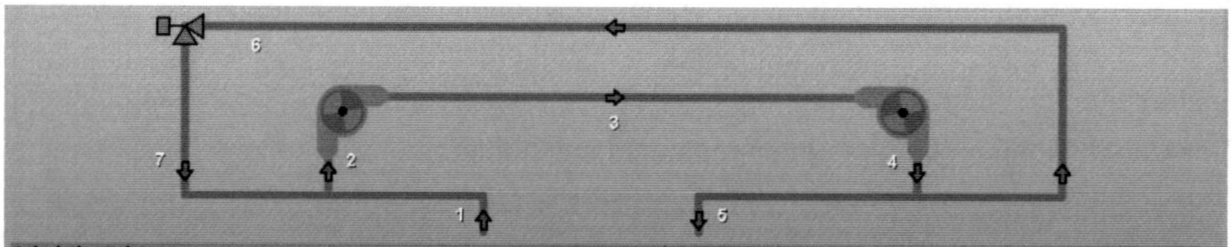

Calculation Point	1	2	3	4	5	6	7	Unit
Oil flow rate (Std Conditions)	0.345	0.345	0.345	0.345	0.345	0.345	0.345	stb/d
Water flow rate (Std Conditions)	0.086	0.086	0.086	0.086	0.086	0.086	0.086	stb/d
Gas flow rate (Std Conditions)	1903435	1903435	1903435	1903435	1903435	1903435	1903435	MMscfd
Pressure	7.14	7.14	7.14	7.14	7.14	7.14	7.14	psia
Temperature	45.4	45.4	45.4	45.4	45.4	45.4	45.4	degF
Oil flow rate (In-Situ Conditions)	0.40	0.40	0.40	0.40	0.40	0.40	0.40	bbl/d
Water flow rate (In-Situ Conditions)	0.09	0.09	0.09	0.09	0.09	0.09	0.09	bbl/d
Gas flow rate (In-Situ Conditions)	0.37	0.37	0.37	0.37	0.37	0.37	0.37	MMcfd
Total flow rate (In-Situ Conditions)	0.49	0.49	0.49	0.49	0.49	0.49	0.49	bbl/d
GVF	0.96	0.96	0.96	0.96	0.96	0.96	0.96	%
Oil density	49.58	49.58	49.58	49.58	49.58	49.58	49.58	lb/ft3
Water density	59.47	59.47	59.47	59.47	59.47	59.47	59.47	lb/ft3
Gas density	5.35	5.35	5.35	5.35	5.35	5.35	5.35	lb/ft3
Mixture density	51.96	51.96	51.96	51.96	51.96	51.96	51.96	lb/ft3
Oil viscosity	1.34	1.34	1.34	1.34	1.34	1.34	1.34	cP
Water viscosity	0.25	0.25	0.25	0.25	0.25	0.25	0.25	cP
Gas viscosity	0.02	0.02	0.02	0.02	0.02	0.02	0.02	cP
Mixture viscosity	1.08	1.08	1.08	1.08	1.08	1.08	1.08	cP

Figure 1. Operating Parameters

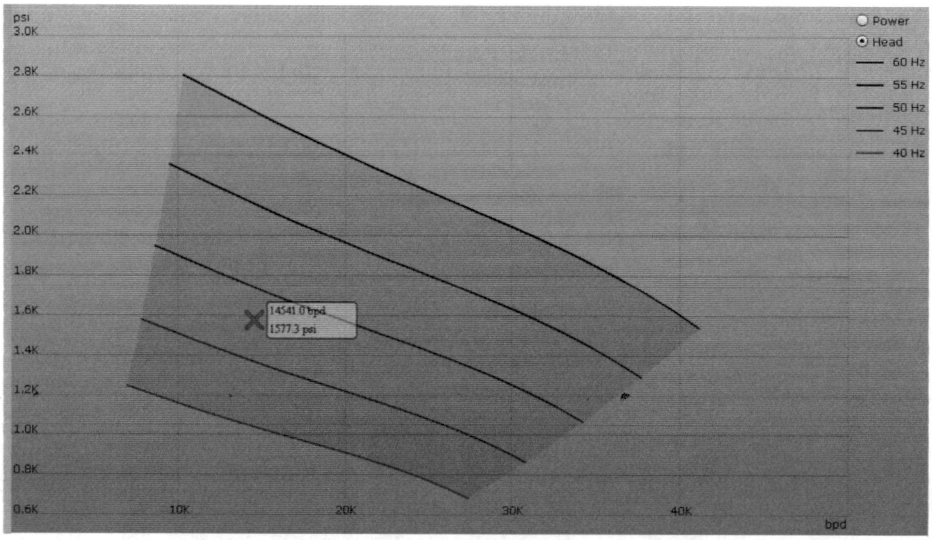

Figure 2. Operating Point on the Head Curves

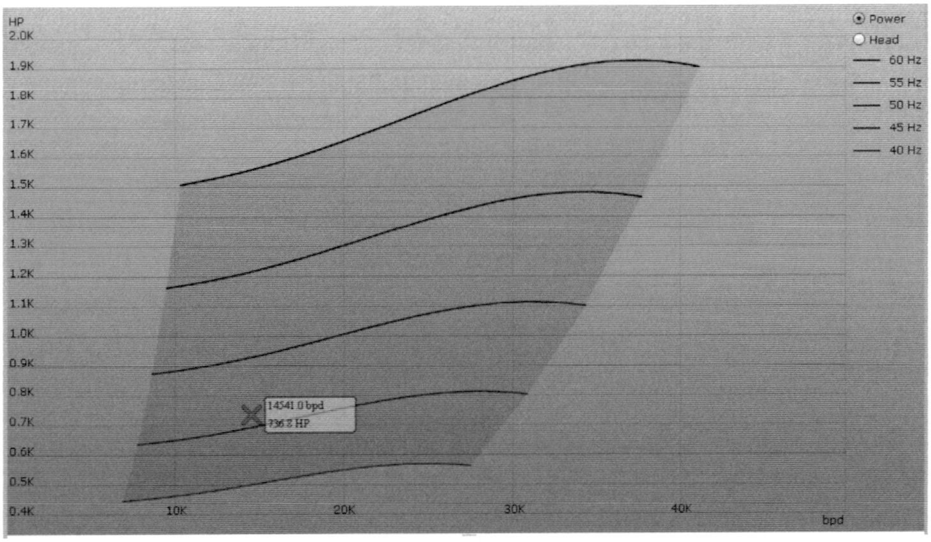

Figure 3. Operating Point on the Power Curves

(56a) Quantifying Flow Assurance Risks During Front End Engineering Design

Allan Rydahland Richard Shea, Flow Business Unit, SPT Group, Houston, TX

Engineering disciplines traditionally focus on providing the most accurate solutions based on the best available information at the time a study is undertaken. Further, it is common practice to refine and update such studies as data changes, based on changes in assumptions, decisions that are made and so forth.

This paper presents a new workflow and methodology which will extend the*one data set in, one exact solution out*engineering approach to a quantifiable, comprehensive risk analysis of the whole engineering solution space being investigated. Effectively, this methodology transitions the flow assurance engineering discipline from being deterministic to becoming probabilistic.

By utilizing a combination of sensitivity analysis, uncertainty analysis and optimization, it is possible to quantify the impact of input assumptions on key flow assurance decisions. This starts at a high level with the conceptual field layout and may progress into details such as field deliverability, pipeline insulation requirements and equipment sizing. The methodology has already been adapted in other engineering disciplines such as reservoir studies and long term production forecasting. It provides a consistent way of quantifying results from a comprehensive analysis of multiple data input scenarios and will lead to better results from FEED projects.

(56b) Multiphase Flow - the Flowing Phases Know What They Are Doing, but Do We?

Sam Kashou, Chevron, Houston, TX

Multiphase Flow in Wells and Pipelines is a complicated science. The production systems are not complicated if the flow is only two phase, however, that is never the case. As the oil and gas industry keeps pushing for deeper and longer production subsea tie backs, multiphase flow gets more and more complicated and harder and harder to predict. At high pressures and low temperatures and various flow rates, many Flow Assurance issues will develop. These include hydrates, wax, asphaltenes, and slugging tendencies, just to name a few.

Steady State models are no longer enough to predict flow behavior. In this presentation we will look at the nature of multiphase flow and shed some light on why it is important to predict what actually happens in the pipelines at flowing conditions. We will look at why multiphase flow is sometimes difficult to predict and what is the industry doing and the tools they use to solve the flow problems that develop.

Integrating Dynamic Simulation and Training for Efficient Design, Startup and Operation

Insuring Against Future Disasters

Norm Stewart, RSI Simcon, Inc., Houston, Texas

Biography

Norm Stewart is the VP, Sales & Marketing, for RSI Simcon, Inc., in Houston, Texas. He has over 25 years of international and US professional experience in the process industries and earned his degree in Chemical Engineering at Mississippi State University. He began his career at The Dow Chemical Company in Freeport, Texas, where he held numerous roles in research, engineering and operations. Norm went on to work in sales, marketing and international business development in the United States, Latin America and most recently, the Middle East. He works with clients to understand their underlying business challenges and how the creative application of simulation technologies and training methodologies can be applied to solve them. He is married with 4 children, 5 grandchildren and 8 great grandchildren.

Introduction

Oil and gas exploration and production companies invest heavily in sizeable assets that must be delivered at the lowest possible cost and then optimized to operate efficiently and safely. While much attention is paid to physical assets and the implementation of the latest technical innovations, operating personnel continue to be trained and developed using techniques perfected in, and suited for, the last century. In fact, training is often an afterthought; something that needs to be ticked off some unit-commissioning checklist. In light of this dichotomy we must ask ourselves – why is the training of old not suited for today, and what are the consequences of not training to competency, and what should we be doing differently?

Statement of Problem

If we take some time and look below the surface, we see some interesting trends. The often referenced baby boom generation has reached the time in their careers where they are retiring or finding some creative way to spend their lives. A direct consequence of this demographic shift is that young, inexperienced employees are being moved into positions of responsibility before they are fully prepared. Technologies being deployed today for finding and producing oil and gas are becoming more complex, so more brain power is needed on an individual basis just to keep up. On an international front, National Oil Companies (NOC) are developing quickly and need more resources for their own domestic and foreign assets.

Extending the life of fields with increasing rates of recovery means there are more and more installations and reservoirs to manage putting unseen pressure on the existing workforce. Combining all of these simply means we need more brains out there with higher capabilities, so experienced personnel are being petitioned to stay on for just a bit more while companies scramble to attract and develop new talent.

So what are the trends and what are we doing to address them? Instead of dealing with major risks and mitigating their effects, companies now have to consider more seriously how to avoid or contain disasters. New technologies and processes promise to deliver the results we seek, but we have to ask "but what about the training?" NOCs are growing beyond their borders, and we have to ask "but what about the training?". Since one solution does not fit all situations, increasingly content is being localized, so once again we ask "but what about the training?". To go where no man has gone before (to paraphrase a well-known movie line), the oil & gas companies demand new technologies, but the question still remains "but what about the training?". Training often becomes the responsibility of HR and that is a logical thing to do, or as companies have an eye toward reducing costs, procurement takes the lead so that a company gets what is available at the lowest possible cost. Because results are not where we would like to them to be, we now ask the universal question "but what about the training?' while emphasizing that it is now a serious issue. This paper positions the premise that a new way to develop people is needed. We propose that the marriage of sound adult learning methods with dynamic simulation technology provides the right platform for accelerating competencies and addressing a shrinking available workforce.

Competency and Certification

Over the past 5 years or so, managers and those responsible for HR have woven 2 words into their vocabulary to where now they just roll off into normal everyday speech, i.e., competency and certification. In some instances, these terms provide cover and show that "we are doing something". In some cases, the rational for competency development can be backed up by proofs and case studies, even guarantees of sorts. But what do we mean by competency and what kind of certificates? Let's look at some definitions.

The European mindset looks at competence in terms of trades.

- **French Wikipedia:** Competence is the **whole of** *knowledge – knowhow – experience* necessary to exert a **trade**

Whereas Americans tend to define competence in terms of what is needed for a particular job, hence the need for "job descriptions" and "job task analyses".

- **English Wikipedia:** Competence is a **standardized requirement** for an individual to properly perform a specific **job**

So no matter where you start in your thinking, we all have some idea of what competency means and even ideas on how to get there. The staff we have manning our platforms and rigs have all achieved a level of competency and certainly are seen as capable. So when a disaster happens, the natural reaction is simply to ask "Why?" knowing at least superficially that our people are competent. But are they? Could there be more below the surface when we look at the link between training and competence? We know these are indeed linked when we consider that poor training cannot deliver competent people.

Link between Training and Competence

Looking at the two definitions from above, we know that competence is built from 3 primary sources: Knowledge, Skills or Know How, and Behaviors. Each of these competency characteristics are obtained from 3 sources. Knowledge comes through formal training at schools or universities, or even some type of professional training. Skills develop through professional training and OJT, or On the Job Training. Finally, behavior comes about by adding to the previous mix some targeted soft skill training, coaching and most importantly, experience. The latter many times being "the best teacher", but often the most expensive when wrong decisions are made.

A 'new way to develop people'?

What is the best way to train? We can only answer that question when we realize what training is not. It is not "giving good habits and reflexes" as that approach emphasizes and builds exactly the behavior we are hoping to avoid. If we train only by having operators practice a pre-determined set of actions and reactions without understanding the underlying process fundamentals and without a defined way to analyze a situation, we are only encouraging repetition of the disasters we have all seen the past decade or so. A new way of improving performance of an asset focuses on the human side of the equation by combining proven Adult Training Methods with Dynamic Simulation.

Dynamic simulation tools provide a powerful operations and training platform. Using modern technology, one dynamic process model provides value all the way through a production asset's life that ranges from conceptual engineering to long-term operations. The same model provides the foundation and core of an effective training program.

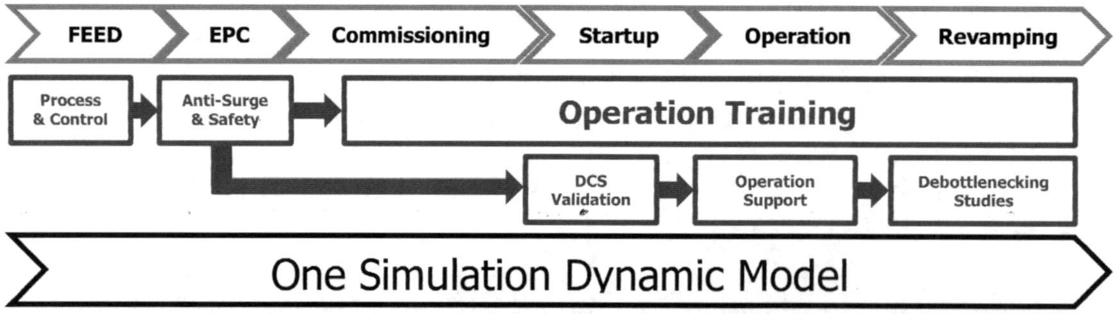

Figure 1 – One Dynamic Model from FEED through Operation

Figure 1 shows the logical progressive use of simulation tools based on dynamic models. Process and control studies uncover unknowns that often reflect the final design, testing of anti-surge strategies and safety systems often reveal additional design considerations, and validating control logic and configuration of a DCS early can avoid costly field expenses later. From there, the same system will often be used to improve processes by proving out ideas and debottlenecking applications. Positioned in the middle of this figure is the traditional use of simulators, i.e., as Operator Training System; however, the tool is of important, but limited, use without a sound training program built around it

Figure 2 graphically shows how each of the various aspects, or views, of a dynamic modeling system interact with each other. A burgeoning evolution of the dynamic model is 3 dimensional (3D) presentations. As shown here, the 3D view gives a sense of space so additional training can be delivered for such things as field operations required during startup.

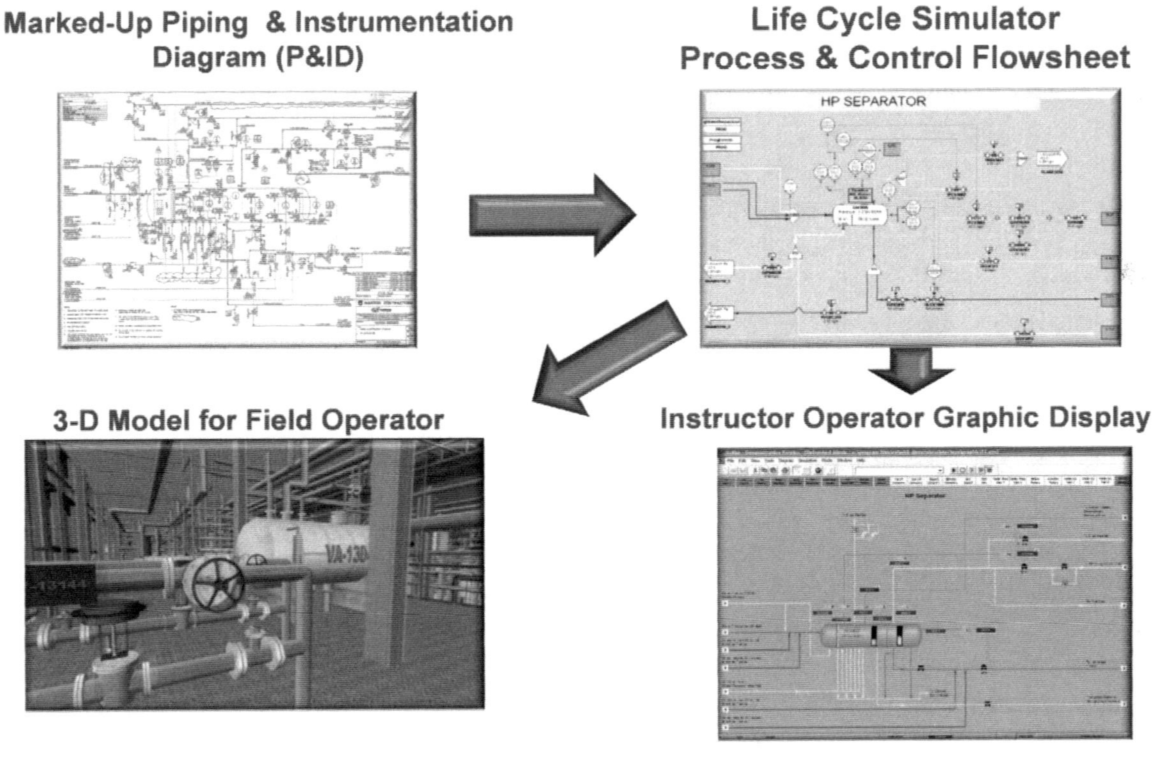

Figure 2 – Virtual Plant from Design to Operation

What is the new way?

High performance, safely operated assets have a common trait; the operations workforce receives structured training that incorporates continuous and creative uses of simulators. Traditional operator training techniques in use for decades continue to dominate the industry. As a result, operator performance achieves only minimal levels of competency.

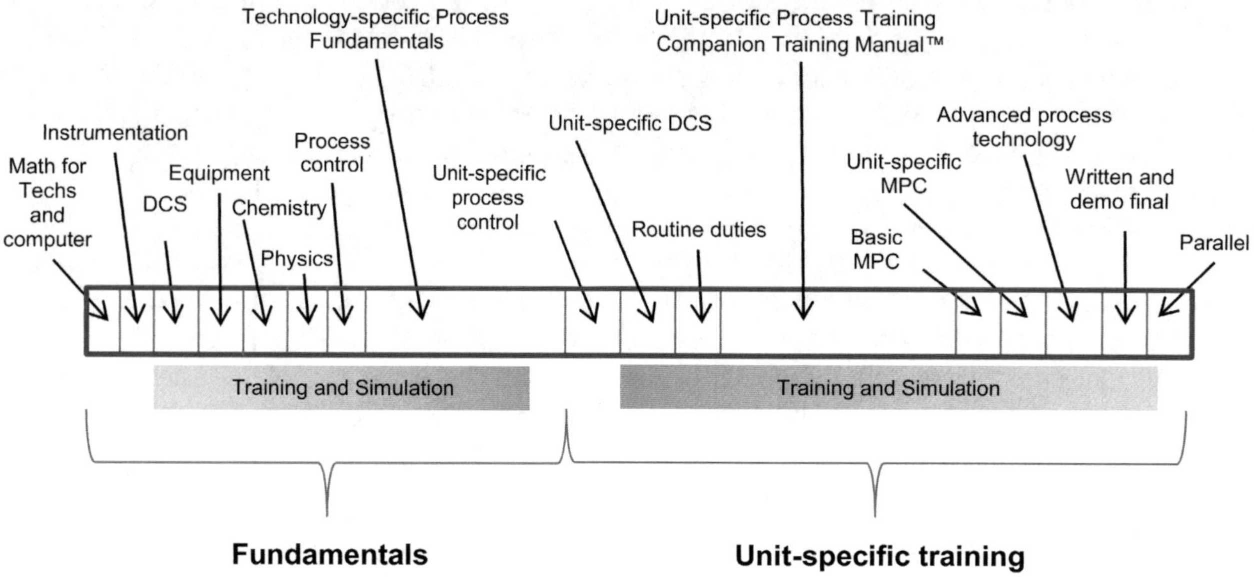

Figure 3 – Comprehensive Training Curriculum

A comprehensive training curriculum builds upon training methods that center upon learning the underlying process fundamentals and how to think. That is, learning how to analyze a situation by logically walking through the causes and effects of primary operating variables, then acting upon the most reliable conclusion. Compare this approach to one where the operator acts because he or she believes they have seen something similar before. A program such as described goes further by integrating dynamic simulators into the training so that the proper experiences are continually invoked to build knowledge and confidence in the operating staff.

What about certification?

At first blush, the word "certification" seems fairly straight forward, but, interpretation of the word varies from company to company. Some companies that equate certification to training have been known to award a certificate simply because the student attended a class; however, that is not exactly a good way to measure success and is only an internal recognition of attendance. An improvement is to award a certificate upon successful completion of an exam at the end of a course. When courses are isolated to the classroom and evaluations are only internal, we are still missing something. The best way is to award a certificate based on successful completion of a simulator based training program that measures performance in real world scenarios given and controlled by an independent third party. Only then can we declare an operator as truly qualified to run our particular unit operation or plant.

So what can we do to avoid the next disaster?

Industry demands that operators safely achieve high levels of production in an optimal and safe manner. This gap between demands and the capabilities imparted with outdated training methods must be addressed quickly and confidently. When simulation technologies and training programs are creatively integrated we can accelerate the acquisition of competencies to desired levels. When we can develop all individuals to think on their own and make decisions based on accurate observations and analyses, we can have a workforce that can safely operate a plant, optimize its production and recover more quickly from upsets. When our people know and understand the risks based on sound scientific principles, then they can dare to act and do so appropriately. Those who know they can dare to do right will care to do right.

RSI and IFP Training are examples of an adaptation to this changing market and the evolution of technology. These two sister companies that have historically marketed simulation systems and training independently have now merged to better respond to the demands of their customers. Together, as RSI Simcon, we are bringing to market integrated modeling technology and training methodologies to address concerns voiced by our industry.

(56d) Quantifying Asphaltene Impairment Risks During Front End Engineering Design

Stefan Smuk and Ian Roberts, Schlumberger, Sugar Land, TX

Asphaltenes precipitation and deposition occur within both reservoirs and production systems, posing serious operational and economic risks in numerous developments worldwide. Traditionally, these risks are assessed strictly on a thermodynamic basis. Laboratory tests of varying degrees of sophistication are used to identify asphaltenes-onset conditions, these are compared with the thermal-hydraulic and compositional path of the fluid system, and chemical intervention or process selection are specified accordingly. To the extent that such data are valid, this approach is "correct" in the sense that it provides for prevention or remediation downstream of the point of asphaltenes occurrence within a system. However, this approach cannot provide quantification of impairment rates or severity, as required for a rigorous, probabilistic cost-benefit analysis of competing management strategies. To help address this deficiency, deposition-kinetics analysis tools have recently come into use; however, their application tends to be semi-quantitative (as in chemical screening), and issues of scale-up and validation limit their direct applicability and generalization to FEED-scale system analysis. Finally, it is generally untenable to use a purely comparative/experiential approach to asphaltenes-management planning, because asphaltenes behavior is highly system-specific, and because historical operational data for asphaltenes impairment are generally very limited in quantity, quality and/or detail.

This paper presents a workflow methodology by which novel mechanistic modeling tools for asphaltenes deposition, within both porous-media and pipe/tubing flow environments, can be used to address the existing gaps in technology and field experience in planning for asphaltenes management. Using a combination of laboratory analysis, reservoir- and production-simulation tools, and sensitivity analysis, it is possible to quantify the range of probable levels of productivity impairment due to asphaltene phenomena, and their economic significance. This in turn provides a basis for optimization of the asphaltenes management strategy – which might include process design considerations, such as modifications to recovery and lift methods, along with direct intervention/remediation protocols such as periodic or continuous chemical inhibition, solvent treatments, or other stimulation, cleaning and workover operations. This approach can provide for greatly reduced risk and uncertainty associated with the FEED analysis of asphaltene-prone developments.

(56e) Designing and Delivering Integrated Control, Monitoring and Optimization Systems for Subsea or Topside Systems

Espen Storkaas, Olav Slupphaug and Bjorn Bringedal, ABB AS Norway, Oslo, Norway

For efficient operation of any remote production asset, be it an oil platform or a subsea tie-in with wellhead and potentially some subsea processing, advanced control and monitoring functionality is getting more and more important as the oil companies moves to more remote locations, deeper waters and more complex reservoirs. In addition, the pursuit of reduced operational expenditures drives the trends towards more of the operational support and monitoring being done from a centralized, onshore support facility using an Integrated Operations approach. These developments lead to new challenges for the design and delivery of control and monitoring systems

Sophisticated monitoring and control applications for wells and pipelines have been available from several vendors for some time. However, these applications have generally been stand-alone expert applications connected to a single subsystem, for example, a slug control solution for pipelines located in the plant control system or a virtual flow metering system for wells located in the subsea (control) system. The usage and benefit of these systems have therefore been limited. This paper outlines how monitoring and control applications for gathering networks should be structured in an integrated operations framework, and which benefits this will give for operators.

Structuring of the different applications ensures that data from monitoring applications are easily available for a large group of users while ensuring that closed-loop control applications retains the robustness and security that is required. Furthermore, recent developments of the applications itself, partially made possible by the modern integrated operations system topology with increased data availability, provides additional functionality not only for expert users, but for generalists as well. Finally, synergies between different monitoring and control applications can give additional value to the users.

(57a) Interfacially-Active Nanohybrids for Improved Oil Recovery

Daniel E. Resasco[1], Jeffrey Harwell[2], Santiago Drexler[3], Jimmy A. Faria[3], M. Pilar Ruiz[3], J. Baez[4]and Bor-Jier Shiau[5], (1)School of Chemical, Biological and Materials Engineering and Carbon Nanotube Technology Center, University of Oklahoma, Norman, OK, (2)Chemical Engineering, University of Oklahoma, Norman, OK, (3)School of Chemical, Biological and Materials Engineering, University of Oklahoma, Norman, OK, (4)University of Oklahoma, Norman, (5)School of Petroleum and Geological Engineering, University of Oklahoma, Norman, OK

Development of enhanced oil recovery (EOR) techniques is an important challenge for engineers and researchers working in the oil industry. In recent years, the use of nanoparticles for application of chemical EOR has attracted significant attention. In this contribution, a novel technique for potential enhancement of oil. The technique consists in injecting amphiphilic nanoparticles into the water. These hybrid nanoparticles can simultaneously act as emulsion stabilizers as well as carriers for catalytic species such as a metal that can be active for partial oxidation and hydrogenation reactions. Partial oxidation of organic compounds can lower the water – oil interfacial tension. Partial hydrogenation can enhance the viscosity of the oil phase in the emulsion, thus improving the mobility ratio. In addition, partial hydrogenation can be an effective pre – treatment of the oil to favor the subsequent partial oxidation.

We have investigated the application of nanohybrid particles that preferentially locate at the water/oil interface and catalyze oxidation and hydrogenation, and pondered how these reactions could impact EOR operations. In a previous study, we demonstrated the stabilization of Pickering emulsions by nanohybrid particles composed of Single-Walled Nanotubes (SWNT) and silica. We found that the volume fraction of emulsion and droplet size can be controlled by adjusting the water/oil (W/O) ratio and the amount of nanohybrids. Here, we describe the applications of the concept by incorporating active species on the surface of nanohybrids, which catalyze reactions at the oil-water interface. We found that the oxidation of aromatic compounds and hydrocarbons at the interface reduces the oil – water IFT since the species with higher dipole moment (oxygenates) dominate the interface. Also, the complete hydrogenation of the polyaromatic molecules increases the viscosity of the resulting products. The selective hydrogenation could be an appropriate strategy to increase the viscosity of the sweeping fluid (emulsion), improving the oil mobility.

In addition, we have conducted mobility studies on crushed Berea sand columns and demonstrated that the synthesized nanohybrids successfully flow through the porous media, particularly with the aid of small amounts of polymers.

(57b) Effect of Nanoparticle As Interfacial Modifiers in EP Operations

Ramanan Krishnamoorti, Chemical & Biomolecular Engineering, University of Houston, Houston, TX

Particles have been shown to be outstanding interfacial agents and in particular nanoparticles have been shown to drive the interfacial tension between oil and water to values below 15 mN/m at concentrations of the order of a few ppm. We have developed a comprehensive program to examine and understand the ability of nanoparticles to act as interfacial agent. Specifically, we have attempted to understand the role of particle chemistry (metal, metal oxide, inorganic, carbon) along with surface functionalization chemistry and that of the particle size and to a limited extent shape on the interfacial tension reduction achieved by such nanoparticles. Additionally, to understand the origins of these interfacial properties we are also using grazing incidence x-ray scattering from oil (alkane) water interfaces with particles with tailored surface properties to examine the interfacial structuring of the particles. Recent studies using "Janus" spherical gold nanoparticles (diameter 10 nm) with a hydrophobic polymer on one side and a hydrophilic polymer on the other have shown a significant enrichment of the surface concentration with the incipient formation of hexagonally ordered domains at the surface coinciding with the saturation of the interfacial tension reduction caused by the particles.

(57c) Rock-Based Micromodel and CFD Studies of Nanomaterial Transport for Reservoir Applications

Karsten E Thompson, Craft & Hawkins Department of Petroleum Engineering, Louisiana State University, Baton Rouge, LA, Saade Bou-Mikael, Chemical Engineering, Louisiana State University, Baton Rouge, LA, Daniel S. Park, Center for BioModular Multi-Scale Systems (CBMM), Louisiana State University, Baton Rouge, LA, Dimitris Nikitopoulos, Mechanical Engineering, Louisiana State University, Baton Rouge, LA and Clinton S. Willson, Civil and Environmental Engineering, Louisiana State University, Baton Rouge, LA

Both physical micromodels and pore-scale computational models have a long history of providing insight into pore-scale transport mechanism relevant to the oil and gas industry. Additionally, during the past decade microcomputed tomography (microCT) has become widely used for pore-scale analysis and modeling because it can provide nondestructive images of the interior geometry, composition, and fluid distribution in porous rocks. In this work, we integrate microCT with micromodel and pore-scale modeling technology to help understand the transport of micro- and nano-particle transport in rock samples.

A new technique has been developed in which a 3D microCT image is used to create a pattern for fabrication of 2.5D micromodels. The terminology "2.5D" means that the micromodel has a varying floor elevation (in addition to the heterogeneous areal pattern). However, it does not have intertwined channels that pass over or under one another as in a truly 3D structure. The basis of the micromodel pattern is a 3D microCT image of an outcrop sandstone. To construct the 2.5D pattern, an optimization procedure was used in which a series of computer simulations were performed to match both the pore structure and single-phase fluid mechanics in the 2.5D micromodel to the 3D rock as closely as possible.

Micromodel experiments as well as 3D computer simulations (on the original rock) are being performed to better understand the transport and retention of microparticles and nanoparticles. Single-particle computer simulations are used to help understand the effect of the pore geometry on particle trajectories, locations of retention, and the forces on the particles. Micromodel experiments can be performed with many particles and include physics that cannot yet be incorporated into the computational modeling. They provide insight into factors that are contributing to transport versus retention of the particles.

(57e) New Nano-Particle Based Drilling Fluids

Mukul M. Sharma, Petroleum Engineering, University of Texas, Austin, TX, Martin Chenevert, Petroleum Engineering, University of Texas - Austin, Austin, TX, Chung-min Jang, University of Texas, Austin, TX and James Friedheim, M-I Swaco, Houston, TX

Data is presented in this paper that shows for the first time that it is possible to formulate water-based muds using inexpensive non-modified and commercially available silica nanoparticles and that these muds significantly reduce the invasion of water into shale. The addition of silica nanoparticles to water-based muds may offer a powerful and economic solution when dealing with wellbore stability problems in troublesome shale formations. Fluid penetration from water-based muds into shale formations results in swelling and subsequent wellbore instability. Particles in conventional drilling fluids are too large to seal the nano-sized pore throats of shales and to build an effective mud cake on shale surface and reduce fluid invasion. This paper presents laboratory data showing the positive impact of adding silica nanoparticles (particle sizes from 5 nm to 22 nm) to water-based drilling muds and their effect on water invasion into shale.

(57f) Designing Brine-Stable, Transportable Nanoparticles for Downhole Sensing

Michael S. Wong, Chemical and Biomolecular Engineering, Rice University, Houston, TX, James M. Tour, Chemistry, Rice University, Houston, TX and Mason B. Tomson, Civil and Environmental Engineering, Rice University, Houston, TX

The continuing increase in worldwide energy demand has motivated recent efforts of designing nanomaterials with signaling, sensing, and detection capabilities for oil exploration.

Contrast-enhancing agents help detect specific targets at threshold levels and through nanoscale properties, enhance response and resolution of existing x-ray, electromagnetic and seismic-based detection methods. We recently synthesized carbon-based nanoparticles with hydrocarbon sensing capability, in which a hydrophobic compound is released when transporting through oil-containing rock (Berlin et al., Energy Environ. Sci., 4:2, 505-509, 2011). I will discuss the design and synthesis of this material, and present evidence for their colloidal stability and transport in briny fluids.

(58a) Flow Assurance: Modeling Wax Deposition in Subsea Pipelines

H. Scott Fogler, Chemical Engineering, University of Michigan - Ann Arbor, Ann Arbor, MI

As drilling in the Gulf of Mexico has moved further and further off shore, wax deposition in subsea oil pipelines has become a billion dollar problem. As the wax builds up in the pipelines it hinders (and in some cases, stops) oil production and transport The fundamentals of fluid flow, heat and mass transfer, thermodynamics and kinetics have been used to accurately predict single phase wax deposition and buildup without the use of any empirical turning factors. Knowing the location and hardness of the deposit, we can design remediation methods such as pigging frequency or fused chemical reactions to remove the wax deposits in subsea pipelines.

(58b) Asphaltenes Deposition

Jill Buckley, New Mexico Institute of Mining and Technology, Socorro, NM

Although asphaltenes are always present in crude oils, they present problems only when they become insoluble in the remainder of the oil. A change in asphaltene solubility can occur because of changes in pressure, composition, or temperature. Whether asphaltene insolubility constitutes a problem, however, depends primarily on when and where in the oil production, transportation, and processing stream it occurs. Instability leading to deposition in a deepwater well is high on the list of potentially problematic situations.

While insolubility is a necessary condition, it is not sufficient to predict the buildup of an asphaltene deposit in a production well. Flocculated asphaltenes may deposit or be carried out of the well in the flowing oil phase. To improve understanding and hence prediction of asphaltene deposition, we have used a capillary deposition test to investigate the tendency of a wide variety of oils to create deposits. Observations, in addition to pressure increases during flow through a capillary, include changes in morphology of asphaltene flocs as they slowly separate from unstable oils, the amount of asphaltene produced as a function of time, and chemical analysis of deposits recovered from capillary tests. As insight grows, the data produced are being used by our collaborators at Rice University and Chevron to develop a simulator to predict the rate of growth of arterial asphaltene deposits under wellbore conditions.

(58c) Review of Paraffin Deposition Under Multiphase Flowing Conditions

Cem Sarica, Petroleum Engineering, University of Tulsa, Tulsa, OK and Ekarit Panachareonsawad, Petroleum Engineering, The University of Tulsa, Tulsa, OK

Paraffin deposition in subsea pipelines is one of the flow assurance problems for oil and gas production. A reliable paraffin deposition prediction tool is needed to develop mitigation strategies. Most of the studies address single-phase paraffin deposition. However, in practice, multiphase flow is a common occurrence during oil and gas production. Paraffin deposition prediction under multiphase flowing conditions is largely ignored.

The current practice is to use single-phase paraffin deposition models in conjunction with multiphase flow hydrodynamic and thermal models. Thermal models used do not consider the flow pattern effects rigorously. The use of a single-phase paraffin deposition models for multiphase flow is not proper, and may lead to significant uncertainty in predictions. Thus, there is a need to better understand paraffin deposition under multiphase flow conditions.

Limited number of research studies have been conducted. This paper reviews the current state-of-the-art in multiphase paraffin deposition. Gas-oil and oil-water paraffin deposition models, experimental data, deposition testing techniques, and deposit characterization methods are summarized and discussed. This review also discusses the key challenges and the areas for the advancement of multiphase paraffin deposition research.

(58d) Asphaltene Solubility and Fluid Compatibility

Irv Wiehe, Soluble Solutions, Gladstone, NJ

Asphaltenes are held in petroleum by dispersion/solution in a delicate balance that is easily upset, causing plugging of well bores during production and fouling of heat exchangers and furnace tubes during refining. Methods will be reviewed for predicting the onset of asphaltene precipitation on reduction of the pressure during production or on the blending of crude oils after production. Since petroleum contains only a small amount of atoms other than carbon, hydrogen, and sulfur, the major force of interaction is of the van der Waals type. As a result, practical prediction methods for petroleum macromolecules, including asphaltenes, can be used that were previously found to be only a rough approximation when applied to organic polymers.

(58e) Wax Deposition and Control in North Sea Oil and Gas Production Systems

Hans P Ronningsen, Multiphase and Flow Assurance Department, Statoil, Stavanger, Norway

An overview of experiences with production of waxy petroleum fluids on the Norwegian continental shelf (NCS) is provided.

Although wax deposition is normally avoided by good thermal design (insulation, heating) and operational practice, there are several examples of production and transport systems where wax deposition has been experienced.

A long condensate export pipeline which started to experience severe wax deposition, and corresponding increase of pressure drop, when a waxy condensate was introduced, is given special attention. Wax deposit thickness measurements by a pressure pulse technique and tracer injection has provided useful information about the pipeline status, allowing controlled operation of the line. Attempts have been made to clean the line by chemical treatment, unsuccessfully. A new, highly sophisticated wax removal pig for thick deposits has recently been qualified by Statoil in order to clean the pipeline.

Other examples of wax deposition and wax control in export pipelines, multiphase flow lines and risers, are described. Special design and operational issues related to wax pigging of complex multi-node export pipeline systems, pigging of multiphase gas dominated flow lines, production of low temperature reservoirs and production of high pour point fluids are mentioned. Examples of ongoing activities to improve the prediction of wax pigging frequency will be shown.

An integrated cold flow concept for long distance multiphase transport of waxy oils, based on technology patented by Statoil, is described. This may be a candidate for long tie-backs of remote reservoirs to shore or a central hub.

(58f) New Asphaltene Nanoscience and Its Impact on Reservoir Characterization

Oliver Mullins, Schlumberger, Houston, TX

Asphaltene science has undergone a renaissance in recent years with many of the advances being subsumed in the "modified Yen model" which consists of asphaltene molecules, nanoaggregates and clusters of nanoaggregates. Various molecular diffusion and mass spectral measurements show that asphaltene molecular weights are ~750 (range 500-1000) Daltons. Molecular diffusion measurements originally established the predominance of one polycyclic aromatic hydrocarbon (PAH) per molecule; this is now supported by unimolecular decomposition studies. This molecular architecture yields nanoaggregates with <10 aggregation number. High-Q ultrasonic studies were first to show that asphaltene nanoaggregates form at 10^{-4}mass fraction. SANS and SAXS studies show small stable asphaltene nanoaggregates; DC-conductivity studies support all these findings of the size and aggregation concentration. Clusters form at ~10^{-3}mass fraction as shown by a flocculation kinetics and other measurements. Surface studies of molecular orientation and surface morphology dependence on aggregation are consistent with the modified Yen model. The combination of this new asphaltene nanoscience with new technology of performing chemical measurements in situ in oil wells is proving of enormous value in the oilfield reservoir evaluation. Depending on asphaltene concentration, all three forms, molecules, nanoaggregates and clusters are found in reservoirs. The new asphaltene nanoscience has given rise to the industry's first predictive equation of state for asphaltene gradients; in turn, this aids in charactering reservoir connectivity, heavy oil, tar mat formation and fluid disequilibrium. New science combined with new technology provides powerful new solutions to key problems.

Available know-how in transforming an emulsified drilling fluid to be removed from unwanted location into a low-viscosity single phase system.

Jean-Louis Salager*[,1], Ana María Forgiarini*[,1], Raquel Elena Antón[1] and Lirio Quintero[2]
[1]Laboratorio FIRP, Universidad de Los Andes, Mérida Venezuela
[2]Baker Hughes, Houston TX, USA

This invited paper was prepared for presentation in the 1st International Conference on Upstream Engineering and Flow Assurance, a part of the 2012 AIChE Spring Meeting, held in Houston, Texas, USA, 1-5 April 2012.

ABSTRACT

Drilling fluids are generally emulsions, either O/W or W/O, with an aqueous phase containing various electrolytes and an oil phase ranging from hydrocarbons to polar oils. Such multiphasic systems are stabilized by surfactants and cosurfactants whose nature is related to the emulsion morphology. Stable emulsions should have an unbalanced generalized formulation which is also related to the phase behaviro at equilibrium. If the drilling fluid tends to form a viscous multiphase in the casing void or to enter the porous medium and to plug it by capillarity, it has to be removed as a single phase system of the mesophase type. This usually requires the formation of a bicontinuous microemulsion with a surfactant system exhibiting the highest solubilization of both oil and water phases, a situation that takes place when the formulation is exactly balanced. Moreover this should also occur with an amount of surfactant as small as possible and this is yet an unsolved problem. Recent developments indicate that there are several trends toward performance improvement that can be used in parallel or together, and this is what has been used in practical cases.

▪ INTRODUCTION

Drilling fluids for oil and gas exploration-production, often called drilling muds, are complex multiphase systems of the suspemulsion type, i.e. O/W or W/O emulsions with solid particles that modify the rheology as clays, or the density as barite or carbonate. The amount of oil and water is variable from one case to the other, but both immiscible phases are always present because of their fundamental roles in cooling and lubricating. The external phase of the emulsion should be relatively efficient in suspending drops and particles, hence it should contain a surfactant system which is either hydrophilic (respectively lipophilic) to insure a protection against coalescence of O/W (respectively W/O) emulsions. Used surfactants can be extremely variable. Their main emulsifier role is not directly linked with their nature but rather with the physicochemical formulation in which they contribute, specifically the relative hydrophilicity-lipophilicity which has to be matched with the formulation. The nature of the surfactant is however important because it actually influences second order effects dealing with solubilization and emulsion stability (1). It also should satisfy restrictions and compatibility problems in particular with the aqueous salinity and the temperature.
Oil phases contain hydrocarbons such as petroleum cuts like diesel, olefins or mineral oils, as well as natural oil derivatives when green restrictions are imposed. The aqueous phase contains electrolytes like salts of mono and divalent cations, often similar to the connate water in the reservoir.

The drilling fluids tend to produce some cake at the formation surface wall and to penetrate in the porous medium with some plugging inside the pores. Cake properties such as thickness, thoughness, slickness and permeability are important for the expected roles of the drilling fluid, which are sometimes conflicting. The cake build up tends to avoid the filtration of the liquid phase into the formation, but it also can cause stuck pipe and other drilling problems. Reduced oil production can results from formation damage produced by a poor filtering cake that allows deep filtrate invasion. The following discussion particularly deals with the worse situation in which an "internal" cake has been formed inside the formation and must be removed to increase the oil production. For the sake of simplicity the discussion will apply to oil based mud, i.e. W/O emulsions containing solid particles, stabilized with lipophilic surfactants. The production is reduced by two phenomena: the first one is the presence of interfacial meniscus between oil and water produced by the trapping of the water drops in the pores. At each meniscus a Laplace capillary pressure drop takes place, thus resulting in a considerable pressure opposition to flow. The fact that the contact angle exhibits an hysteresis phenomenon when a oil/water boundary is displaced, further increases the plugging effect. On the other hand, the presence of solid particles reduces the porosity and thus tends to reduce the permeability.

It may be said that an improved permeability implies the elimination of the interface, or at least the considerably reduction of the oil/water interfacial tension, i.e. formulation issues, as well as the partial or total removing of solid particles, i.e. their dissolution.

The present short review will focus on surface phenomena to be generated, managed and manipulated, without dealing with specific surfactant species and process details that could be quite different depending on specific cases, as could be found in patents.

- FORMULATION OF SURFACTANT-OIL-WATER SYSTEMS AT EQUILIBRIUM

Winsor's R ratio to describe formulation effects

Surfactant-oil-water (SOW) systems are used in many application as detergency, home care products, foods, paints etc. However it may be said that the most significant advances have been produced during the second half of the 1970's with the first research drive in enhanced oil recovery by surfactant flooding. At this time the main issue was to produce a so-called optimum formulation to attain oil/water miscibility, or at least an ultralow interfacial tension below 0.001 mN/m (2). The condition to attain such a situation has been presented by Winsor in the early 1950's according to his R ratio formulation description, a theoretical presentation which is still the best pedagogical approach in a most simple case in which a surfactant molecule is adsorbed at the oil/water interface (3).

Fig 1 indicates the interaction between an adsorbed surfactant molecule and oil (Aco) and water (Acw) at interface. R = Aco/Acw, i.e. the ratio of these two interactions, is the formulation criterion that determines the phase behavior of the SOW system in the multiphase zone, when not enough surfactant is available to produce a single phase.

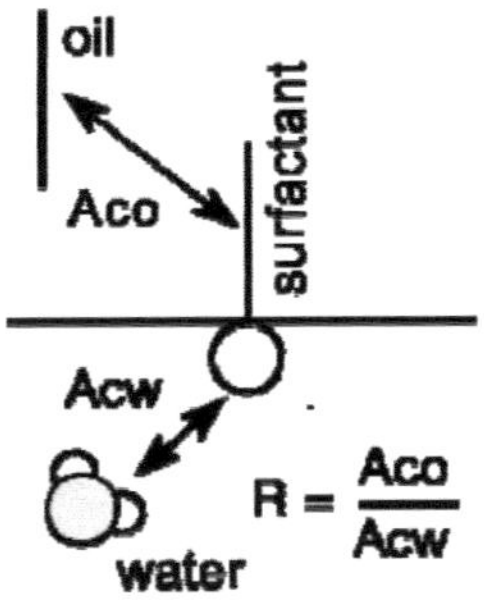

Fig 1 Adsorbed surfactant interactions with oil and water at interface and R ratio

As the R parameter changes, the multiphase behavior exhibits three different aspects. When the amount of surfactant is small, i.e. a few %, the system typically exhibits two separated phases, one of which contains most of the surfactant as a micellar solution sometimes containing structures swollen by solubilization, and the other one is an excess phase, essentially with no surfactant (4). Fig 2 test tubes on the left and right extremes indicate the aspect of the two-phase systems depending on the R value. For R > 1 (respectively R < 1) a surfactant rich oil (respectively water) phase is in equilibrium with water (respectively oil) phase. The whitish liquids in the corresponding test tubes of the Winsor I (WI) and Winsor II (WII) phase behavior types contain swollen micelle systems that are often called microemulsions when the micelles are numerous and large enough to be percolated, i.e. to essentially touch one another. When the value of R is unity (center), the system separates into three phases, i.e. a surfactant rich middle phase which is in equilibrium with excess water and oil phases. In this Winsor III (WIII) phase behavior, the surfactant rich phase is a bicontinuous structure, either a lamellar liquid crystal or a microemulsion depending of the degree of disorder. The microemulsion structure was presented by Winsor as a percolation of normal and inverted micelles (see Fig 2) in which the average curvature is zero, a depiction which is consistent with the currently accepted structure.

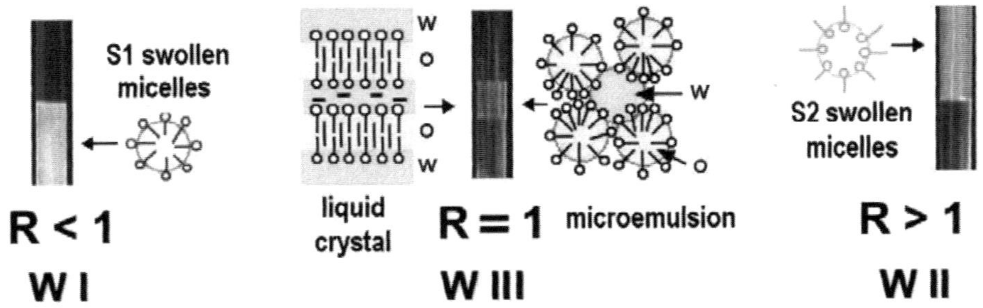

Fig 2 phase behavior in multiphase systems

When the surfactant amount is large, says 20-30 %, in most cases a single phase is exhibited in essentially all systems. Although this case is a welcome single phase, it is not in general the sought situation for applications, since it is not necessarily associated with the best surfactant performance.

When a formulation variable is changed, one or both interactions of the surfactant with oil and water are altered, and the R ratio could increase or decrease. For instance, if the aqueous phase salinity increases as in Fig 3 from left to right, the ion shielding between the surfactant head and water results in a reduction of Acw. On the other hand, an increase in surfactant tail length tends to increase the interaction with oil Aco. Both formulation changes produce an increase in R. If the formulation change range is appropriate, R would increase from $R < 1$ to $R > 1$, and the phase behavior would evolve along the WI → WIII → WII transition, as shown in Fig 3 test tubes from left to right, where the colored phase contains most of the surfactant.

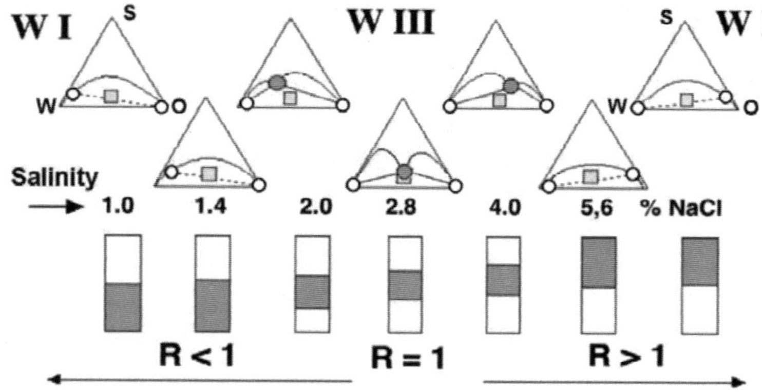

Fig 3 Phase behavior variation along a formulation scan

This sequence, so-called a formulation scan, corresponds to the SOW Winsor's diagram series illustrated in Fig 3, in which the square dot indicates the composition of the SOW system and the shaded round dot represents the composition of the surfactant rich phase, which is a microemulsion (shaded in the test tubes). The white round dots in Fig 3 diagrams represent the phases in equilibrium with the surfactant rich phase, i.e. one (or two) excess phases. Fig 3 scan sequence indicates that the microemulsion or swollen micelle phase continuously varies from one side to the other, i.e. from water to oil when the salinity is increased (4).
Fig 3 indicates that the height of the ternary diagram multiphase region above the square point (i.e. at WOR = 1 in this case) is minimum at R =1. This height, which is the minimum concentration of surfactant C^*_S required to attain a single phase behavior (at WOR=1), is a measurement of the performance of the surfactant formulation (5). The C^*_S value is essentially inverse to the solubilization parameter that measures the amount of oil and water solubilized by unit mass of surfactant, usually referred to as SP*, which is the maximum at R=1 (2).

Effect of formulation of SOW system properties

Fig 4 indicates the variation of the interfacial tension γ and multiphase height C_S (at WOR=1) as a function of the formulation along a salinity scan. The minimum interfacial tension γ^* and maximum solubilization, i.e. minimum C^*_S, is attained at a salinity that corresponds to R=1. Since an ultralow interfacial tension was the principal goal in enhanced oil recovery studies in the mid 1970's, this physicochemical situation (R=1) was called optimum formulation, a term which is still used in all applications so far, even if it does not correspond to the best case sought. The two examples (with hexane and octane oils) in fig 4 indicate that the lower the minimum interfacial tension γ^* is, the lower is the surfactant concentration C^*_S to generate a single phase microemulsion at optimum formulation. This is consistent with Huh's relationship that relates the tension with the solubilization parameter (6). It is also seen in Fig 4 that the formulation to attain a low tension or high solubilization is extremely precise at the center of the three phase region, and that a slight departure from optimum formulation would result in a considerable performance penalty. As a consequence, it is extremely important in practice to determine the optimum formulation with a good accuracy. This was the main driving force behind the research effort in physicochemical formulation in the 1970's.

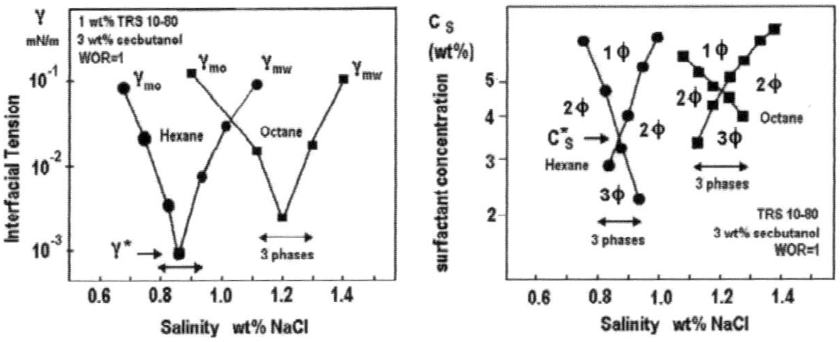

Fig 4 Formulation performance according to the interfacial tension and multiphase region (7)

Similar scans are taking place from $R < 1$ to $R > 1$ as some formulation variables are increased (brine salinity, surfactant tail length, temperature with nonionic surfactants, addition of lipophilic alcohol) or decreased (alkane carbon number of the oil phase, ethoxylation degree of nonionics, temperature with ionics, pH in fatty acid formulation). However, the actual effect of these variables on Aco and Acw interactions cannot be calculated with any accuracy, just as the HLB number proposed by Griffin 60 years ago (8). Consequently, a substitute of R having an exact numerical value had to be elaborated.

Correlations for the attainment of optimum formulation

Since different formulation variables result in opposite phase behavior effects, it was possible to carry out systematic studies to estimate the compensation between the different contributions to the formulation variation. Thousands of experiments were carried out to collect data and generate a generalized correlation between the different variables to attain an optimum physicochemical situation. The first studies reported a numerical correlation between common characteristics of simple systems, i.e. sodium chloride brine and n-alkane, surfactants characteristic similar to HLB but more precise, and eventually alcohol cosurfactant separated effect.

The condition to attain an optimum formulation was expressed as:

$$\ln S - K\,ACN + \sigma + f(A) - a_T\,(T - 25) = 0 \qquad [1]$$

for anionic (9) or cationic (10) surfactant systems

$$b\,S - K\,ACN + \beta + \phi(A) + c_T\,(T-25) = 0 \qquad [2]$$

for polyethoxylated nonionic surfactants systems (11).

where S is the aqueous phase salinity expressed as wt% NaCl, σ and β are characteristic parameters of ionic and nonionic surfactant, ACN is the alkane carbon number a characteristic of the oil, f(A) and ϕ(A) are effects that depend on the alcohol co-surfactant type and concentration, and T is the temperature in °C. Symbols b, K, a_T and c_T are positive constants.
Surfactant parameters σ and β increases linearly with the number of carbon atom added in a linear tail group. $\beta = \alpha$ - EON where EON is the average number of ethylene oxide groups in the head and α is the contribution of the surfactant tail that also increases with the number of carbon atoms in the tail (11-12). Some values of the characteristic parameters have been published elsewhere (4,13).

Generalized formulation concept SAD and HLD dimensionless expression

These correlations were found to express the standard chemical potential change when a surfactant molecule passes from the oil phase to the water phase, which was called surfactant affinity difference (SAD), i.e. for an ionic surfactant (14, 15):

$$\Delta\mu = \mu^*_O - \mu^*_W = \text{SAD} = \text{RT (correlation term [1] or [2])} \qquad [3]$$

SAD was expressed in a dimensionless way as the Hydrophilic-Lipophilic Deviation (HLD), a generalized formulation value characteristic of the physicochemical situation of the system at equilibrium. This is the same general concept than HLB, but it is much more accurate and takes into account the contribution of all variables (16).

$$\text{HLD} = (\text{SAD}-\text{SAD}_{ref})/\text{RT} = \ln S - k\ \text{ACN} + \sigma + f(A) - a_T\,(T-T_{ref}) \quad \text{for ionic surfactant} \qquad [4]$$
$$\text{HLD} = (\text{SAD}-\text{SAD}_{ref})/\text{RT} = bS - k\ \text{ACN} + \beta + \phi(A) + c_T\,(T-T_{ref}) \quad \text{for nonionic surfactant} \qquad [5]$$

An optimum formulation is described by HLD = 0, whatever the actual value of the terms representing the contribution of the brine (S), the oil (ACN), the surfactant (σ or β), the cosurfactant f(A) or ϕ(A), and the temperature.
For the best performance as far as a high solubilization is concerned, the system to be produced down hole should be at HLD = 0, when all the ingredients present in the drilling mud and the injected formula are taken into account.
It means that in practice the different terms in the HLD expression have to be calculated not for a single component but for mixtures, sometimes very complex. For the sake of simplicity the present review will emphasize the technique to select a proper surfactant/cosurfactant mixture to achieve the purpose, i.e. what corresponds to the terms σ + f(A) or β + ϕ(A) in HLD. It is assumed that the injected system has an oil equivalent alkane carbon number (EACN) and a brine salinity S identical to those of the drilling fluid. If it is not the case, there is no significant variation in the formulation technique, just a simple correction to be introduced in the HLD expression, with the average values for EACN and S calculated by the mixing rules techniques for oil and brine which have been reported elsewhere (17,18).
In what follows, the formulation technique will be discussed as the selection of the surfactant/cosurfactant ingredients.

Emulsion properties

The formulation has been found to be the most important factor in determining the emulsion properties (1). Fig 5 indicates that the emulsion type is directly related with HLD sign. In numerical terms a deviation of about 0.5 units from HLD zero value is enough to change the emulsion morphology. The most important feature of this figure is indicated by the central graph, that shows that the emulsion is very unstable at HLD = 0 (19, 20), while it becomes stable when a deviation from HLD = 0 becomes significant, typically 1-2 units. The symbol DF indicates the location of a stable W/O emulsion like a typical drilling fluid.
The right graph indicates that even as an emulsion, the system located at HLD = 0 exhibits a low viscosity, because of the associated low tension that facilitates the drop elongation (21). As a single phase microemulsion it will exhibit a low viscosity.

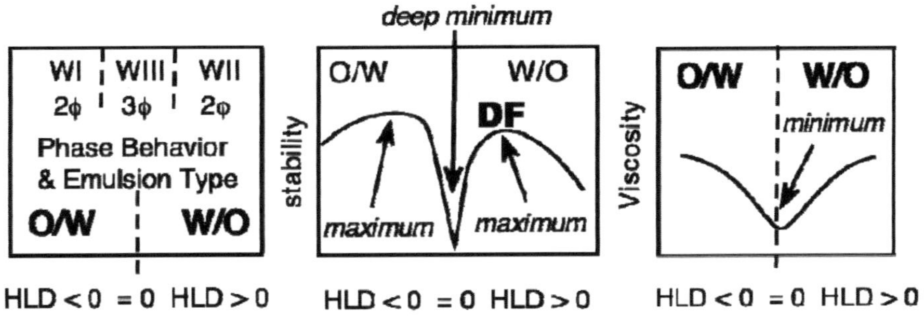

Fig 5. Phase behavior and emulsion properties as a function of generalized formulation

- ATTAINEMENT OF AN OPTIMUM FORMULATION BY MIXING AN INJECTED FORMULA WITH THE DRILLING FLUID

The vicinity of HLD = 0 indicates the presence of a three phase behavior region in a ternary diagram at constant formulation for a simple WIII type diagram. In real practice a single amphiphile specie is not sufficient. In most cases, some alcohol cosurfactant is added to avoid the formation of a liquid crystal. In the currently discussed case, a fourth component is required even without any eventuality about the presence of alcohol. There is a surfactant (S_1) in the drilling fluid. There will be another surfactant (S_2) introduced in the injected formula to modify the drilling fluid. This requirement implies a fourth component in the system, which will be symbolized a the S_1S_2OW quaternary.
For the sake of simplicity, the same O and W liquid phases are supposed to be used everywhere, i.e. in the initial formulation of the drilling fluid (S_1OW) and the in the injected formula (S_2OW). Hence, the mixture of the two systems down hole will have four components S_1, S_2, O and W, in its most simple description. Since the W/O emulsion drilling fluid has a HLD > 0 formulation, the injected one should have a HLD < 0 formulation, so that their mixing could end up in a HLD = 0 situation, which is compulsory to attain the best solubilization performance in a scan. Although each of the two systems can be described in a ternary diagram, a tetrahedral diagram is required to describe the quaternary S_1S_2OW system phase behavior.
Since such representation cannot be laid down on a sheet of paper, the usual solution to this practical problem is to make a two-dimensional cut at constant S_1/S_2 through the tetrahedron as indicated in Fig 6 (22).
In this particularly simple case A = S_1 is a lipophilic cosurfactant (n-pentanol) and S = S_2 a hydrophilic ethoxylated nonionic surfactant. The ternary AOW has a WII (HLD > 0) diagram,

whereas the ternary SOW exhibits a WI type (HLD < 0). Different mixtures of A and S result in an amphiphilic pseudocomponent with a term ($\beta + \phi(A)$) that can change the sign of HLD, and thus swap the phase behavior. The figure indicates the different cuts obtained when the S/A ratio of the two amphiphiles is changed. When there is 80% or more of S, a WI diagram is exhibited, and when there is more than 50% of A it is a WII diagram. In between, a WIII phase behavior diagram is found with a displacement of the middle phase microemulsion as in Fig 3. It is worth noting that he three phase triangle is not completely shown because the excess phases are not in the cutting plane. Nevertheless, the general aspect of a WIII diagram is conserved.

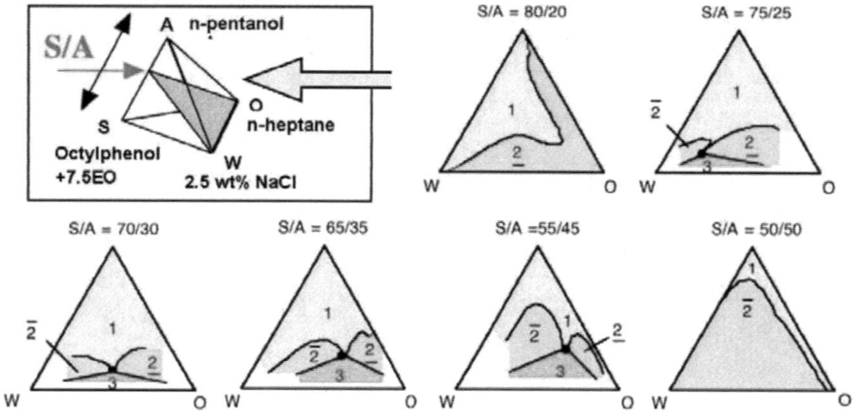

Fig 6. Tetrahedral diagram phase behavior content represented through a sequence of two dimensional cuts at S/A constant (22).

If the same tetrahedral diagram is cut at constant water/oil ratio as in Figure 7, the same phase behavior content appears in a quite different aspect, as in a gamma or fish three-phase region. There is no doubt that this cut does not look like a WIII diagram, and this is the main reason why the navigation in multidimensional space to observe the phase behavior characteristics and to find a formulation path is not an easy task.

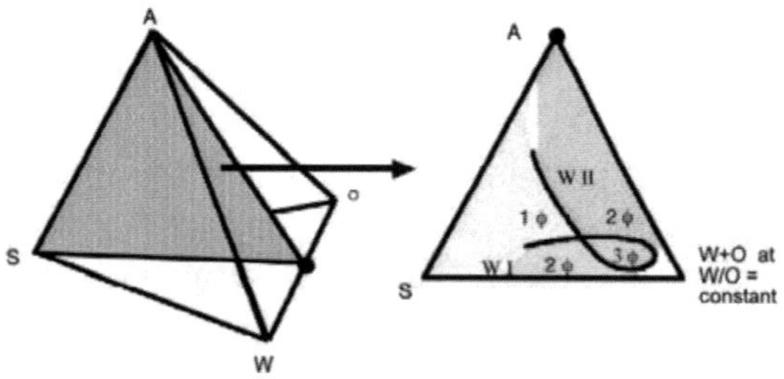

Fig 7. Tetrahedral diagram phase behavior content represented in a two dimensional cut at O/W constant (22)

It is worth recognizing that Winsor's diagram aspect for a simple SOW ternary is not a very common one, not even as a cut of S1/S2-OW simple quaternaries. This is mainly because the presence of two surfactant/cosurfactant species generally results in a predominance of one of them in some areas, and not necessarily in other. For instance in Fig 8, S (sodium dodecyl sulfate) is a very hydrophilic surfactant, whereas A (n-butanol) is really lipophilic only when it is in large amount, being hydrophilic or neutral at low concentration. As a consequence, at low alcohol content the A/S mixture is hydrophilic, i.e. the S amphiphile dominates, while it is the opposite at high alcohol concentration. The cut at a A/S ratio 50/50 exhibits the consequence of this situation. As the A/S mixture concentration increases from bottom to top, the phase behavior changes as WI → WIII → WII → microemulsion single phase. Such behavior has been found to be quite common and to correspond to the so-called twisted gamma or fish diagram (5, 23).

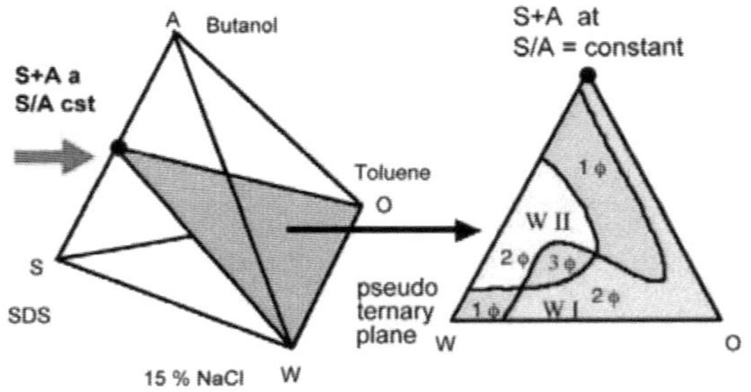

Fig 8. Two dimensional cut in a tetrahedral diagram showing a formulation effect of the variation of the surfactant/cosurfactant total concentration, in spite of a constant S/A ratio.

Actual systems with mixtures of surfactants, cosurfactants, oils and electrolytes could be even more complex. The present discussion only deals with the effect of a two surfactant/cosurfactant mixture on the phase behavior map, but it should be stressed that other ingredients susceptible to alter HLD have a substantial effect on the phase behavior map (24, 25), for instance the salinity (23, 26) or the temperature (27).

In systems close to the HLD = 0 condition, there is a WIII region that has a touching point with the single phase region that pinpoints the best performance conditions to be attained. This point is the top of the three phase triangle in Fig 3 or 6, or the crossing point of the gamma in Fig 7. The proper path to reach this point by mixing the drilling fluid in situ with the injected formula is the challenge of the treatment process. In practice the path would have to be represented and followed in the proper cut, for instance at S_1/S_2 constant ratio as in Fig 6.

The phase diagram (a) in Fig 9 is similar to the one previously shown in Fig 6, this time with a S_1 lipophilic surfactant to stabilize the W/O emulsion drilling fluid whose composition is indicated by a square dot. The injected formula indicated in diagram (b1) as a white circle dot is an aqueous solution of hydrophilic surfactant S_2, eventually with oil if the dot is in the O/W 50/50 plane cut. The mixing between the drilling fluid in situ and the injected formula takes place along the arrow until the middle of the three phase zone in Fig 9 (b1). This is an optimum formulation, with particularly low tension that considerably reduces the trapping capillary phenomena produced by the W/O emulsion, but it is not a single phase system. A better injected formulation is shown in diagram (b2). It is contains not only water and S_2, but also oil and some S_2 surfactant, and by mixing with the drilling fluid, it ends up in a three phase behavior zone with a large amount of microemulsion. In diagram (b3) the injected formula is a single phase microemulsion that has the capacity to solubilize all the drilling fluid and to stay as a single phase. This situation could be however quite expensive because of the large amount of surfactant required in (b3) injected formula.

The (b1) situation in which the mixed system exhibits a three phase behavior may be improved by injecting afterward a formula containing a S_3 surfactant with a formulation close to HLD = 0. The path along the arrow attains the single phase microemulsion at the best location as far as the

cost is concerned. The S_3 surfactant does not appear in the diagram since it is in another dimension, but a path equivalent to (b1) + (c1) may be produced by mixing S_2 and S_3 in the injected formula.
An even better trick is shown in diagram (d1). What has been altered here is the phase boundary by changing something else like the salinity or the temperature, or adding a new surfactant that boosts solubilization as discussed next. In this case the mixing of the injected formula with the drilling fluid ends up at the less expensive single phase point.

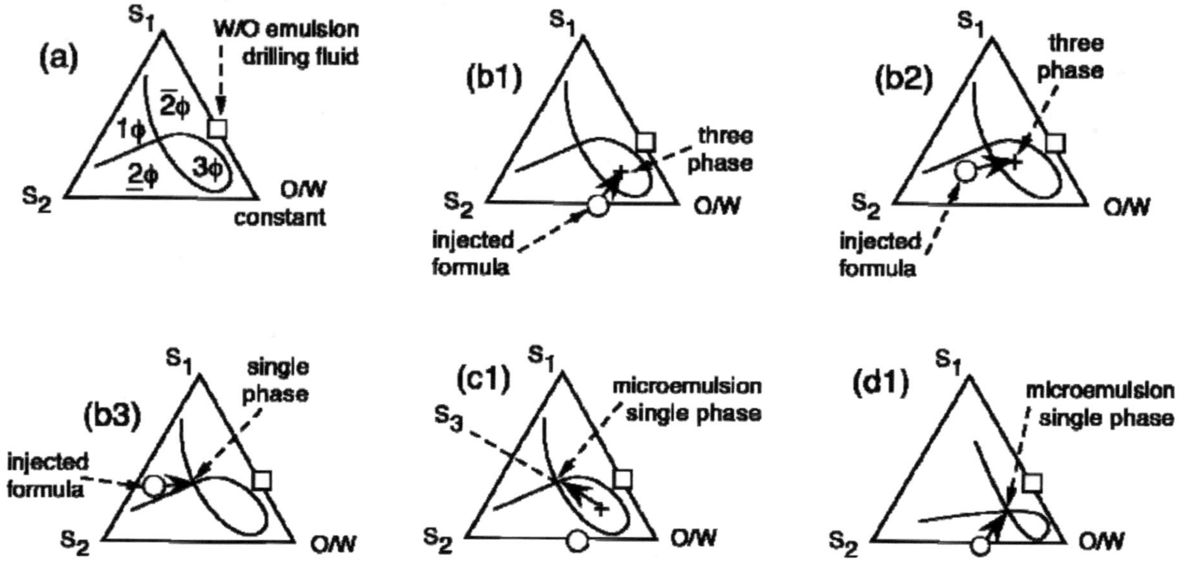

Fig 9. Formulation paths to transform a W/O drilling fluid emulsion in a single phase microemulsion by mixing it with an injected formula.

- IMPROVEMENT OF THE SOLUBILIZATION PERFORMANCE

The solubilization performance of a given SOW system depends on many variables, and its optimization is not easy. However, since a performance boosting has been sought for numerous applications of microemulsions, many systematic research efforts have been carried out from petroleum recovery to cosmetics and pharmaceuticals. As a consequence several tendencies have been discovered by comparing cases in which HLD = 0 has been attained in different ways. The present section is just a short sum up of the different trends available to improve the solubilization in microemulsion; more details may be found elsewhere (28). The first guideline has been introduced by Winsor 60 years ago, who first stated that an optimum formulation with R = 1 is a situation which should be compared with another optimum formulation also with R = 1. However, since R = 1 may be attained with different interactions on both sides of the interface, i.e. as R = 2/2 or R = 10/10, Winsor proposed that the best solubilization performance was associated to higher interactions on the two sides, i.e. to R = 10/10 in this comparison (4).
Such an improvement is attained by increasing the interaction of the surfactant with oil, e.g. by increasing the surfactant tail length on the oil side, as well as decreasing the salinity on the water

side, so that the optimum formulation is maintained. In ethoxylated alcohol surfactants such an improvement on both sides of the interface could be produced by increasing both the lipophilic alkyl chain length in the tail, and the ethylene oxide average number (EON) in the head to stay at an optimum formulation. This kind of change is equivalent to use bigger surfactant molecules, i.e. alcohol ethoxylates with longer tail and longer head, as shown in Fig 10 (a) graph, and corroborated by several studies (29, 30).
However, there is a limit to the increase of the surfactant size, i.e. its solubility in the system, particularly in the water phase. This limit approximately corresponds to a C16 linear hydrocarbon tail. The presence of double bond, branching or ramification in the surfactant tail illustrated in Fig 10 (b-c-d), tends to allow a significant larger size in the tail without precipitation, and also provides a geometrical disorder that prevents the formation of liquid crystals (31). This tendency pushes further the precipitation limit beyond C16-C18 carbons atoms in the tail. However, there is still a limit, in particular with high salinity.

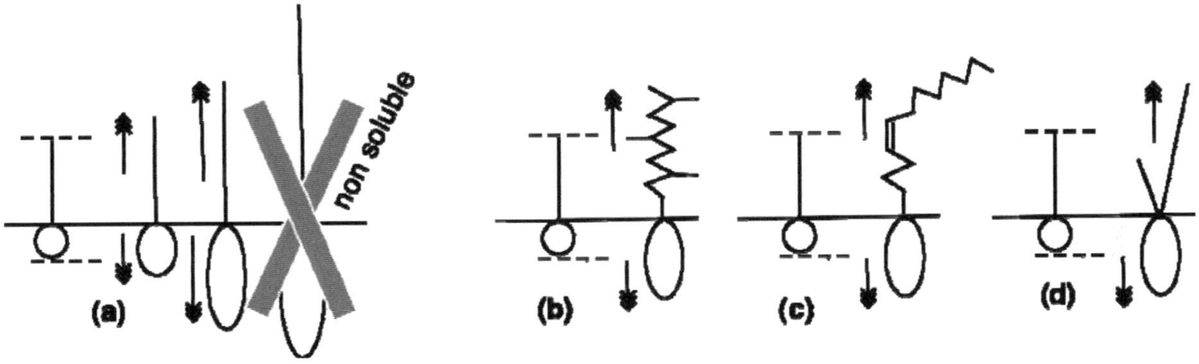

Fig 10. Improving solubilization by (a) increasing both sides of the surfactant with a size limit for a linear tail. Further increase in tail size without precipitation with (b) branching, (c) double bound, and (d) ramification.

The second feature to be effective in improving the solubilization performance is the utilization of surfactant intermolecular mixtures, with some hydrophilic and some lipophilic species adsorbing together at interface in some kind of collective behavior. This combination illustrated in Fig 11 (a) increases the interaction of the surfactant mixture on the oil side thanks to the lipophilic species, and on the water side thanks to the hydrophilic ones, sometimes with a higher compactation that boost the adsorption density, as indicated by the horizontal arrows in fig 11 (a) This mixture concept works quite well with very complex systems and allow to solve other problems as the liquid crystal formation, the high salinity tolerance (32-34), or the temperature insensibility (35). However, there is also a limit which is the fractionation, i.e. the unequal partitioning of the different species between interface and phase bulks. The result is a loss of surfactant/cosurfactant into the bulk phases, which results in a decrease in the solubilization performance, i.e. an increase in C^*_S, unless a single phase system is attained at the end. The presence of polar oil species that tend to segregate close to interface has also been a way to improve solubilization, beyond Winsor's rules, by prolongating the tail and head of the surfactant as shown in Fig 11 (b). Such effects called lipophilic linking (LL), and similar hydrophilic linking (HL), have been found to improve solubilization, but also to result in the loss of the species which are not located at interface (36-39).

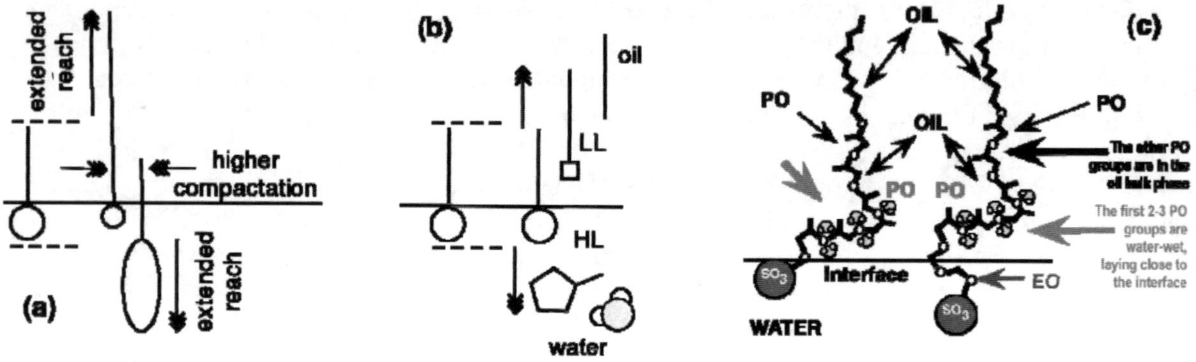

Fig 11 Improving solubilization by (a) intermolecularly mixing a hydophilic and a lipophilic surfactants, (b) by adding a lipophilic linker (LL) and a hydrophilic linker (HL), (c) by using an intramolecularly mixed extended surfactant.

The anchoring of such features resulting from intermolecular mixtures, in the same amphiphilic molecule is a clever way to avoid non-collective behaviors and species fractioning. Such an intramolecular mixture is attained by designing the so-called extended surfactant (40), which is a molecule with an intermediate polarity polypropyleneoxide spacer between the head and tail usual groups, as shown in Fig 11 (c). This extension enables the surfactant to increase its reach in both phases and thus improve solubilization, without the inconvenients of precipitating or fractioning. Such an intramolecular mixing is the third trend to favor improved solubilization, not only of hydrocarbons oils, but also of polar oils, particularly natural triglycerides and derivatives, which is an exceptional feature in surfactant science. (40-43). Consequently such extended surfactant may be considered as a compulsory ingredient if the oil phase contains polar substances, particularly esters.

Extended surfactants with a propoxylated intermediate chain present other interesting properties like a better tolerance to salinity because of some nonionic character, and a less hydrophobic tail which may be longer and thus could provide an increased interaction with oil without precipitation. As indicated in Fig 11 (c) the first 2-3 propylenoxide units close to the head group are generally slightly wetted, hence the surfactant tail is twisted. The surfactant is thus less likely to produce liquid crystals (44, 45). Consequently it is not necessary to add alcohol as a cosurfactant when such extended species are included in the mixture.

The introduction of a few ethylenoxide groups between the propoxylated intermediate and the anionic head group, is likely to improve the salt tolerance even with a high divalent cation content (46). Of course a better salt tolerance is attained in this and other inter and intramolecular surfactant mixture by using glucidic head groups instead of anionc like carboxylate, sulfate or sulfonates.

- ADDITIONAL FORMULATION ISSUES

As mentioned in the introduction, drilling fluids contain suspended particles of different types and for different purposes. The total dissolution of such particles would of course increase the porosity and the permeability for production. An only partial dissolution would reduce the viscosity of the fluid, and facilitate the particle removal. Some particles could be dissolved by an acid brine, e.g. carbonate, while other may be solubilized by a divalent ion sequestrant or chelatant like ethylendiamine tetraacetate or nitrilo acetate. These additives to be incorporated in the aqueous phase does not produce particular worries, and just requires some checking of incompatibility with the surfactant/cosurfactant species included in the drilling fluid and the injected formula. Many surfactants are compatible with acid pH and most of the chelatant may be considered as weak electrolytes with no strong effect. In any case some preliminary checking will enable to evaluate the effect of these additive and the dissolved substances on the formulation, particularly the surfactant mixture and salinity alteration. A slight variation in salinity, alcohol or surfactant is generally sufficient to compensate the aditive effect on HLD.

▪ CONCLUSION: APPLICATION TO WELL TREATMENT IN PRACTICE

The previous sections have presented a few phase behavior cases and a few alternatives for building up a process to treat a formation damage by transforming a viscous emulsified drilling fluid into a low viscosity single phase microemulsion that can be pumped out easily. The actual variety of phase diagrams is considerable, in particular because of the actual dimension (much above 3 or 4) and number of degrees of freedom, that make some of them exhibit curious features with advantages and drawbacks. The experience has shown that a slight change in the formulation can considerably modify the phase behavior frontier. As a consequence a previous experience with many system is a precious know-how.

The few general tendencies to boost the solubilization performance are an outstanding tool to improve the phase diagram perspective to offer a single phase microemulsion in a relatively simple and inexpensive way. In any case the formulation path to be taken to reach the solution has to be designed on the corresponding phase behavior chart, and since the chart is essentially unique with each case, it may be said that there is no general formula, but good and better ones to be tested in details once a first HLD = 0 formulation system has been obtained from basic techniques, particularly surfactant mixing (18). Examples of actual cases are found in some patents (47-49), but taking them as a starting point is not necessarily a good or a fast technique, since the approach of a HLD = 0 formulation is different from one case to another, just as it is for designing the surfactant slug in enhanced oil recovery, or the demulsifier in crude oil dehydration. It seems that the formulator performance is associated with his comprehension of the physicochemical framework and with a long experience in interpreting fragmented phase behavior maps, that allow him to reduce the huge number of required trial and error experiments.

▪ AUTHOR INFORMATION

Corresponding Authors

*Tel. ++582742402954 Emails: salager@ula.ve, anafor@ula.ve

REFERENCES

1) Salager, J.L.; Bullón, J.; Pizzino, P.; Rondón-González, M.; Tolosa, L.I. In *Encyclopedia of Surface and Colloid Science.* P. Somasundaran Ed., Taylor & Francis **2010,** 1:1, 1-6
2) Reed, R. L.; Healy, R. N. In *Improved Oil Recovery by Surfactant and Polymer Flooding,* D. O. Shah, R. S. Schechter, Eds. Academic Press . **1977** pp 383-347.
3) Winsor, P. *Solvent Properties of Amphiphilic Compounds,* Butterworth London, **1954**.
4) Bourrel, M.; Schechter R.S. *Microemulsions & Related Systems,* Marcel Dekker New York **1988.**
5) Bourrel, M.; Chambu, C.; Schechter, R. S.; Wade, W. H. *Soc. Petroleum Eng. J.* **1982**, *22*, 28-36
6) Huh, C. *J. Colloid Interface Sci.* **1979**, *71*, 408-426.
7) Salager J.L. PhD dissertation. University of Texas at Austin, USA (**1977**).
8) Griffin, W. C. *J. Society Cosmetic Chemists* **1949**, *1*, 311-326
9) Salager, J. L.; Vasquez, E.; Morgan, J.; Schechter, R. S.; Wade. W. H. *Soc. Petroleum Eng. J.,* **1979**, *19*, 107-115
10) Antón, R. E.; Garcés, N., Yajure A., *J. Dispersion Sci. Technology,* **1997,** *18,* 539-555.
11) Bourrel, M.; Salager, J. L.; Schechter, R. S.; Wade ,W. H. *J. Colloid Interface Sci.,* 1**980**, *75*, 451-461
12) Hayes, M.; Bourrel, M.; El-Emary, M.; Schechter, R. S.,; Wade, W. H. *Soc. Petroleum Eng. J.,* **1979**, *19*, 349-356
13) Salager, J. L.; Antón, R. E.; Andérez, J. M.; Aubry, J. M. In *Techniques de l'Ingénieur,* **2001**, *Vol. Génie des Procédés J2,* Chapter 157, 1-20
14) Salager, J. L. In *Handbook of Detergents - Part A: Properties,* G. Broze Ed. M. Dekker New York **1999;** pp 253-302
15) Salager, J. L.; Márquez, N.; Graciaa, A.; Lachaise, J. *Langmuir,* **2000**, *16*, 5534-5539.
16) Salager, J. L.; Antón, R. E.; Forgiarini, A.; Márquez, L. In *Microemulsions — Background, new concepts, applications, perspectives.* C. Stubenrauch, Ed. Blackwell Pub. Oxford UK **2009**. pp 84-121,
17) Salager, J. L.; Bourrel, M.; Schechter, R. S.; Wade, W. H. *Soc. Petroleum Eng. J.* **1979**, *19,* 271-278.
18) Antón, R. E.; Andérez, J. M.; Bracho, C.; Vejar, F.; Salager, J. L. *Adv. Polymer Sci.*, **2008,** *218* : 83-113
19) Bourrel, M.; Graciaa, A.; Schechter, R.S.; Wade, W.H. *J. Colloid Interface Sci.* **1979,** *72*, 161-163
20) Anton, R.E.; Salager, J.L. *J. Colloid Interface Sci.* **1986**, 111, 54-59
21) Salager, J.L.; Miñana-Perez, M.; Andérez, J.M.; Grosso, J.L.; Rojas, C.I. *J. Dispersion Sci. Tecnology* **1983**, *4*,161-173
22) Antón, R.E., MS thesis, Universidad de Oriente, Puerto La Cruz, Venezuela **1981**
23) Bellocq, A.M.; Bias, J.; Clin, B.; Gelot, A.; Lalanne, P.; Lemanceau, B. *J. Colloid Interface Sci.* **1980,** *74*, 311-321
24) Forgiarini, A.M.; Esquena, J.; Gonzalez, C.; Solans, C. *Langmuir* **2001**, *17,* 2076-2083
25) Forgiarini, A.M.; Esquena, J., Gonzalez, C.; Solans, C. *Progress Colloid Polymer Sci.* **2000,** *115,* 36-95
26) Bellocq, A.M.; Gazeau, D. *Progress Colloid Polym. Sci.* **1988,** *76*, 203-210
27) Kunieda, H.; Shinoda, K. J. Dispersion Sci. Technology. **1982**, *3,* 233-244

28) Salager, J. L.; Anton, R. E.; Sabatini, D. A.; Harwell, J. H.; Acosta, E. J.; Tolosa L. I. *J. Surfactants Detergents* **2005**, *8,* 3-21
29) Bourrel, M.; Chambu, C. *Soc. Petroleum Eng. J.* **1983**, *23*, 327-338
30) Sottmann, T.; Strey, R. *J. Chem. Phys.* **1997**, *106,* 8606-8615
31) Sunwoo, Ch. K.; Wade, W. H. *J. Dispersion Sci. Technology* **1992,** *13,* 491-514.
32) Bansal, V.K.; Shah, D.O. *Soc. Petroleum Eng J.* **1978**, *18:* 167-172
33) Bansal, V. K.; Shah, D. O. *J. Colloid Interface Sci.* **1978**, *65*: 451-459
34) Abe, M.; Schechter, R. S.; Selliah, R. D.; Sheikh, B.; Wade, W. H. *J. Dispersion Sci. Technology* **1987**, *8,* 157-172
35) Antón, R.E.; Graciaa, A.; Lachaise, J.; Salager, J.L. *J. Dispersion Sci. Technology* **1992,** *13,* 656-579
36) Graciaa, A.; Lachaise, J.; Cucuphat, C.; Bourrel, M.; Salager, J.L. *Langmuir* **1993**, *9,* 669-672
37) Graciaa, A.; Lachaise, J.; Cucuphat, C.; Bourrel, M.; Salager, J.L. *Langmuir* **1993**, *9,* 3371-3374
38) Acosta, E.; Uchiyama, H.; Sabatini, D.; Harwell, J. H. *J. Surfactants Detergents* **2002,** *5,* 151-157
39) Uchiyama, H.; Acosta, E.; Tran, S.; Sabatini, D. A.; Harwell, J. H. *Ind. Eng. Chem. Res.* **2000**, *39,* 2704-2708
40) Miñana-Pérez, M.; Graciaa, A.; Lachaise, J.; Salager, J. L. *Colloids Surf. A* **1995**, *100,* 217-224.
41) Miñana-Pérez, M.; Graciaa, A.; Lachaise, J.; Salager, J. L. *Progress Colloid Polymer Sci.* **1995**, *98,* 177-179
42) Do, L.D.; Witthayapanyanon, A.; Harwell, J.H.; Sabatini, D.A. *J. Surfactants Detergents* **2009,** *12:* 91-99
43) Phan, T.T.; Attaphong, C.; Sabatini, D.A. *J. Am. Oil Chem. Soc.* **2011,** *88,* 1223-1228
44) Velasquez, J.; Scorzza, C.; Vejar, F.; Forgiarini, A.; Antón, R.E.; Salager, J.L. *J. Surfactants Detergents* **2010**, *13*: 69-73
45) Forgiarini, A.M.; Scorzza, C.; Velasquez, J.; Vejar, F.; Zambrano, E.; Salager, J.L. *J. Surfactants Detergents* **2010**, *13,* 451–458
46) Aoudia, M.; Al-Shibli, M.N.; Al-Kasimi, L.H.; Al-Maamari, R.; Al-Bemani, A. *J. Surfactants Detergents* **2006**, *9,* 287-293
47) Jones, T.A.; Clark, D.E.; Quintero, L.; Rae Ewen, B. US Patent 7,134,496 B2 (Nov. 14, **2006**)
48) Quintero, L.; Clark, D.E.; Jones, T.A.; Salager, J.L.; Forgiarini, A.M. US Patent 8,091,645 B2 (Jan. 10, **2012**)
49) Quintero, L.; Jones, T.A.; Clark, D.E.; Gabrysch, A.D.; Forgiarini, A.M.; Salager J.L. US Patent 8,091,646 B2 (Jan. 10, **2012**)

Aqueous Polymer-Surfactant Mixtures for Wellbore Treatment and Completion

Mario R. Rojas, Ajay Raj, Jeffrey Harwell and Bor-Jier Shiau
University of Oklahoma, Norman OK

This invited paper was prepared for presentation in the 1st International Conference on Upstream Engineering and Flow Assurance, a part of the 2012 AIChE Spring Meeting, held in Houston, Texas, USA, 1-5 April 2012.

Abstract

The rheological properties of mixtures of aqueous soluble biopolymers and mixtures of these polymers with ionic surfactants were studied in natural and artificial brines for their application in wellbore cleaning. Polymers mixtures that show synergistic effects offer enhanced rheological properties for high solid suspension, minimum invasion to the formation and removal of the filter-cake. Different surfactants were tested for their ability to interact with the polymers to increase viscosity and improve the cleaning process while reducing the amount of polymer needed in the formulation.

Introduction

Natural polysaccharides and celluloses are attractive biopolymers for viscosity enhancement in oil exploration and production operations due to their relatively low cost, high resistance to mechanical and thermal degradation, and long-term biodegradability. In oil production, biopolymers have been employed in enhanced oil recovery, and as part of the formulation of drilling muds and completion fluids. In drilling operations, it is important to remove and suspend the cuttings produced by the drill in the well bore, which is thought to be aided by the high viscosity provided by the polymers.

Xanthan gum (XG) is an extracellular polysaccharide produced by the bacteria *Xanthomonas Campestris*. It consists of a glucose backbone substituted by a trisaccharide side chain (1). Guar gum (GG) is a galactomannan produced in the endosperms of seeds of *Cyamopsis tetragonolobus*. The chain is formed by a mannose backbone with galactose side chains. The mannose to galactose molar ratio is approximately 1.8 (2).

Xanthan, guar gum and their mixtures are extensively used in food, cosmetics, and oil recovery applications. The synergistic effect exhibited by these two polymers is attributed to direct interactions between the xanthan gum chain and exposed mannose segments in the backbone of the guar gum macromolecule (3). The intermolecular binding occurs between the ordered Xanthan and un-substituted regions of the Guar backbone. The synergic effect is strongly dependent on the amount of galactose content. Other authors suggest that intermolecular attractions occur between the side chains of Xanthan in the helical form and the backbone of the Guar (4). However, synergistic effect on viscosity of polymer mixtures is not a common phenomenon. Most polymers mixtures exhibit logarithmic additive increases in viscosity when mixed in solution, as the only interaction present is intermolecular overlapping.

Scleroglucan (SC) is another polysaccharide produced by microbial fermentation. Initial application of scleroglucan for oil recovery has shown greater efficiency and stability than xanthan over a wide range of temperature and pH. On the other hand, presently, scleroglucan is not as economical as xanthan gum, but the potential use of scleroglucan could soon become economical for oil recovery and drilling (5).

Well drilling and completion applications rely on turbulent flow so that the fluid can effectively transport solid particles that are product of the drilling operation. However, there is a compromise between turbulent forces and viscous forces in the transport of solids. Low shear rate viscosity of the fluid has been identified as a major contributor to the suspension of concentrated solids suspension in the wellbore (3).

In this work, the rheology of xanthan gum/guar gum mixtures, and mixtures of scleroglucan/hydroxyethyl cellulose (HEC) in high salinities brines has been studied for their potential use in cleaning near wellbore operations. In addition to reporting the synergistic effects on viscosity exhibited by these two binary polymeric systems at high salinities, the interaction of the polymers with ionic surfactants is shown. Some of the polymer-surfactant mixtures show increases in viscosity of up to one order of magnitude with respect to the pure polymer solutions.

Materials and Methods

Materials

Commercial industrial grade polymers were used: Xanthan Gum (CP Kelco), Guar Gum (Frac-Chem), Scleroglucan CS-11 (Cargill), and a Hydroxyethyl cellulose drilling grade (DOW). Sodium chloride and calcium chloride (Sigma) were used to prepare artificial brines of high TDS. Anionic surfactants sodium laureth sulfate with 3 EO groups (Steol CS460 from Stepan), and Dioctyl sodium sulfosuccinate (Aerosol OT, Fisher), were used without further purification. Additional surfactants tested were alkyl propoxy alcohols sulfates of different chemical structures (8 to 12 C; 3 to 8 PO, from Sasol).

Natural brine samples were collected from an oilfield in northeast Oklahoma. The brine samples were vacuum filtered with 8 microns glass fiber filters. A total of 250 ppm (mg/L) sodium gluconate was added to the filtered brine to prevent iron precipitation, while 200 ppm of formaldehyde was added to prevent polymer biodegradation. Field brine sample contained total dissolve solids (TDS) concentration of 16.4%. Standard tap water was used as a control to compare with the different formulations at higher TDS. Standard tap water was prepared by adding 1.0 grams of NaCl and 0.15 grams of $CaCl_2$ to 1 L of deionizer water. Effect of salinity was also studied with solutions of 5% NaCl and 10% NaCl.

Preparation and Measurements

Polymer solutions were prepared by adding the components to a high speed blender (Hamilton Beach) for 8-10 minutes. Solutions were then left at room temperature for at least 12 hours before viscosity measurements were taken.

A Brookfield rotational viscometer (DV II Pro Digital) equipped with a LV Couette geometry was used to measure viscosity as a function of shear rate at 25ºC. Temperature was controlled by the use of a water jacket attached to the viscometer and a water bath. For highly viscous solutions, a Rheosys (Merlin II) equipped with a viscometer was used.

Results and Discussion

Xanthan gum and guar gum interactions and their effects on the rheology of the two polymers have been studied and reported before (3). The addition of guar gum in small fractions increases the viscosity with respect to that of the pure guar gum, and reduces the amount of xanthan gum necessary to reach the same viscosity (Figure 1). In this particular case, when the total polymer concentration is below 3,000 ppm, the interaction is not strong enough to induce an increase in viscosity for the mixtures that exceeds both, the viscosity of the pure XG, and GG at the same total concentration. At 4,000 ppm, the intermolecular interactions reach their maximum level, and the viscosities of the mixtures are significantly higher than those of the pure polymers.

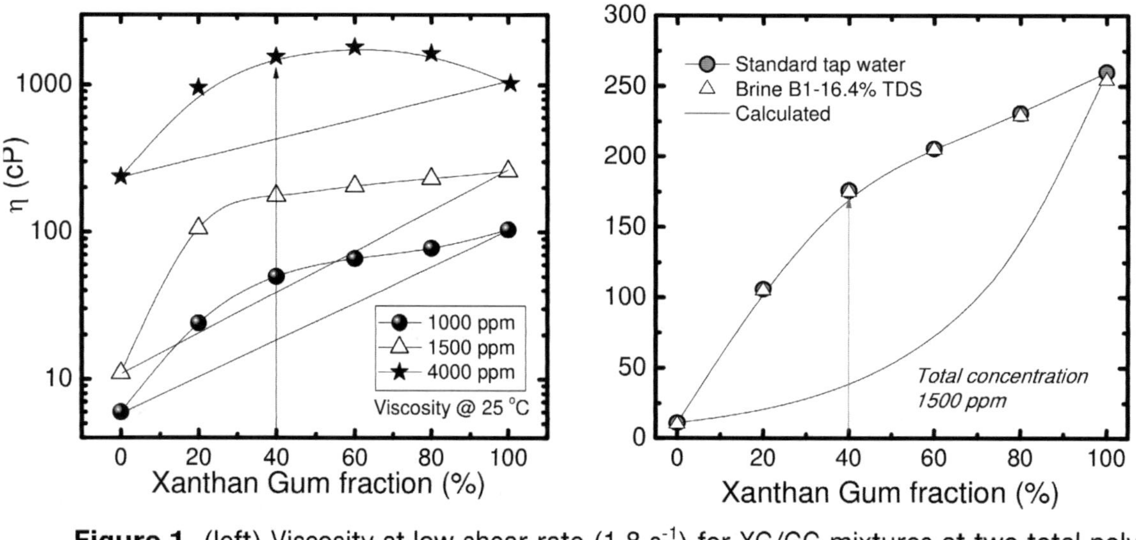

Figure 1. (left) Viscosity at low shear rate (1.8 s^{-1}) for XG/GG mixtures at two total polymer concentrations in standard tap water. (right) Comparison of the experimental performance of the mixture and the calculated viscosity if no synergy was present (solid lines). Red line shows the optimum in synergy for the mixture. T=25 ºC.

The synergy is such that total polymer concentration could be reduced by using the mixture. Furthermore, on average, GG has a third to a half of the cost of XG, which means that the use of GG in the formulation near the optimum (40/60 XG/GG), may reduce even more the cost of the overall formulation. According to the results, the ionic strength of the solution had no major effect on the performance of the system (Figure 1, right).

Different mixtures of the four polymers studied (XG, GG, Sc, HEC) were tested for synergistic rheological effects. Besides XG/GG, the only other combination that showed important synergistic features was the scleroglucan-hydroxyethyl cellulose system (Figure 2). Similarly, the mixture exhibit low-shear viscosity that exceeds the values of an ideal additive system (solid line). The effect is enhanced at higher total polymer concentrations as the entanglements and intermolecular interactions for both polymers increase. However, the nature of this rheological synergy has to involve more complex interactions. The synergy could be related to physical interactions between the cylindrical triple-helix conformation of the scleroglucan molecules (6), and the highly expanded chains of HEC stabilized by hydrogen bonds. The interconnected network of polymer molecules provides a high viscosity and increases the elasticity of the system.

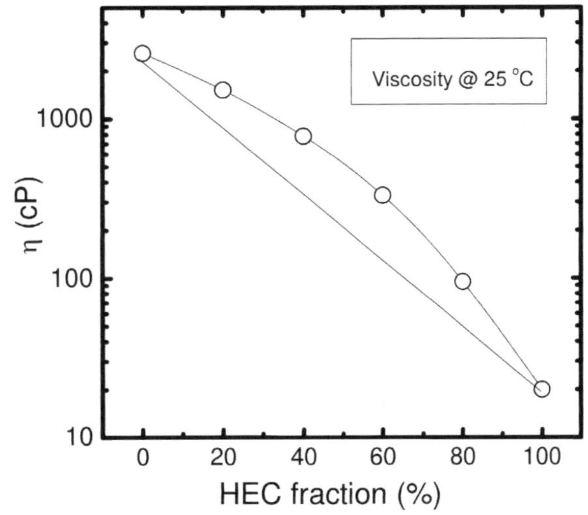

Figure 2. Viscosity at low shear rate (1.8 s^{-1}) for SC-HEC mixtures (2000 ppm total polymer concentration) in standard tap water. T=25 ºC.

Scleroglucan showed a strong dependence of viscosity upon increasing the ion strength of the solution. In standard tap water, viscosity at 0.6 s^{-1} and 25 ºC, of a 2,000 ppm SC solution was 785 cP; in 5% NaCl solution viscosity increased to 1,010 cP; and in 10% NaCl solution to 1,350 cP. In the natural high-TDS brine, viscosity of the SC was 1,605 cP.

The benefits of formulating fluids with these polymer mixtures are noteworthy. On the other hand, one common additive of wellbore cleaning and completion fluids are surface active chemicals. In addition to acting as a displacement fluid, the casing cleaning fluid should remove, or dissolve the remaining drilling fluid and possible debris remaining after the drilling operation (7). These functions are usually achieved by the addition of surfactants. In some cases, surfactants can interact with viscosifying polymers in the formulation to produce highly viscous fluids with unique properties. Surfactants that are able to create rod-like micelles at low concentrations and high salinities are of special interest, since these structures can promote strong interactions with associative and non-associative polymers which lead to significant increases in viscosity and elasticity of the fluids (8). However, most surfactants able to generate these rod-like micelles are relatively expensive cationic surfactants, and for anionic surfactants there are fewer reported in the literature (9).

From all the surfactants and their combination with polymers tested in this work, only one system showed promising results in terms of interactions and stability at high salinities: HEC-only, and HEC-SC mixtures combined with sodium laureth sulfate (Figure 3). At relatively low salinities (below 5%), the surfactant does not change the polymer viscosity in solution (Newtonian profiles). As salt concentration increases, the wormlike micelles start to create an interconnected network that results in one order of magnitude increase in viscosity with respect to the pure HEC solution, along with the appearance of non-Newtonian behavior. On the other hands, the surfactant-only solution (no polymer), showed a viscosity of 35 cP in a 10% NaCl solution.

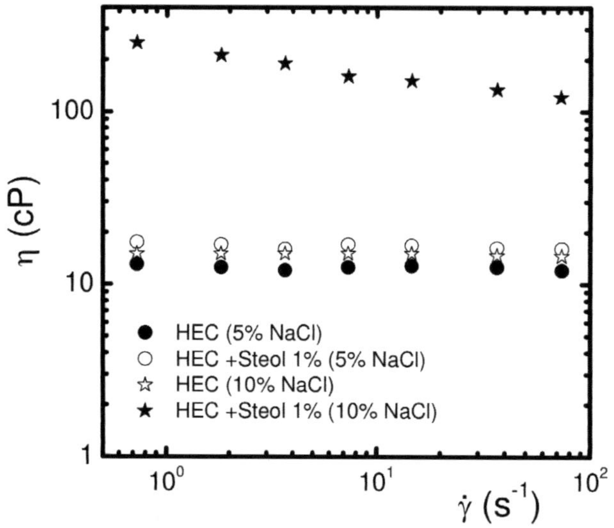

Figure 3. Viscosity as a function of shear rate for HEC and HEC-Steol 460 mixtures at two salt concentrations. Polymer concentration: 2000 ppm. T=25 ºC.

SC-surfactant mixtures did not show any significant changes in rheological properties (data not shown). However, given the synergy between SC and HEC, binary mixtures of these polymers in conjunction with the surfactant are particularly interesting. The interaction of the polymer mixture (SC/HEC) with anionic surfactant was studied using the same approach described above (Figure 4). The results demonstrate that the presence of small amounts of HEC in the mixture (20 wt% in a 2,000 ppm total polymer concentration), are enough to induce noticeable change in the viscosity of the polymers/surfactant solution. At low shear rates (2.9 s^{-1}), the viscosity of the system increases from 197 to 910 cP upon the addition of the surfactant at relatively high salinities (10% NaCl).

Non-associative polymers, such as HEC do not form molecular complexes with surfactants; i.e., the rod-like micelles do not attach directly to the polymer chains. Surface tension measurements have been used to confirm that the synergy in the viscosity seems to be a physical effect product of cooperative entanglements between the polymer chains and the wormlike micelles (8).

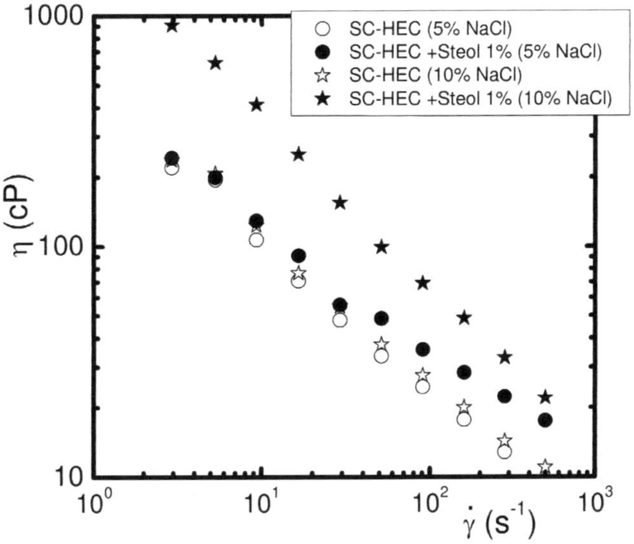

Figure 4. Viscosity as a function of shear rate for SC/HEC 80/20 and SC/HEC 80/20 with Steol 460 mixtures at two salt concentrations. Polymer concentration: 2000 ppm. T=25 ºC.

Concluding Remarks

Mixtures of natural biopolymers can lead to a reduction in the total amount of polymer needed to formulate drill-in and completion fluids by taking advantage of the synergistic effects exhibited by these systems, particularly xanthan gum/guar gum and scleroglucan/hydroxyethyl cellulose.

The interaction between the polymers is directly related to their total concentration in solution. High salinities do not affect performance for the XG/GG mixtures, while increases in salt content enhance rheological properties of SC/HEC mixtures in solution. The last system shows a positive interaction with the addition of sodium laureth sulfate at high salinities (above 5% NaCl). The polymer-surfactant mixture exhibits viscosities up to one order of magnitude above their counterpart pure polymer and pure surfactant solutions.

References

1. Lundin, L; Hermansson, A.M (1995), *"Supermolecular aspects of xanthán–locust bean gum gels based on rheology and electron microscopy".* Carbohydrate Polymers 26, 129-1402.

2. Schorsch, C; Garnier C, Doublier, JL (1995), "Microscopy of xanthan/galactomannan mixtures". Carbohydrate Polymers 28, 319-323.

3. Amundarain, J.L; Castro, L.J.; Rojas, M.R.; Siquier, S; Ramírez, N.; Müller, A.; Sáez, E.A. (2009), *"Solutions of xanthan gum/guar gum mixtures: shear rheology, porous media flow, and solids transport in annular flow"*. Rheological Acta 48, 491-498.

4. Dea, I.C.; Morris, E.R.; Rees, D.A.; Welsh, E.J.; Barnes, H.A.; Price, J. (1977), *"Association of like and unlike polysaccharides: mechanism and specificity in galactomannans, interacting bacterial polysaccharides, and related systems"*. Carbohydrate Research, 57, 249-272.

5. Survase, Shrikant, A.; Saudagar, Parag S.; Bajaj, Ishwar B. and Singhal, Rekha S. (2007), " Scleroglucan: Fermentative Production, Downstream Processing and Applications". Food Technol. Biotechnol. 45, 107-118.

6. Palleschi A, Bocchinfuso G, Coviello T, Alhaique F. (2005), *"Molecular dynamics investigations of the polysaccharide scleroglucan: first study on the triple helix structure"*. Carbohydrate Research 340, 2154–2162.

7. Berg, Erik; Nabati-Shoghl, Farzad; Saasen, Arild; Sedberg, Stian (2002), "*Emulsion-Based Fluid used To Optimize Casing-Cleaning Operations"*. IADC/SPE Drilling Conference, Dallas, Texas, 26–28 February 2002.

8. Rojas, Mario R.; Müller, Alejandro; Sáez, Eduardo (2008), *" Synergistic effects in flows of mixtures of wormlike micelles and hydroxyethyl celluloses with or without hydrophobic modifications"*. Journal of Colloid and Interface Science 322, 65–72.

9. Mu, Jian-Hai; Li, Gan-Zuo (2001), "The formation of wormlike micelles in anionic surfactant aqueous solutions in the presence of bivalent counterion". Chemical Physics Letters 345, 100-104.

New Method For Removing Formation Damage With Nanoparticle Dispersions

By Paul McElfresh, David Holcomb, and Daniel Ector, FTS International, Inc.

This invited paper was prepared for presentation in the 1st International Conference on Upstream Engineering and Flow Assurance, a part of the 2012 AIChE Spring Meeting, held in Houston, Texas, USA, 1-5 April 2012.

Abstract

A new surface energy mechanism, based on recent joint industry-university research using nanoparticle dispersions (NPD) is now available to the oil and gas industry. Nanoparticle dispersions (or colloidal particle dispersions in the classic surface chemical sense) provide a unique enabling mechanism to improve the efficiency of fluids and additives in the performance of their intended actions during a number of reservoir intervention applications. This novel fluid chemistry has shown potential for improving stimulation fluid recovery, remediating wellbore damage issues such as paraffin, water block and deep induced imbibition, as well as enhancing the recovery of oil, gas and water from a wide range of reservoir types. Also being readied for implementation is the use of NPDs to provide increased hydrocarbon production and injection efficiency from water flooding and other improved hydrocarbon recovery operations. Increasing injectivity into saltwater disposal wells (SWD) has been accomplished in beta test field trials. Graphic experimental demonstrations of these mechanisms will be shown and discussed, with emphasis on their current and potential field treatment applications, and with comparisons to another successfully utilized additive technology.

This paper will focus on the colloidal dispersion of solid particles. The nanoparticles in NPD utilize a relatively new application for the mechanism of disjoining pressure by self-assembling into a wedge film once the nanoparticles encounter a discontinuous phase. Innovative, aqueous dispersed, amorphous silica nanoparticle (4 - 20 nm) fluids have been developed and tested which uniquely demonstrate disjoining pressure functioning with respect to wetting agent and nanoparticles dispersed in various type aqueous phases. This wedge film acts to separate formation fluids (oil, paraffin, water, and/or gas) from the formation's surface, thereby recovering more fluids than previously possible with conventional additives or fluids. Several examples of the use of disjoining pressure to improve fluid efficiency will be shown along with case studies from beta testing.

Introduction

Nanoparticle dispersions (or colloidal particle dispersions in the classic surface chemical sense) provide a unique enabling mechanism to improve the efficiency of fluids and additives in the performance of their intended actions during a number of reservoir intervention applications.

A wide range of applications for nanotechnology utilizing the mechanism of disjoining pressure have been investigated. Custom designed nanoparticle fluid systems enable solutions that range from typical work-over squeeze treatments for mitigation of paraffin and asphaltene blockage, to recovery of residual oil from depleted fields in water flood applications. The mechanism works to remove a discontinuous phase like oil, gas, or damaging material such as paraffin, polymer, biofilm, or scale from the rock substrate. Fluid systems can be specifically designed for each of these applications by selecting various particles, with or without surface modifications, or with alterations to the electronic nature of the carrier fluid.

Currently, research has been focused on enhancements for use in remedial treatments. In these systems, nanofluids act as an enhancer to established methods of dissolution. The benefits the new mechanism provides is the removal of foreign matter from reservoir surfaces to be produced from the well, rather than established practices of dissolution alone that can displace damage from the near wellbore region further into the reservoir. These traditional methods of remediation provide a temporary solution that will have to be treated continually with further interventions. The complete removal of the damage from the well results in a treatment that gives longer lasting results, providing for a higher ROI as well as a lowered chemical footprint. For use in water flooding applications, nanofluids have been studied in conjunction with conventional surfactants. The addition of nanoparticles to the surfactant systems improves the detergency of the flood, leading to higher recoveries of residual oil.

Disjoining Pressure

Nanofluids are stable colloidal dispersions or micellar dispersions that accelerate recovery of hydrocarbon from oil and gas reservoirs, and were developed by the authors in conjunction with the Illinois Institute of Technology's Dr. Darsh Wasan and Dr. Alex Nikolov. This paper will focus on the colloidal dispersion of solid particles. The nanoparticles in NPD utilize a relatively new application for the mechanism of disjoining pressure, **Figure 1**, by self-assembling into a wedge film, **Figure 2**, once the nanoparticles encounter a discontinuous phase. This wedge film acts to separate formation fluids (oil, paraffin, water, and/or gas) from the formation's surface, thereby recovering more fluids than previously possible with conventional additives or fluids.

Disjoining pressure is a function of surface energy, interfacial tension, capillary pressure, and zeta potential. These forces act together to cause a self-assembly of particles which act energetically at an interface in such a manner that causes displacement of material on that surface[3]. This force can be calculated and modeled[1]. In many laboratory experiments, pressures of 50,000 Pa have been observed.

Because of the inherent low stability of colloids in high salt environments, modification of the nanoparticles is necessary to maintain the disjoining pressure functionality. Stability is achieved by controlling the surface charge density through either surface modification of the particles, control of the ionic environment of the fluid via surfactants, or a combination of both. This method of stability control enables beneficial secondary effects such as decreased capillary pressure and formation wettability. In the examples to follow, the use of surfactants is enhanced by the driving force of the disjoining pressure to initiate removal of material and then with the subsequent function of surfactants to cause separation of the phases involved.

Example Paraffin Treatments

Most treatments performed to date have been on vertical wells with extremely low production rates. Many of these wells had organic blockage problems associated with paraffin, or asphaltene. In these wells, the most successful treatments have been performed in conjunction with a traditionally used solvent, such as xylene or a terpene. Treatment sizes range from 75-200 gal/ft of pay depending upon the severity of the problem. The ratio of solvent to the aqueous phase, containing the NPD additive, varies from 25% to 50% - also depending on the nature of the treatment. It is hypothesized this two-fold approach provides superior results to using each additive alone because each chemical has its own unique advantages: a solvent will dissolve a paraffin or asphaltene, but as mentioned, can displace the damage over a wide area further into the reservoir. A nanofluid, is able to effectively remove damage from the rock surface, but will not dissolve the organic matter. The use of both fluids in tandem denotes the beneficial properties unique to each system, and provides superior results.

This approach was applied to two Nacotach Formation salt water disposal (SWD) wells. A 5-stage treatment was performed at 5 bbl/min where each stage consisted of 2,000 gal of load water, 2,000 gal of 10 – 15% HCl, 215 gal paraffin solvent, 660 gal 15% NPD mixture, and 1,000 gal of diverter in 2% KCl brine. One job was pumped with ball sealers and one used rock salt in between stages. Results immediately after treatment and over the next six months showed both wells improved from 500 bpd to 1500 bpd injection volume at 3,700 psi injection pressure. At $1/bbl. disposal cost, the treatments paid for themselves in 30 days. Previous treatment procedures involving paraffin solvent /acid jobs showed only a 200-300 bbl/day increase in injectivity. The only change to the old procedure was the addition of the NPD fluid.

In another field example of NPD usage, 2,000 gallons of treatment consisting of 60% (17% active) NPD fluid and 40% paraffin solvent were placed in the annulus of a Sprayberry formation well and shut in overnight. When the well was returned to pumping, bulk paraffin was produced intermittently for several days and final oil production was increased from 6 to 12 BOPD. Treatment lifetime also improved from 30 to 90 days.

Fracturing Polymer Example

This mechanism can also be applied to scale and polymer blocks, usually caused by intervention processes during completion, or increasing brackishness of produced water. Again, a traditional treatment was used in conjunction with the disjoining pressure mechanism provided by the nanofluid. In this case, oxidizing breaker performance was enhanced to more effectively remove polymer damage from linear guar in proppants. In severe cases, oxidizers will not sufficiently remove all of the polymer damage. Experiments performed have shown a nanofluid/oxidizer combination enhances removal of polymer skin damage compared to oxidizer alone.

In this example, filter cakes were deposited onto ceramic disks at a concentration of 200 lbs/1000 gallons[2]. The filter cakes were treated with a 1.25 lb/1000 gallon concentration of ammonium persulfate. A second sample was treated with the same fluid with the addition of 10% (v/v) surface modified NPD fluid. Both fluids were digested at 165°F for 16 hours. The results in **Figure 3** show partially reacted guar residue that tends to settle from solution. However, in **Figure 4**, the guar residue is completely lifted from the ceramic disk as illustrated by the floating residue in the fluid. Inclusion of the modified NPD has enhanced the removal of the filter cake even though the ammonium persulfate has been completely reacted. The conclusion from this example indicates that fracturing treatments with this additional additive will enable removal of more polymer residual and lead to better productivity.

EOR Examples

The most promising application of the technology is in the area of EOR. Nanoparticles can be added with a surfactant package, enhancing its detergent properties utilizing the same technical mechanism. Core testing has shown nanofluids are an effective additive in water flood applications whereby residual oil is removed from the substrate for production. Additional recoveries of 25% OOIP after simulated primary production have been observed in lab settings. Fluid systems are designed to be compatible with downhole chemistries, and geological parameters. Basic parameters in this regard were high salt brines and compatibility with both sandstone and limestone formations.

In order to estimate the effectiveness of the NPD, a series of fluid imbibition tests were conducted in API brine (8% NaCl and 2% $CaCl_2$ w/v) at room temperature and at 195°F. These tests were run with Berea sandstone core samples, 160 mD, saturated with San Andres crude which had been cured in a 120°F oven overnight. Emulsion tests were performed with the NPD fluid to ensure compatibility with the oil as well as compatibility with API brine. Typical imbibition results are shown in **Figure 5**.

Core flow testing was also conducted on oil-saturated sandstone and limestone cores. Generally, the procedure was as follows:

1. Core saturated with San Andres crude oil
2. Baseline established with 2% KCl
3. Followed immediately with 10 pore volumes of API brine/10% NPD additive
4. Shut in for 16 hours then injected 10 pore volumes of API brine/10% NPD additive
5. Produced using 2% KCl water drive

Figure 6 shows this procedure in diagrammatic form. Initial flow of 2% KCl brine removed 65% of the oil in the core. After the introduction and shut-in of the NPD fluid, 35% of the remaining oil was removed, leaving only 12% of the original oil in the core.

The same procedure was applied to Indiana limestone cores, 6.9 mD, that were saturated with Eagle Ford oil. The effectiveness of the formulation was screened with imbibition tests to find a suitable candidate for core flow testing. **Figure 7** shows a typical pressure chart for the most effective NPD formulation developed from the screening tests. The same NPD fluid used for sandstones showed proved to also be effective in limestone, showing no damage and slight stimulation of the core. The fluid removed 9.6% additional oil compared to 2% KCl alone. Owing to the lower permeability, this result was considered an improvement in oil recovery.

Conclusions

Nanoparticle dispersions have been shown to be a versatile and effective fluid to enhance the removal of hydrocarbons from mineral surfaces by utilization of a unique mechanism of disjoining pressure. Combined with more conventional surfactant technology, fluids can be formulated that will enhance performance and improve lifetime. In both experimental laboratory evaluations and field trials NPD fluids have been shown to provide improvements in the remediation of paraffin, hydrocarbon production improvement, as well as increased injectivity compared to conventional treatments. Through utilization of the strong surface interactions with NPD, it is possible to develop similarly effective fluids to address many other flow assurance and formation damage problems.

Summary of Nano-Fluid spreading to Form a wetting wedge which creates a spreading force that removes oil, gas and/or water from solid surfaces relative to three phase contact angle and the surface tension of an adsorbed film of oil, gas and/or water.

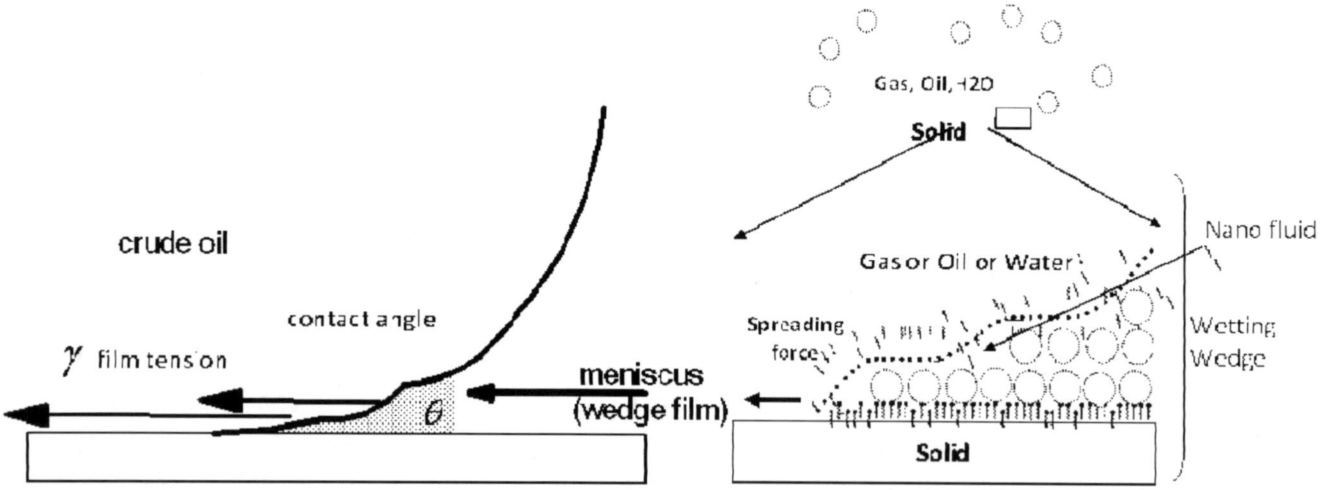

Gas Oil and/or Water Removal Mechanism in the Presence Of Nano-Fluid (Colloidal Particle Dispersion)

Figure 1. Nanofluid Spreading to Form a Wedge Film[1]

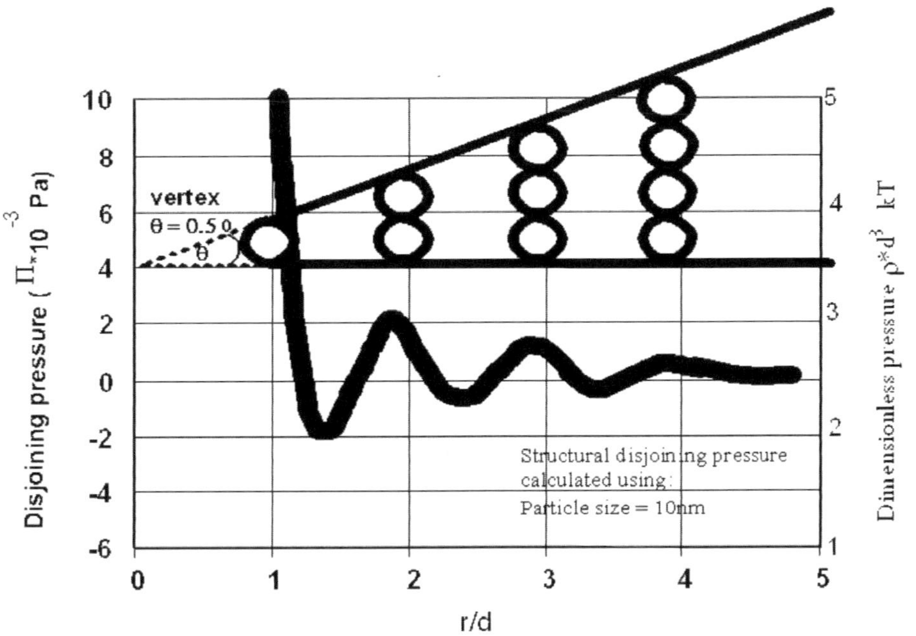

"Pressure profile on the wedge walls for 0.5 ° contact angle at the vortex as a function of radial distance"

Figure 2. Particle distance scaled by diameter, r/d; pressure near the vortex is approximately 50,000 Pascals which is dependent on the effective particle volume fraction and particle size. d=particle diameter; r= particle to center (radial) distance

Figure 3. A 200 lb linear gel filter cake treated with 1.25 pptg ammonium persulfate – elapsed time 16 hours

Figure 4. A 200 lb linear gel filter cake treated with 1.25 pptg ammonium persulfate and 10% NPD fluid – elapsed time 16 hours

Figure 5. Typical Imbibition Test Results

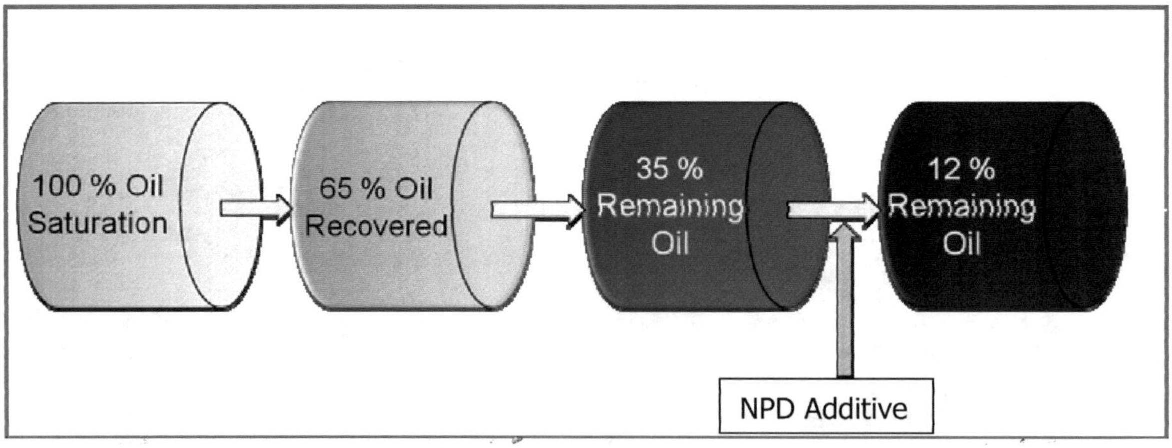

Figure 6. Core Testing - Recoverable Oil With San Andres Oil in Berea Sandstone

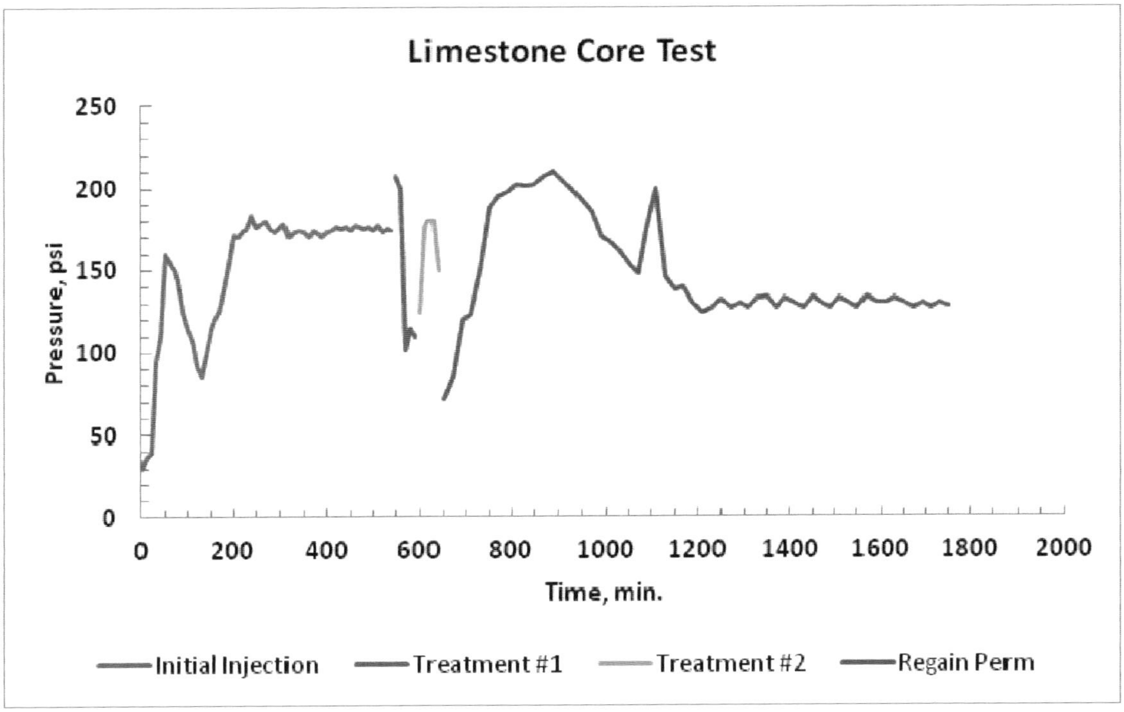

Figure 7. Pressure Chart for Limestone Core Flow Test

References

1. A. Trokhymchuk, D. Henderson, A. Nikolov, and D.T. Wasan, "A Simple Calculation of Structural and Depletion Forces for Fluids/Suspensions Confined in a Film", **Langmuir,** 17:4940-4947 (2001).
2. P. McElfresh, Carla Olguin and D. Ector, SPE 151848, "The Application of Nanoparticle Dispersions to remove Paraffin and Polymer Filter Cake Damage", SPE International Symposium and Exhibition on Formation Damage Control , Lafayette, Louisiana, USA, 15–17 February 2012.
3. D.T. Wasan and A.D. Nikolov, "Spreading of Nanofluids on Solid", **Nature,** 423: 156-159 (2003).

Further Reading

D. Boda, K. Y. Chan, D. Henderson, D. T. Wasan, and A. D. Nikolov, "Pressure profile on the wedge walls for 0.5 ^{o} contact angle at the vortex as a function of radial distance", **Langmuir,** 15, 4311-4313, 1999

A. Chengara, A.D. Nikolov, D.T. Wasan, A. Trokhymchuk, and D. Henderson, "Spreading of Nanofluids Driven by the Structural Disjoining Pressure Gradient", **Journal of Colloid and Interface Science,** 280:192-201 (2004).

Temperature Control of Drilling Fluid with Phase-Change Materials

Othon R. Monteiro and Lirio Quintero, Baker Hughes, Houston, Texas, and*
Martin Bates, Duncan Bruce, Elisabeth Wilcock, Antonina I. Smirnova and Alastair J. Logan,
University of York, York, England.
**Address correspondence to Othon.Monteiro@bakerhughes.com*

This invited paper was prepared for presentation in the 1st International Conference on Upstream Engineering and Flow Assurance, a part of the 2012 AIChE Spring Meeting, held in Houston, Texas, USA, 1-5 April 2012.

Abstract

The trend toward drilling deeper and consequently hotter wells places increasing demands on the electronics in the bottomhole assembly. This article introduces and discusses the benefits and limitations of a new technology to reduce the temperature of circulating fluids in wellbores. This technology makes use of phase-change materials (PCM), which utilize the latent heat of phase change to absorb the energy transferred to the drilling fluid from the formation or the drilling string. A combination of mathematical and physical modeling is used to determine the temperature reduction in drilling fluids that can be achieved with conventional PCMs, and new forms of PCMs are discussed.

Introduction

The increasing world's demand for energy and the favorable economics drive oil companies to prospect and drill for oil in deeper and more challenging reservoirs, which usually translates into higher pressures and higher temperatures.[1] This places more stringent requirements on the properties of drilling fluids. Stable rheological properties at these elevated temperatures are important for the fluid to perform its main functions such as cleaning holes, suspending solids, controlling sub-surface pressure and minimizing formation damage. An additional important function of drilling fluids is to cool the tools in the drillstring, especially the heat-sensitive instrumentation for measuring while drilling (MWD) in the bottomhole assembly (BHA), which contains multiple electronics. Conventional electronic devices are designed for continuous operations at temperatures up to 75°C and peak temperatures up to 150°C (302°F). Recent developments in Silicon-on-Insulator (SOI) technology have expanded this range by allowing maximum temperatures up to 225°C, but this technology is still costly and of limited use. Other benefits of limiting the temperature of the drilling fluid include enhancing borehole stability and reducing bit temperature and subsequently, bit wear. Finally, the cooling action of the drilling fluid helps maintain the temperature of elastomers below their design failure limits.

A conventional method to lower the temperature of the drilling fluid involves the use of mud-coolers, which are basically heat-exchangers located at the surface of the rig[2] that reduce the mud temperature closer to ambient temperature before the fluid is re-injected into the drill string. These coolers have been primarily used for geo-thermal drilling due to the elevated temperatures encountered in said systems. This paper presents a new method for lowering the temperature of circulating fluids in the wellbore through the use of phase-change materials (PCM) added directly to the drilling fluids.

PCMs have been used for energy storage or thermal stabilization in several applications. PCMs can absorb or release latent heat when the temperature of the material increases or decreases beyond its melting point.[3] The heat absorbed or released during a phase change (per unit mass) as latent heat is significantly higher than the sensible heat, and equally as important, the heat absorption and release can be designed to take place at a temperature of interest. PCMs have been developed for various applications due to their different phase change intervals – e.g., stabilizing low temperatures in air-conditioning applications, and preventing temperatures surges that can lead to spontaneous ignition.[4] The variety of PCMs investigated to date can be subdivided into organic, inorganic and eutectic PCMs.[5]

The criteria governing the selection of PCMs for heat storage applications are: phase-change temperature in the desired temperature range of the application; high latent heat per unit mass; high thermal conductivity; thermal stability; small volume changes during phase transition; little or no sub-cooling; chemical stability; and availability at low costs and high quantities. A new and improved type of PCM (in microcapsules) has been recently developed to mitigate the challenges typically encountered by PCMs such as thermal stability, high super-cooling effect and low thermal conductivity. Micro-capsule PCMs improve the heat transfer efficiency because of the increase in surface area, reduce the PCM reactivity with the environment by physically isolating the phase-change material core from direct contact with the environment, and control the change in volume during the phase change.[6]

This article analyzes the effect of different types of PCMs in drilling fluids. A mathematical model of the circulating fluid in a wellbore is developed and used to investigate the potential gains in temperature of at the BHA. The article also presents the results of a series of experiments with different materials highlighting the challenges in using them in wellbore applications.

Experimental:

a. Materials

The PCMs tested for this work are listed in Table 1. Due to the temperature range of interest, the focus here was on materials with a liquid-crystal phase and other organic materials such as waxes and thermoplastics. Inorganic PCMs were out of scope because their melting temperatures are outside the range of interest. Two materials with liquid crystal phases were tested in this study: cholesteryl stearate and zinc stearate from Sigma-Aldrich (product numbers C79409 and 26428, respectively). These materials undergo solid-liquid crystal-isotropic liquid phase transition. The transition temperatures to isotropic liquid, and the enthalpy of phase change corresponding to the total latent heat including intermediate phase changes are shown in Table 1. The selection of these liquid crystal materials followed an extensive analysis of latent heat and densities of over 3,000 compounds using data from a recently published compendium of materials.[7] The materials selected were those that led to the theoretical maximum change in temperature induced by the phase change. Another criterion to narrow down the list of candidates was the cost and availability of the material. Figure 1 shows one example of this analysis that led to the choice of zinc stearate. The other materials in Table 1 have been used as PCM in other applications (wax A and the thermoplastic polymer) or have the potential for the particular set of conditions in the wellbore (wax B).

b. Methods

Latent heat and temperature of phase change were determined by differential scanning calorimetry in a DSC Q20 by TA Instruments (New Castle, DE) with a heating rate of 5°C/min up to 150°C, and similar cooling rate. Thermal stability of the tested PCMs was also studied by carrying out consecutive heating-cooling cycles in the DSC and measuring the transition temperatures and the phase change enthalpies. The calorimetry experiments were carried out with the pure substance (not dispersed in any fluid).

The stability of the PCMs suspended in fluids was investigated by cycling the temperatures of the suspensions in mineral oil and in water up to 150°C, in a pressure cell. Morphology of the PCMs was monitored before and after the treatment.

Table 1: Phase Change Materials

Material	T_{PC} *(°C)	ΔH (J g^{-1})
Cholesteryl Stearate	83	109
Zinc Stearate	128	150
Themoplastic Polymer	65	
Wax A	55	200
Wax B	129	250

*: Phase change temperature

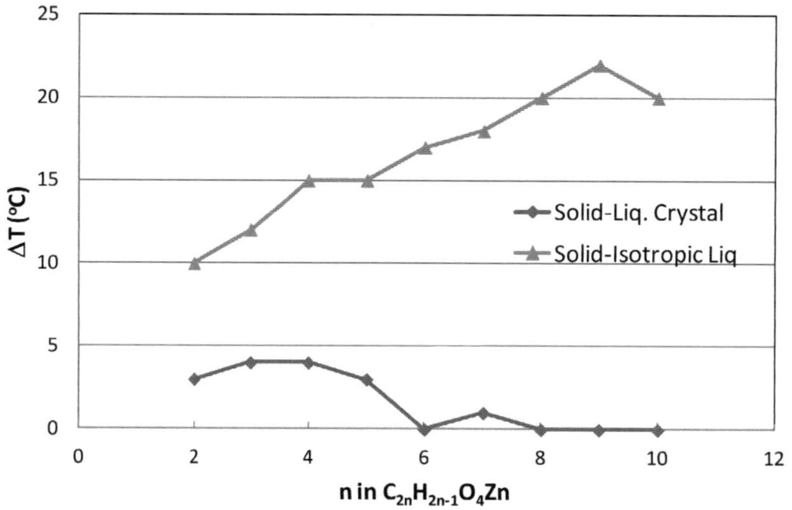

Figure 1: Maximum temperature change resulting from the addition of 5%w/w LC2 series. The phase change temperature and enthalpy data were obtained from Acree[7] and the maximum temperature was calculated from those values.

Mathematical Modeling

A one-dimensional mathematical model of the circulating fluid through the drill-pipe and annulus was developed to estimate the impact of adding PCMs on the temperature of the fluid. A schematic representation of the geometry considered is shown in Figure 2. For the sake of simplicity, the hole diameter was assumed to vary linearly with depth, i.e., we did not consider stepwise changes in the well diameter. Due to the extreme diameter-to-length aspect ratios encountered in wellbores (e.g., only a few cm in diameter and thousands of meters deep), the error introduced by having the wellbore diameter to vary linearly with depth instead of having discrete changes in diameter at certain depths is negligible. This aspect ratio makes the fulcrum geometry virtually cylindrical, except for the small impact of recirculation at locations of changing casing diameter. The temperature of the surrounding rock is assumed to vary linearly with depth. This is a simplification with respect to more sophisticated models that assume the temperature of the rock at an infinite distance from the wellbore to be constant and make use of a thermal conductivity of the rock.

The wellbore of depth (L) is divided into N vertical slices of thickness Δz, such that L = N Δz. Slices are numbered starting from the surface. The circulating fluid is split into two regions; region I inside the drill string and region II outside, in the annulus. At the bottom of the wellbore, we assume that the circulating fluid flows from region I to region II, without considering any specific geometry / shape. The model is one-dimensional and does not consider radial variation of temperature.

Considering initially the circulating fluid in region I inside the drill string, the temperature of the fluid inside each cell (or slice) *i* is T_i^I. In the flowing fluid, the total energy of the fluid in cell *i* varies due to fluid entering from cell $i-1$, leaving to cell $i+1$ and the convective exchange of heat with the drill string. The fluid and temperature inside each cell is assumed constant and homogeneous, and for the purposes of simplification, heat generated inside the cell is neglected. The temperature of the fluid entering cell *i* is T_{i-1}^I and the fluid leaving is at temperature T_i^I.

The change in energy of a cell *i* inside region I, ΔQ_i^I, over a short time period Δt is

$$\Delta Q_i^I = F\Delta t\rho C T_{i-1}^I - F\Delta t\rho C T_i^I + h\Delta t\,\pi d_1\Delta z(T_i^P - T_i^I) \qquad (1)$$

where *F* is the flow rate, ρ is the mass density of the fluid and C is the specific heat capacity. The first two terms correspond to the fluid entering and leaving the cell. The third term originates from the heat supplied to the fluid from the drill pipe wall, and T_i^F is its temperature in slice *i* and *h* is the heat transfer coefficient of the fluid and surface of the drill pipe wall. The change in temperature ΔT_i^I of the fluid in cell *i* is determined and updated by using Equation 2:

$$\Delta Q_i^I = \frac{\pi d_1^2\Delta z}{4}\rho C\Delta T_i^I \qquad (2)$$

A similar set of equations is written for the returning fluid in region II, but this time the interaction with the external rock surface is include as is the change in diameter with depth of the cells in region II.

$$\Delta Q_i^{II} = F\Delta t\rho C T_{i+1}^{II} - F\Delta t\rho C T_i^{II} + h\Delta t\,\pi d_2\Delta z(T_i^{II} - T_i^P) + h`\Delta t\,\pi A(T_i^R - T_i^{II}) \qquad (3)$$

where

$$A = \sqrt{(d_{iu}^2 - d_{il}^2)^2 + \left(2\Delta z(d_{iu} + d_{il})\right)^2} \qquad (4)$$

The first term in equation 3 represents the flow of the circulating fluid into and out of the cell, as before. The second term is the energy lost to the drill string; this is again modified to take into account the external diameter of the drill string d_2. The final term is the energy gain by convection from the rock surface. Here *h*` is the heat transfer coefficient of the fluid and surface of the rock and T_i^R is the temperature of the formation at this depth; d_{iu} and d_{il} are the diameters of the upper and lower faces of the frustum shaped volume slice *i*. As before, the change in temperature ΔT_i^{II} of the circulating fluid in the slice *i* of the annulus is calculated from

$$\Delta Q_i^{II} = \frac{\pi\Delta z}{4}\left(\frac{1}{3}(d_{iu}^2 + d_{il}^2 + d_{iu}d_{il}) - d_2^2\right)\rho C\Delta T_i^{II} \qquad (5)$$

The last equation for the model stems from the assumption that the drill string is thin enough so that it can be described as having constant temperature in each slice. In this case, heat conduction along the length of the drill string can be ignored, and

$$\Delta Q_i^P = h\Delta t\, \pi d_2 \Delta z (T_i^{II} - T_i^P) - h\Delta t\, \pi d_1 (T_i^P - T_i^I) \tag{6}$$

For multi-component fluids, ρC should be replaced with $\sum x_j \rho_j C_j$ in equations 1 through 5 where the subscript *j* here represents the components in the mixture and x_j is its volumetric fraction.

In the case of phase changes, the calculations consider the flow of heat rather than temperature in each slice, for in regions undergoing a transition, the temperature is constant but the heat content will change. For simplicity, we assume that the fluid is a two component system with component 2 having a phase transition, and in region I the equation equivalent to Equation 1 becomes

$$\Delta Q_i^I = F\Delta t \frac{Q_{i-1}^I}{V_{i-1}} - F\Delta t \frac{Q_i^I}{V_i} + h\Delta t\, \pi d_1 \Delta z (T_i^P - T_i^I) \tag{7}$$

where Q_i and V_i are the heat content and volume of cell *i*, respectively. It is still necessary to determine the temperature of cell *i* to be able to calculate the heat exchange with the drill string. In order to calculate the temperature, we determine two thresholds; the heat (per unit volume) required to raise the temperature from some arbitrary reference point to the transition temperature, and this threshold plus the specific heat of melting (or enthalpy of fusion, again per unit volume). By checking whether the energy content of the cell is below, between or above these thresholds the temperature in the volume element is calculated. Once we know which regime we have, the temperature is determined as before from the relevant heat capacity.

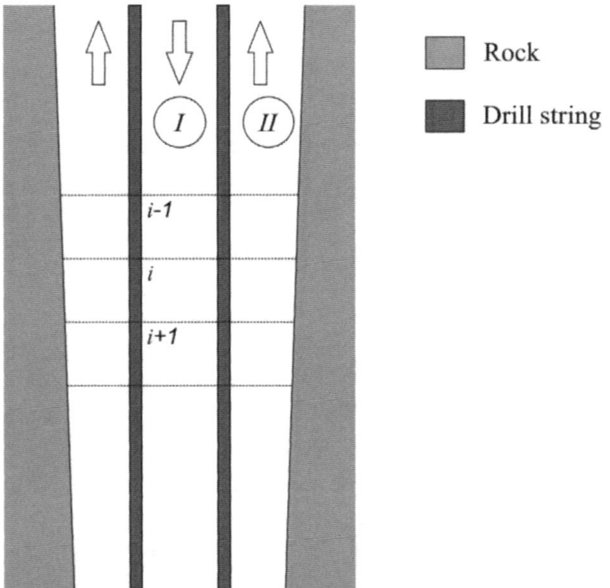

Figure 2: Schematic representation of the geometry used in the mathematical model of the wellbore temperature

Results and Discussions

a. Modeling:

Figure 3(a) shows the steady state temperature profile of a circulating water-based drilling fluid with and without PCM. The temperature of the drilling fluid at the entrance of the well is assumed to be 25°C, and the return temperature of the fluid in the annulus at the surface is left to float to its final value, which results from the energy balance through its path. The phase-change temperature of the PCM is easily seen as the horizontal plateau in the dashed curve. Figure 3(b) shows a magnified view of the final region of the drill pipe. The particular PCM used for the calculations in Figure 3(b) was a wax with melting point of 90°C and latent heat of $3\text{x}10^5$ J/kg. Introduction of this PCM results in lowering the temperature by 2.5°C and 5.5°C for a 5% and 10% addition, respectively.

The calculated decrease in temperature ΔT near the BHA is not as large as what would be expected if one considers only the latent heat of fusion of the PCM. This results from the differences in the specific heat of the base fluid, and the distinct phases of the PCM; organic PCMs have a lower specific heat than water, and the specific heat of the solid PCM is larger than that of the liquid PCM. Metal-base PCMs have higher specific heat as well as higher density that the base fluid, and therefore could bring greater benefit.

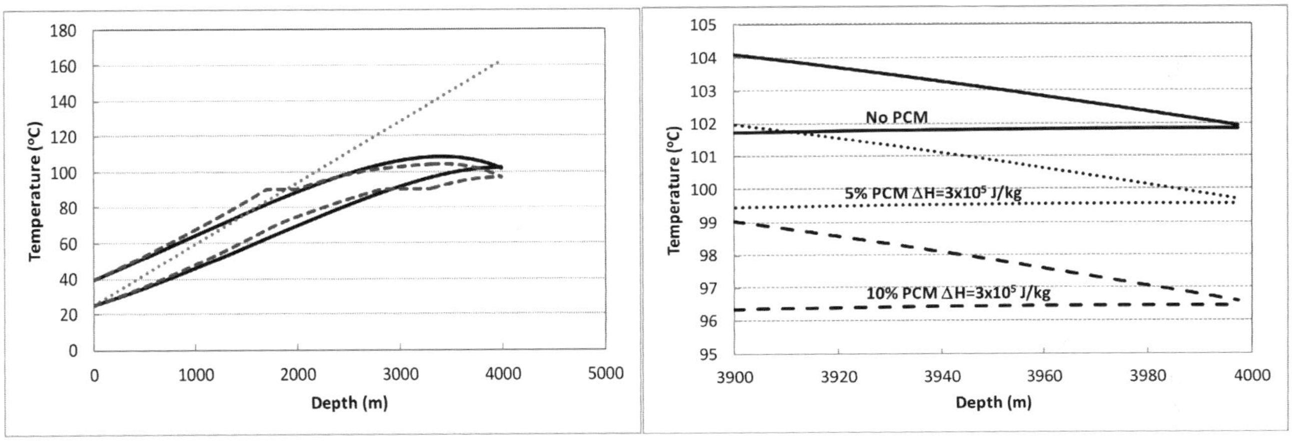

Figure 3: (a) Temperature profile for a circulating drilling fluid without PCM (solid) and with a PCM with melting point of 90°C (dashed). The dotted straight line is formation temperature. (b) Zoomed-in temperature near the bottom for a drilling fluid without PCM and with two concentrations of PCM.

The higher specific heat of water when compared to base oils result in higher fluid temperatures at the bottom of the wellbore when oil-based drilling fluids are used. The effectiveness of the PCM in reducing this temperature is virtually the same as indicated in the example shown in Figure 4, where the temperature profiles for the case of fluids with similar amount of PCMs. The actual temperature change due to the addition of 10% PCM is virtually the same when one utilizes water-based or oil-based fluids although the maximum temperature in the case of an oil-based drilling fluid is higher than that of a water-based drilling fluid.

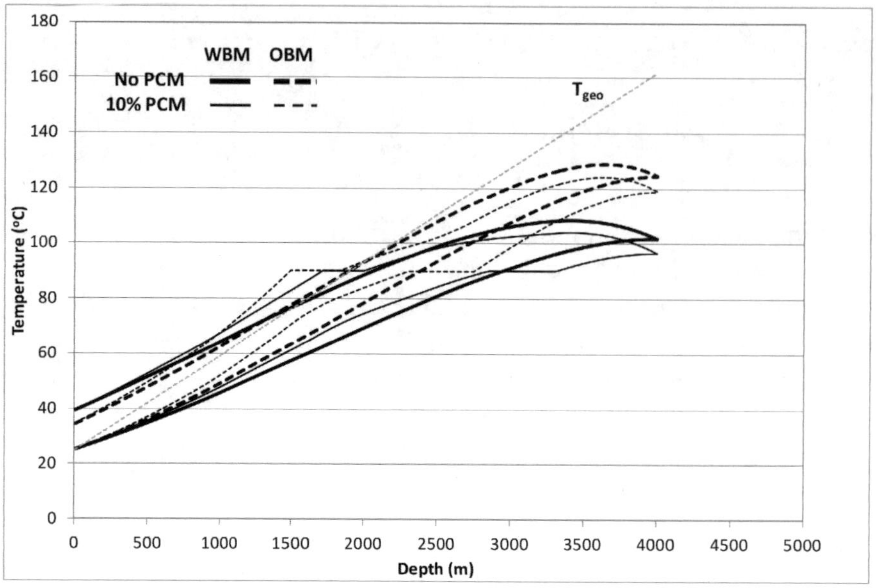

Figure 4: Calculated temperature distribution in the drill string and the annulus for the cases of oil-based and water-based drilling fluids with and without PCM.

Another important parameter when choosing the proper PCM is the phase-change temperature. Figure 5 shows the impact of the phase-change temperature on the temperature of the fluid at maximum depth. Increasing the phase-change temperature lowers the bottom-hole temperature, as long as the entire volume of PCM undergoes the phase change. Beyond that, the temperature of the fluid at the bottom begins to increase until the phase-change temperature exceeds the bottom-hole temperature that the fluid would have in the absence of PCM. Figure 5(a) shows the temperature of the fluid in the drill string and the annulus when two PCMs with different melting temperatures are added, namely 90°C and 100°C. The bottom-hole temperature decreases from 120°C to 117°C when an oil-based drilling fluid is used, whereas in the case of the water-based drilling fluid, the same increase in melting temperature of the PCM results in a increase in the temperature of fluid at the bottom from 97°C to 99°C. Figure 5(b) shows the temperature at maximum depth as a function of the temperature of the phase change of the PCM for a water-based drilling fluid.

b. *Materials*

Despite the potential large reduction in the temperature of the fluid around the BHA achieved when liquid crystals are used as indicated in Figure 1, the materials tested here have encountered a number of difficulties when exposed to conditions similar to those in the wellbore. Both stearates failed the temperature-cycling tests in water and in mineral oil. Cholesteryl stearate was stable in water but not in mineral oil at elevated temperatures. Zinc stearate had the opposite behavior, i.e., stable in oil but not in water. A common problem both materials was that when they did not dissolve, they formed large clumps upon solidification. Solid and liquid PCMs should be insoluble in the carrying fluid and should not form large agglomerates that can lead to obstruction of fluid flow in the drill pipe or annulus. The PCM should also maintain small dimension so that it is not removed from the fluid by the solids control equipment.

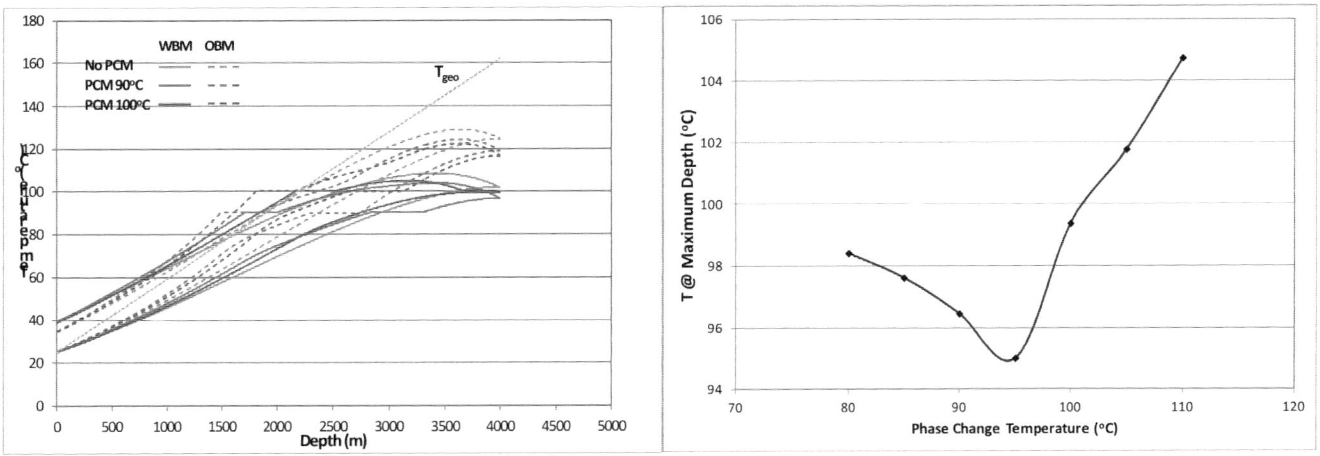

Figure 5(a): Temperature profile for WBM and OBM with PCM with different melting temperature; (b) Temperature at maximum depth in WBM as a function of the temperature of phase change.

Uncoated organic PCMs have encountered similar challenges. The result of a thermal cycling experiment with thermoplastic beads is shown in Figure 6. Even under high shear conditions, the plastic beads agglomerated upon cooling, and over time the particles lose their original appearance.

(a) (b) (c) (d)

Figure 6: Thermoplastic polymer beads: (a) before heating; (b) upon heating in water, the beads become a colorless soft plastic; (c) upon cooling, aggregation leads to network (low density) or clumping (high density) of beads; (d) After 30 minutes of fast stirring, individual particles lost their characteristics.

An approach to make PCMs more resistant to the environmental conditions in the wellbore (thermal and chemical) comprises encapsulating the PCM with a layer that separate the core from the fluid. Wu et al[8] have described a technique to produce microencapsulated n-Octadecane with styrene-divinylbenzene co-polymer shell which are stable up to 260°C. Other forms of core-shell particles have been described in the literature.[9-11] The shell materials can be either organic or inorganic coatings, albeit the latter has been used to a lesser extent in PCMs.[11]

Conclusions

Phase-change materials have been proposed as an approach to improve the cooling capabilities of drilling fluids for HPHT wells. The maximum benefit of PCMs will depend on the latent heat of the phase change as well as the phase change temperature with respect to the temperature profile of the fluid in absence of the PCM. There is an optimal phase change temperature that maximizes the decrease of fluid temperature at the bottom of the well.

A particular challenge among the materials tested here was the form stability, which is critical for PCM to find application in drilling fluids. The formulation of micro-encapsulated or nano-encapsulated materials appear as a promising option to mitigate this problem. Moreover, in a PCM with a shell-core morphology, the shell material may be chosen with a specific (sensible) heat that also contributes for the lowering the fluid temperature.

References

1. Godwin, W.; Ogbonna, J.; Boniface, O. "Advances in Mud Design and Challenges in HPHT Wells" in the Nigeria Annual International Conference and Exhibition, Abuja, Nigeria, 2001 SPE-150737-PP
2. Maury, V.; Guenot, A. "Practical Advantages of Mud Cooling Systems for Drillling", *SPE Drilling and Completion*, 1995 March, p. 42-48.
3. You, M.; Wang, X.; Zhang, X.; Zhang, L.; Wang J. "Microencapsulated n-Octadecane with Styrene-Divinyl Benzene Co-Polymer Shells", *J. Polym. Res.* **2011,** 18, 49-58
4. Fang, G.; Li, H.; Yang, F.; Liu, X.; Wu, S. "Preparation and Characterization of Nano-encapsulated n-tetradecane as Phase Change Material for Thermal Energy Storage" *Chemical Eng. J.* **2009,** 153, 217-221
5. Agyenim, F.; Hewitt, N.; Eames, P.; Smyth, M. "A Review of Materials, Heat Transfer and Phase Change Problem Formulation for Latent Heat Thermal Energy Storage Systems", *Renewable and Sustainable Energy Reviews* **2010,** 14, 615-628
6. Rong, Y.; Chen, H.Z.; Wei, D.C.; Sun, J.Z.; Wang, M. "Microcapsules with Compact Membrane Structure from Gelatin and Styrene–maleic Anhydride Copolymer by Complex Coacervation" *Colloids and Surfaces A: Physicochemical and Engineering Aspects* **2004,** 242, 17–20.
7. Acree, W.E. "Phase Change Enthalpies and Entropies of Liquid Crystals" *J. Phys. Chem. Ref. Data* **2006**, 35, 1051-1330, 2006.
8. Wu, S., Zhu, D., Zhang, X. and Huang, J. "Preparation and Melting/Freezing Characteristics of Cu/Paraffin Nanofluid as Phase Change Material" *Energy Fuels* **2010**, 24, 1894-1898
9. Yi, Y. and Yanqing, L. "Polystyrene Microspheres Coated with Smooth Polyaniline Shells: Preparation and Characterization", *Tsinghua Science and Technol.* **2009,** 14, 546-550
10. Xing, L., Hongyan, L., Shujun, W., Lu, Z and Hua, C. "Preparation and Thermal Properties of Form Stable Paraffin Phase Material Encapsulation" *Energy Conversion and Management* **2006,** 47, 2515-2522.
11. Zhang, P., Ma, Z.W. and Wang, R.Z. "An Overview of Phase Change Material Slurries: MPCS and CHS" *Renewable and Sustainable Energy Reviews* **2010,** 14, 598-614

PHASE BEHAVIOR OF RESERVOIR FLUIDS: COMPARISONS OF PC-SAFT AND CUBIC EOS SIMULATIONS

Sukit Leekumjorn and Kristian Krejbjerg*
Calsep Inc.
11490 Westheimer Road Suite 500
Houston, TX 77077
*Address correspondence to sl@calsep.com

Abstract

Cubic Equations of State (EOS) is an industry standard for simulating the phase behavior of reservoir fluids. Especially the Soave-Redlich-Kwong and the Peng-Robinson equations are widely used. Applied with volume correction, these models can generally capture the phase behavior of reservoir fluids in terms of saturation points and gas-liquid ratios. However for gas compressibility factors and oil compressibilities, cubic EOS do not fully meet today's requirements to accuracy. This paper will investigate whether the PC-SAFT EOS can provide an alternative to the cubic EOS that will improve simulation accuracy within the problem areas and still provide the same reliability as the cubic equations with respect to phase equilibrium calculations. The paper will further analyze whether it is a problem for simulating EOR gas injection processes that the PC-SAFT equation unlike a cubic equation is not bound to match the pure component critical point.

Keywords: Cubic EOS, SRK, PR, Peneloux volume correction, oil compressibility, gas compressibility factor, PC-SAFT EOS, critical point simulation.

Introduction

Over the years, various forms of EOS models have been developed, tested, and modified to simulate the phase behavior of petroleum reservoir fluids. Today, the most commonly used EOS models for reservoir fluids are the Soave-Redlich-Kwong (SRK) equation [1] and Peng-Robinson (PR) equation [2] applied with Peneloux (P) volume correction [3]. Both EOS models are very good in terms of predicting saturation points, gas-liquid ratio, and liquid density of hydrocarbon fluids. The cubic EOS models are, however, less predictive for calculating gas compressibility factors and oil compressibilities. This inaccuracy is associated with inherent mathematical constraints in the cubic equations, which cannot be corrected by using a volume shift parameter [4, 5]. Poorly simulated gas compressibility factors present problems for example when an EOS is used with gas metering systems to convert a volumetric flow rate to a mass or molar flow rate. Inaccurate oil compressibilities would give a false picture of how the pressure in an undersaturated oil reservoir would respond to material being removed from the reservoir during production.

Recently, the Perturbed-Chain Statistical Associating Fluid Theory (PC-SAFT) model has been proposed as a potential next generation EOS model for the oil industry [4]. PC-SAFT was developed to model fluid systems with high differences in molecular weight (MW) including vapor-liquid-polymer systems [6, 7]. PC-SAFT has shown good results for asphaltene precipitation in oil systems [8] and, as can be seen from Table 1 and Figure 1, PC-SAFT is superior to the volume corrected SRK and PR equations for gas compressibility factors of pure gas and oil constituents and for oil compressibilities.

Table 1. Average absolute deviation (AAD) for gas compressibility factors (Z) for defined components for P = 0-1,000 bar and T = 0-250°C. PC-SAFT parameters for pure components were obtained from Gross and Sadowski [6]. The comparisons were made against the data generated from NIST RefProf[9].

EOS	Average Absolute Deviation for Gas Compressibility									
	C_1	C_2	C_3	iC_4	nC_4	iC_5	nC_5	nC_6	CO_2	H_2S
PC-SAFT	0.0112	0.0148	0.0148	0.0124	0.0151	0.0154	0.0127	0.0144	0.0095	0.0175
SRK-P	0.0127	0.0274	0.0379	0.0278	0.0433	0.057	0.0529	0.0479	0.027	0.0242
PR-P	0.0115	0.0137	0.0192	0.0225	0.0238	0.0337	0.0296	0.0425	0.0193	0.0290

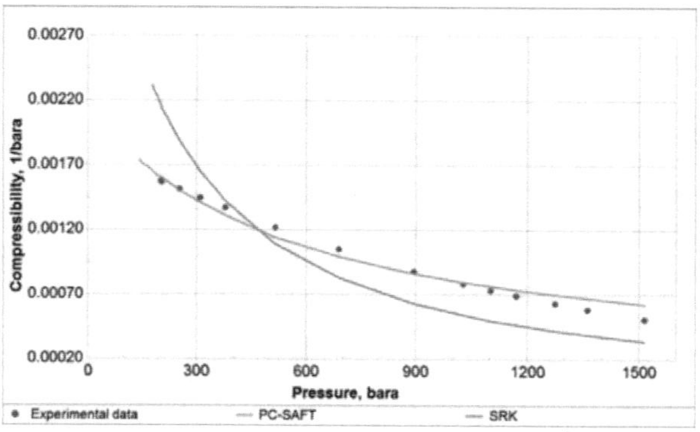

Figure 1. Experimental and simulated liquid compressibility of a highly undersaturated oil at high pressure. PC-SAFT and SRK-P are used for comparison. Reproduced from Larsen et al. [10].

Although PC-SAFT is well recognized in the oil industry for the above mentioned capabilities, it still needs to be tested against more complex fluid systems to prove general applicability for the simulation of the phase behavior of reservoir fluids, including reservoir fluids undergoing gas injection. To simulate the correct asphaltene onset pressures, a model is needed, which will provide the right ratio between the compressibility of the oil and the asphaltene phase. That makes PC-SAFT a strong candidate for this application. Aphaltene precipitation problems are however mainly seen during gas injection. Not only does it require that the applied EOS is good for liquid phase compressibilities, it must also be capable of simulating how the saturation pressure and fluid density develop with gas injection and when the saturation pressure will change from a bubble point to a dew point at which point on the swelling curve the fluid composition is critical. A correct match of the critical point is a precondition for a correct match of the minimum miscibility pressure [11].

It could therefore be of concern that the PC-SAFT equation unlike the cubic equation is not bound to accurately model pure component critical points. Published PC-SAFT parameters for pure components [6] have been estimated to match liquid densities and vapor pressures, but without matching the critical point. The critical pressure (P_c) and critical temperature (T_c) simulated using those parameters are therefore generally too high (1-2% for T_c and 3-23% for P_c for pure hydrocarbons [5]). Nevertheless PC-SAFT has been shown to work well in predicting the critical properties of light gas mixtures (~2% for T_c and ~4% for P_c for light hydrocarbon mixtures with 3-7 components [5]). The lack of matching pure component critical points could however prove to be a problem when modeling fluid systems consisting of an oil mixed with an injection gas at near critical conditions. Unless the critical composition is simulated correctly, the fluid model will be inapplicable to simulate a miscible drive in a reservoir with gas injection. This paper investigates the possibility of using PC-SAFT to model the phase behavior for oil mixtures with gas injection, including matching the critical composition and near-critical phase behavior.

Reservoir Plus Composition and PVT Data

A previous EOS model for a reservoir fluid with extensive enhanced oil recovery (EOR) data from Negahban et al. was selected as a basis for this study [12]. The compositional analysis of the reservoir fluid is shown in Table 2. The experimental PVT data included in this work are saturation point at reservoir temperature (121.1°C), separator test, constant mass expansion (CME), differential liberation, and a swelling test with CO_2.

Table 2. Molar composition to C_{7+} of the reservoir fluid from Negahban et al. [12]. A more detailed composition can be found in the mentioned reference.

Component	Mole%	MW	Density (g/cm^3)
N_2	0.363	28.014	
CO_2	2.991	44.010	
C_1	29.066	16.043	
C_2	7.163	30.070	
C_3	6.577	44.097	
iC_4	1.845	58.124	
nC_4	4.231	58.124	
iC_5	2.274	72.151	

nC_5	2.864	72.151	
C_6	4.104	86.178	0.6640
C_{7+}	38.521	207.648	0.8365

EOS Modeling

PC-SAFT Equation of State

Unlike the SRK and PR models where component critical properties (T_c, P_c, ω) are used as model parameters, PC-SAFT uses a molecular approach, where each component is represented through 3 parameters, its number of segments (*m*), its segment diameter (σ), and its segment energy (ε) [6, 13, 14]. The model concept is shown schematically in Figure 2 for two molecules with *m* number of segments, segment diameters of σ, and an inter-segment radial distance of r. The segment energy, ε, can be understood as the maximum attraction between two segments.

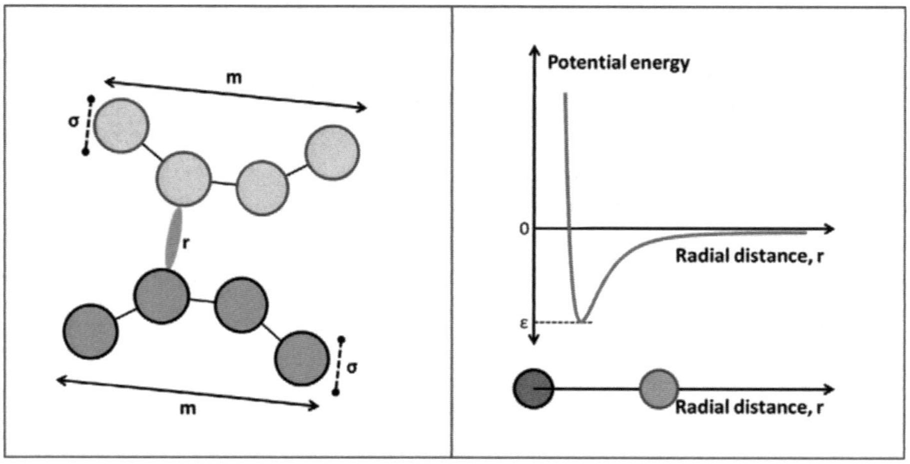

Figure 2. Schematic representation of a PC-SAFT mixture consisting of two components of chain length (*m*) and segment diameter (σ).

As shown in Equation 1, the PC-SAFT expresses the compressibility factor as a deviation from the ideal gas compressibility factor of 1.0.

$$Z = 1 - Z^{hc} + Z^{disp}$$ **Equation 1**

where Z^{hc} is the hard-chain contribution to the compressibility factor and accounts for repulsive molecular interactions and Z^{disp} is an attractive term. Both Z^{hc} and Z^{disp} are functions of the component parameters *m*, σ and ε. For mixtures PC-SAFT makes use of binary interaction coefficients (k_{ij}) along with mixing rules to obtain the mixture values of m, σ and ε.

Initial Consideration about PC-SAFT Component Parameters

Matching of mixture critical points is essential for capturing the correct phase behavior of oils when mixed with gas. It has been shown that PC-SAFT does fairly well in matching the critical points of light gas mixtures [5], which suggest that the PC-SAFT component parameters for light gas components are applicable also for modeling of mixtures. For the heavy end of the fluid (C_{7+}), the PC-SAFT component parameters are as a first estimate generated using a method developed with primary focus on modeling of asphaltene onset pressures for constant composition reservoir fluids [15]. This method uses empirical correlations that cannot be expected to be generally applicable. For the heavy end, model parameters may need to be adjusted to better match the PVT data of a given fluid.

Modeling Results

Default characterization

Table 3 shows the component parameters for 3 fluids generated in this work (denoted PC-SAFT[a-c]). Parameters for PC-SAFT[a] are generated using the default characterization [15]. Parameters for PC-SAFT[b] and PC-SAFT[c] are obtained by tuning PC-SAFT[a] parameters to match the routine PVT data and the routine PVT data + CO_2 swelling data, respectively. As shown, the plus fraction (C_{7+}) has been extrapolated to C_{80} and the hydrocarbon components above C_6 have been lumped to 12 pseudo-components with approximately equal weight percent.

The default characterization honors the trends for σ and ε that were established by Gross and Sadowski [6] for C_1-C_6. This trend shows ε increasing with carbon number, and σ values around 3.8 Å.

Table 3. PC-SAFT component parameters for the reservoir fluid applied in this study. Non-zero k_{ij}'s are shown in Table 4.

Component	Mole%	MW	Density (g/cm³)	PC-SAFT[a]			PC-SAFT[b]			PC-SAFT[c]		
				m	σ(Å)	ε(K)	m	σ(Å)	ε(K)	m	σ(Å)	ε(K)
N_2	0.363	28.014		1.21	3.31	90.96	1.21	3.31	90.96	1.21	3.31	90.96
CO_2	2.991	44.010		2.07	2.79	169.21	2.07	2.79	169.21	2.07	2.79	169.21
C_1	29.066	16.043		1.00	3.70	150.03	1.00	3.70	150.03	1.00	3.70	150.03
C_2	7.163	30.070		1.61	3.52	191.42	1.61	3.52	191.42	1.61	3.52	191.42
C_3	6.577	44.097		2.00	3.62	208.11	2.00	3.62	208.11	2.00	3.62	208.11
iC_4	1.845	58.124		2.26	3.76	216.53	2.26	3.76	216.53	2.26	3.76	216.53
nC_4	4.231	58.124		2.33	3.71	222.88	2.33	3.71	222.88	2.33	3.71	222.88
iC_5	2.274	72.151		2.56	3.83	230.75	2.56	3.83	230.75	2.56	3.83	230.75
nC_5	2.864	72.151		2.69	3.77	231.20	2.69	3.77	231.20	2.69	3.77	231.20
C_6	4.104	86.178	0.6640	3.06	3.80	236.77	3.06	3.80	236.77	3.06	3.80	236.77
C_7	4.203	96.000	0.7369	3.16	3.79	253.72	3.20	3.73	230.20	4.69	3.27	193.19
C_8	3.744	107.000	0.7521	3.50	3.79	254.07	3.54	3.73	230.39	5.24	3.26	194.01
C_9	3.336	121.000	0.7655	3.93	3.78	253.45	4.00	3.72	230.86	5.91	3.26	196.58
C_{10}-C_{11}	5.620	140.125	0.7829	4.52	3.78	253.43	4.61	3.72	231.94	6.77	3.26	197.33
C_{12}-C_{13}	4.461	167.596	0.8027	5.36	3.77	253.63	5.49	3.72	233.94	8.07	3.25	197.80
C_{14}-C_{16}	5.027	204.771	0.8233	6.49	3.77	254.15	6.67	3.71	236.33	9.71	3.25	198.25
C_{17}-C_{18}	2.504	243.596	0.8411	7.66	3.76	255.34	7.89	3.71	239.26	11.61	3.23	198.71
C_{19}-C_{21}	2.822	275.261	0.8561	8.57	3.76	257.55	8.85	3.71	243.34	10.87	3.44	210.00
C_{22}-C_{24}	1.995	317.001	0.8720	9.75	3.76	260.18	10.13	3.71	250.99	11.87	3.49	215.00
C_{25}-C_{29}	2.114	370.314	0.8897	11.25	3.77	263.40	11.72	3.71	257.27	12.19	3.63	220.00
C_{30}-C_{36}	1.500	451.603	0.9120	13.49	3.77	267.89	14.05	3.71	261.65	14.09	3.67	220.00
C_{37}-C_{80}	1.197	624.498	0.9497	18.08	3.77	276.31	18.84	3.72	269.88	28.67	3.20	220.00

[a] Default characterization [15].

[b] Tuned to routine PVT data.

[c] Tuned to routine PVT data and CO_2 gas injection swelling data.

The applied non-zero k_{ij}'s are shown in Table 4.

Table 4. Non-zero binary interaction coefficients for use with the EOS model in Table 3.

Component	PC-SAFT[a]			PC-SAFT[b]			PC-SAFT[c]		
	N_2	CO_2	C_1	N_2	CO_2	C_1	N_2	CO_2	C_1
N_2									
CO_2	-0.032			-0.032			-0.032		
C_1	0.028	0.120		0.028	0.120		0.028	0.120	
C_2	0.041	0.120	0.000	0.041	0.120	0.000	0.041	0.120	0.000
C_3	0.076	0.120	0.000	0.076	0.120	0.000	0.076	0.120	0.000
iC_4	0.094	0.120	0.000	0.094	0.120	0.000	0.094	0.120	0.000
nC_4	0.070	0.120	0.000	0.070	0.120	0.000	0.070	0.120	0.000
iC_5	0.087	0.120	0.000	0.087	0.120	0.000	0.087	0.120	0.000
nC_5	0.088	0.120	0.000	0.088	0.120	0.000	0.088	0.120	0.000
C_6	0.080	0.120	0.000	0.080	0.120	0.000	0.080	0.120	0.000
C_{7+}	0.080	0.100	0.000	0.080	0.100	0.030	0.080	0.080	0.040

Using the PC-SAFT[a] parameters, the routine PVT data was simulated with reasonable accuracy, but further tuning of the PC-SAFT parameters was performed to obtain a better match.

Regression to routine PVT data

A regression was carried out by adjusting the PC-SAFT[a] parameters in Table 3 for the C_{7+} components. The PC-SAFT diameter (σ) and energy (ε) parameters were treated as adjustable parameters in order to match the experimental data including saturation point, oil densities measured in the CME experiment, and oil formation volume factors (FVF) measured in the differential depletion experiment. With σ's and ε's fixed, the m parameters were adjusted to match the pseudo-component liquid densities, which will ensure correct STO densities. As shown in Table 3, both σ and ε parameters for C_{7+} are reduced compared to the default parameters. However, the trend in the σ and ε parameters from the original characterization for C_{7+} component was generally kept.

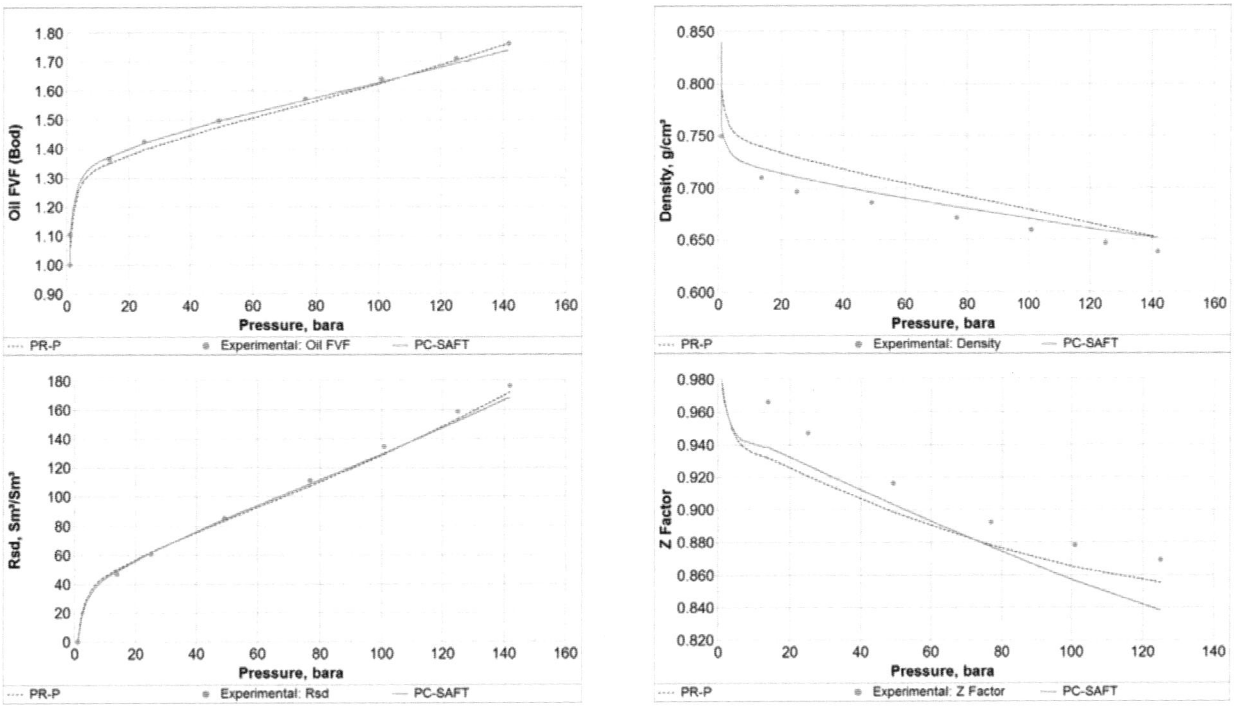

Figure 3. Measured (dots) and simulated (solid line) differential depletion experiments of reservoir fluid tuned to the routine PVT data using the PC-SAFT^b model. Results from Negahban et al. [12] using Peneloux volume corrected Peng-Robinson (PR-P) EOS are also shown in the figure (dashed line) for comparison.

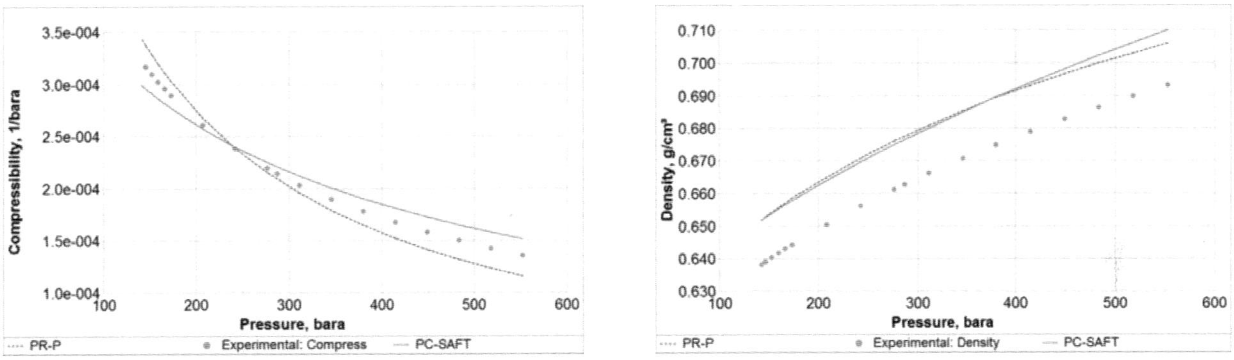

Figure 4. Measured (dots) and simulated (solid line) CME experiments of reservoir fluid tuned to the routine PVT data using the PC-SAFT^b model. Results from Negahban et al. [12] using PR-P are also shown in the figure (dashed line) for comparison.

Table 5. Simulated separator test results using PC-SAFT^b parameters. The errors are relative to the experimental data from Negahban et al. [12].

Total GOR (Sm^3/Sm^3)	% Error	STO Density (g/cm^3)	% Error
124.8	-4.8%	0.8199	0.7%

With the PC-SAFT model created only to match the routine PVT data (PC-SAFT^b in Table 3), a gas injection simulation was carried out to investigate the swelling behavior of the reservoir fluid using pure CO_2. Figure 5 shows the simulation of the saturation pressures and oil densities at different stages of CO_2 injections.

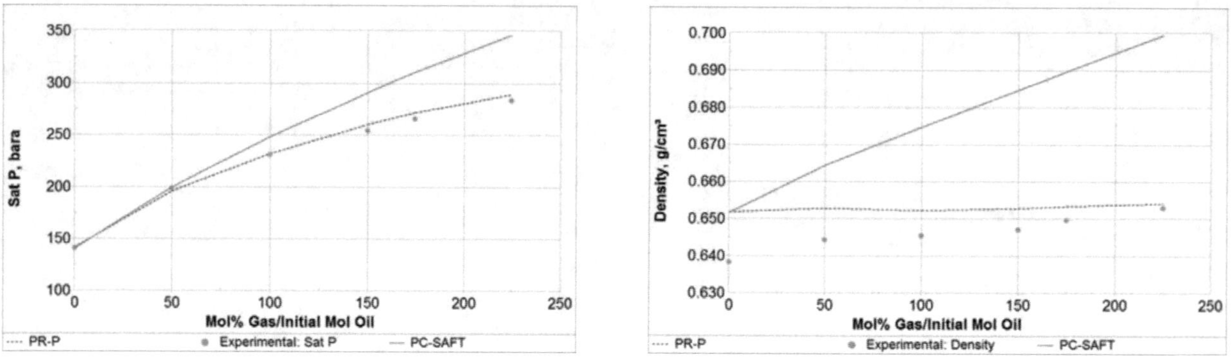

Figure 5. Experimental (dots) and simulated (solid line) saturation pressures/densities of reservoir fluid injected with CO_2 gas using the PC-SAFT^b model. Results from Negahban et al. [12] using PR-P EOS model are also shown in the figure (dashed line) for comparison.

As shown in Figure 5, the simulated saturation pressures and densities significantly deviate from the experimental values. In addition, the critical point on the swelling curve, which was experimentally reported in the range of 175-225 mol gas/mol oil, was not matched with the model, which predicted bubble points for all CO_2 injection stages. In order to match both the routine data and the swelling data with PC-SAFT a new regression that includes the swelling data is required.

Regression to routine PVT and CO_2 swelling data

A new regression was performed to match both the routine PVT data and CO_2 swelling data. The regression was carried out by adjusting the PC-SAFT^a parameters in Table 3 for the C_{7+} components. In this regression scheme, the σ and ε parameters were allowed to deviate from the original trend given by the default characterization. The component parameters of this tuned fluid (PC-SAFT^c in Table 3) shows a significant reduction of both σ and ε parameters as compared to the PC-SAFT^a parameters. Selected simulation results for this fluid are shown in Figures 6-8 and Table 6-7. As can be seen from Table 7, the PC-SAFT^c model provides a good match of the saturation pressures of the swelling experiment and further simulates the shift from bubble point (Pb) to dew point (Pd) for the right mole% of CO_2 injection gas.

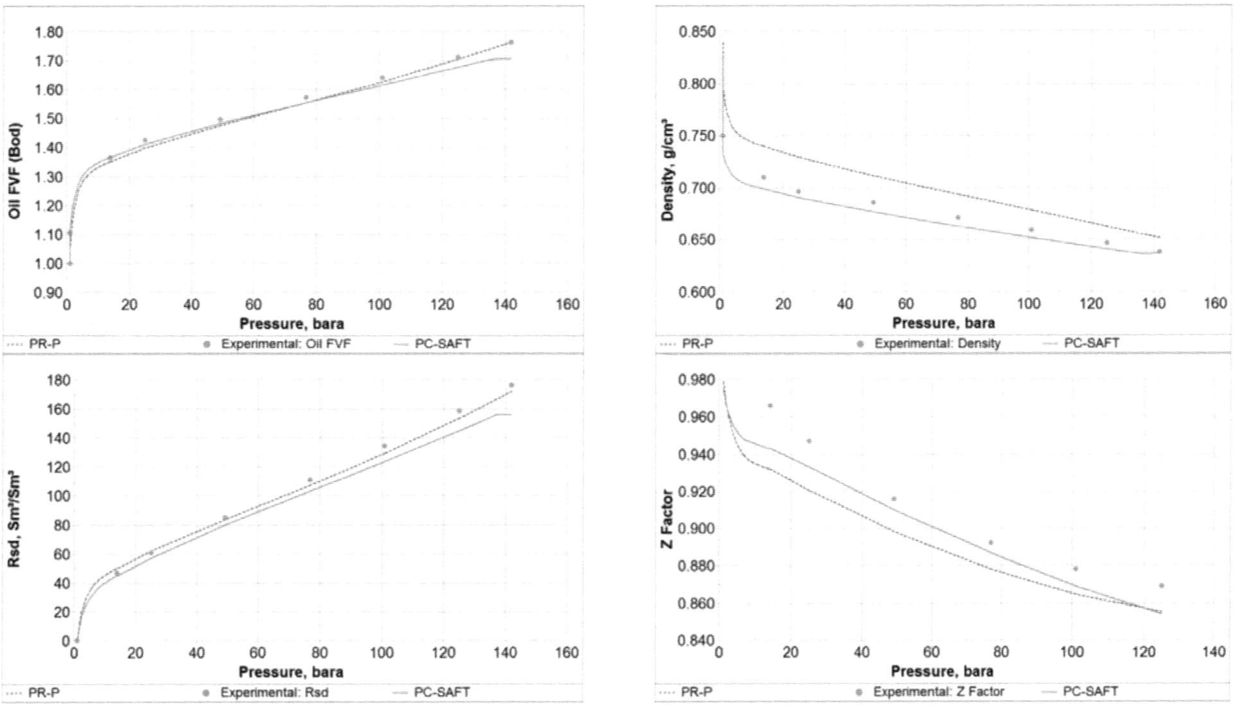

Figure 6. Measured (dots) and simulated (solid line) differential depletion experiments of reservoir fluid tuned to both the routine PVT and CO_2 swelling data using PC-SAFT model. Results from Negahban et al. [12] using PR-P EOS model are also shown in the figure (dashed line) for comparison.

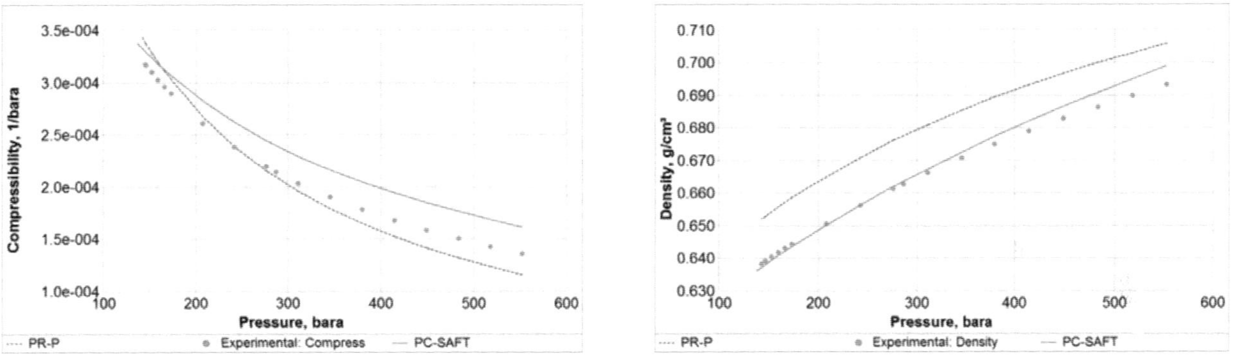

Figure 7. Measured (dots) and simulated (solid line) constant mass expansion (CME) experiments of reservoir fluid tuned to both the routine PVT and CO_2 swelling data using PC-SAFT model. Results from Negahban et al. [12] using PR-P EOS model are also shown in the figure (dashed line) for comparison.

Table 6. Simulated separator test results using PC-SAFT[c] parameters. The errors are relative to the experimental data from Negahban et al. [12]

Total GOR (Sm^3/Sm^3)	% Error	STO Density (g/cm^3)	% Error
117.4	-9.7%	0.809	-0.6%

Table 7. Saturation pressures and phase identification of reservoir fluid injected with pure CO_2 at 121.11°C. Simulated results using PC-SAFT[c] and PR-P [12] models are compared to the experimental values.

Stage	Mole% Gas/ initial Mol Oil	Experiment		PC-SAFT[c]			PR-P		
		Sat P (bar)	Pb/Pd	Sat P (bar)	%error	Pb/Pd	Sat P (bar)	%error	Pb/Pd
0	0	140.91	Pb	136.82	-3.60%	Pb	141.90	+0.70%	Pb
1	50	198.27	Pb	186.05	-4.90%	Pb	195.61	-1.30%	Pb
2	100	230.61	Pb	221.88	-4.30%	Pb	231.85	+0.50%	Pb
3	150	253.91	Pb	250.21	-3.80%	Pb	260.01	+2.40%	Pb
4	175	265.43	Pb	262.90	-3.10%	Pb	271.42	+2.30%	Pb
5	225	283.77	Pd	290.19	+0.20%	Pd	289.55	+2.00%	Pd

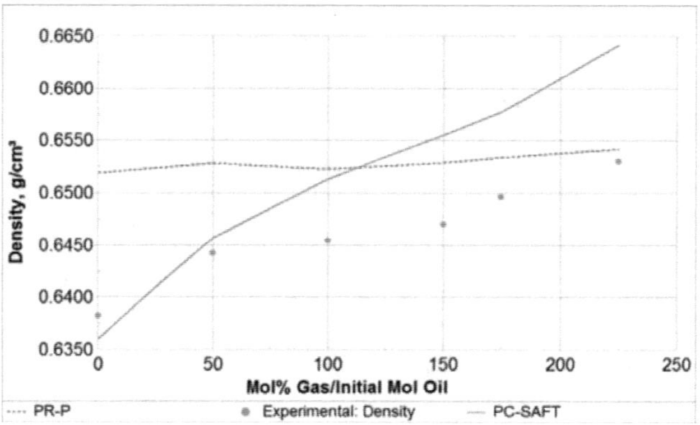

Figure 8. Measured (dots) and simulated (solid line) liquid densities for different swelling stages. Results from Negahban et al. [12] using PR-P EOS model are also shown in the figure (dashed line) for comparison.

From Figures 6-9 and Table 6-7, it is seen that final PC-SAFT[c] model matches both routine PVT data and CO_2 swelling data well and that the match is comparable to the PR-P model. It is noted that the gas compressibility factor calculated using PC-SAFT[c] are more accurate than PR-P, while comparable results are observed for oil compressibilities.

Discussion

For cubic EOS the component parameters, T_c, P_c and ω, are well established for defined components. The critical point can be measured and the ω is defined in terms of critical point and vapor pressure at a reduced temperature of 0.7. The parameters for PC-SAFT are less defined since they relate to molecular structure and interactions between molecular segments which cannot readily be measured. For the C_{7+} fractions, T_c will increase with carbon number, while P_c will decrease with carbon number. A similar trend cannot necessarily be seen with the PC-SAFT component parameters. It further complicates PC-SAFT component parameter estimation that multiple solutions can be found for m, σ, and ε which will give approximately the same T_c, P_c and density.

It is seen that the default characterization, which honors the trend of the defined component parameters from Gross and Sadowski [6], gave a description of the reservoir fluid that match the routine PVT data well. However, further regression was needed to obtain a better match. Two PC-SAFT regression schemes were proposed in this study: 1) tuned to the routine PVT data (PC-SAFT[b]) and 2) tuned to both the routine PVT data and CO_2 swelling data (PC-SAFT[c]). For the PC-SAFT[b] regression, the parameters were constrained such that the trends for σ and ε are kept similar to the default characterization. As shown in Table 3, it was possible to tune PC-SAFT parameters (PC-SAFT[a] to PC-

SAFT[b]) and keep the trend in PC-SAFT parameterization. Despite a good match of routine PVT data, the PC-SAFT[b] parameters were not capable of describing the fluid behavior with CO_2 gas injection.

In order to obtain a match of both the routine PVT data and the CO_2 swelling data, it was determined that it was necessary to allow the σ and ε parameters to deviate from the parameter trends proposed in the default characterization. The results shown in Table 3 indicate that the σ and ε parameters have to be reduced further while the *m* parameters are increased. Moreover, there are also non-monotonic developments for the σ and ε parameters. However, as shown in Figure 9, the T_c and P_c values of the heavy end components keep a monotonic development vs. carbon number as should be the case. Therefore, this non-monotonic trend in PC-SAFT parameters for σ and ε may very well be justified.

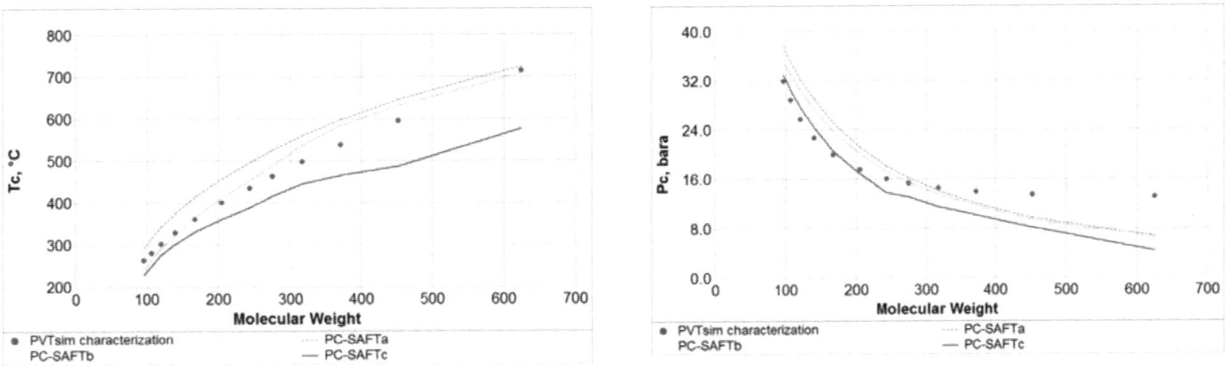

Figure 9. Critical properties of C_{7+} components calculated using PC-SAFT parameters. Critical properties generated using PVTsim default characterization are also shown in the figure.

Conclusion

The PC-SAFT equation is superior to the cubic EOS for prediction of gas phase compressibility factors and oil phase compressibilities. In this work, it was investigated whether the PC-SAFT EOS could be applied to model the phase equilibrium of a reservoir oil with CO_2 gas injection. It was seen that PC-SAFT could be used to model the routine data as well as the gas injection data (including the critical point on the CO_2 swelling curve) and that the results were comparable to those of PR-P. PC-SAFT is therefore a potential candidate as a next generation EOS in the oil and gas industry, and the equation may provide more accurate results for the pressure dependence of phase densities without the well-known reliability of the cubic EOS for simulating saturation points, gas/oil ratio, etc., being jeopardized. Reliable and generally applicable C_{7+} characterization procedures exist for the cubic EOS. Such procedures are still to be developed for PC-SAFT.

Acknowledgements

The authors would like to thank Daniel Kunisch Eriksen and Karen Schou Pedersen from Calsep A/S for their help in calculation of gas compressibility factors and set up of special PC-SAFT regression routines.

References

1. Soave, G. Equilibrium Constants from a Modified Redlich-Kwong Equation of State. *Chem. Eng. Sci.* **1972,** *27,* 1197-1203.
2. Peng, D. Y.; Robinson, D. B. New Two-Constant Equation of State. *Ind. Eng. Chem. Fundam.* **1976,**

15, 59-64.

3. Peneloux, A.; Rauzy, E.; Fréze, R. A Consistent Correction for Redlich-Kwong-Soave Volumes. *Fluid Phase Equilibr.* **1982,** *8,* 7-23.
4. Pedersen, K. S.; Sørensen, C. H. PC-SAFT Equation of State Applied to Petroleum Reservoir Fluids. *SPE ATCE*, Anaheim, CA, 2007; pp SPE-110483-PP.
5. Sørensen, C. H. *Thermodynamic modeling of hydrocarbon mixtures with the PC-SAFT model;* Technical University of Denmark: Kgs. Lyngby, 2008.
6. Gross, J.; Sadowski, G. Perturbed-Chain SAFT: An Equation of State Based on Perturbation Theory for Chain Molecules. *Ind. Eng. Chem. Res.* **2001,** *40,* 1244-1260.
7. Gross, J.; Sadowski, G. Application of the Perturbed-Chain SAFT Equation of State to Associating Systems. *Ind. Chem. Res.* **2002,** *41,* 5510-5515.
8. Ting, D. P.; Gonzalez, D. L.; Hirasaki, G. J.; Chapman, W. G. Application of the PC-SAFT Equation of State to Asphaltene Phase Behavior. In *Asphaltenes, Heavy Oils, and Petroleomics;* Springer, 2007; pp 301-327.
9. Lemmon, E. W. H. M. L.; McLinden, M. O. *NIST Reference Fluid Thermodynamic and Transport Properties;* User's Guide; NIST: Boulder, Colorado, 2007.
10. Larsen, J.; Sørensen, H.; Yang, T.; Pedersen, K. S. EOS and Viscosity Modeling for a Highly Undersaturated Gulf of Mexico Reservoir Fluid. *SPE ATCE*, Denver, CO, 2011; pp SPE-147075-PP.
11. Sah, P.; Pedersen, K. S.; Shaikh, J. A. EOS Modelling, the First Step in Detailed EOR Potential Evaluation - A Case Study. *EORC*, Kuala Lumpur, Malaysia, 2011; pp SPE-144025.
12. Negahban, S.; Pedersen, K. S.; Baisoni, M. A.; Sah, P.; Azeem, J. An EoS Model for a Middle East Reservoir Fluid with an Extensive EOR PVT Data Material. *Abu Dhabi International Petroleum Exhibition & Conference*, Abu Dhabi, UAE, 2010; pp SPE-136530-PP.
13. Chapman, W. G.; Jackson, G.; Gubbins, K. E. Phase Equilibria of Associating Fluids. Chain Molecules with Multiple Bonding Sites. *Mol. Phys* **1988,** *65,* 1057-1079.
14. Chapman, W. G.; Gubbins, K. E.; Jackson, G.; Radosz, M. New Reference Equation of State for Associating Liquids. *Ind. Eng. Chem. Res.* **1990,** *29,* 1709-1721.
15. Calsep A/S. *PVTsim 20: Method Documentation;* Kgs. Lyngby, Denmark, 2011.

Volume-translated Cubic EoS and PC-SAFT Density Models and a Free Volume-Based Viscosity Model for Hydrocarbons at Extreme Temperature and Pressure Conditions

Burgess, W.; Tapriyal, D.; Morreale, B.; Soong, Y.

Baled, H.; Enick, R.

McHugh, M.; Bamgbade, B.; Wu, Y.

This research focuses on providing the petroleum reservoir engineering community with robust models of hydrocarbon density and viscosity at the extreme temperature and pressure conditions (up to 533 K and 276 MPa, respectively) characteristic of ultra-deep reservoirs, such as those associated with the deepwater wells in the Gulf of Mexico. Our strategy is to base the volume-translated (VT) Peng–Robinson (PR) and Soave-Redlich-Kwong (SRK) cubic equations of state and PC-SAFT on an extensive data base of high temperature (278–533 K), high pressure (6.9 – 276 MPa) density rather than fitting the models to low pressure saturated liquid density data. This high temperature high pressure (HTHP) data base consists of literature data for hydrocarbons ranging from methane to C_{40}. Where gaps in the HTHP regions were identified, new data were collected by our team using a windowed, agitated, variable–volume cell. The three new models developed in this work, HTHP VT-PR EOS, HTHP VT-SRK EOS, and HTHP PC-SAFT, yield mean absolute percent deviation values (MAPD) for HTHP hydrocarbon density of ~2.0%, ~1.5%, and < 1.0%, respectively.

An effort was also made to provide accurate hydrocarbon viscosity models based on literature data. Viscosity values are estimated with the frictional theory (f–theory) and free volume (FV) theory of viscosity. MAPD values of up to 5% for the viscosity are obtained when the PR and SRK EOSs, or their HTHP VT analogs, are used to estimate the attractive and repulsive pressure input values for the f-theory. However, the FV theory, which requires fluid density as an input, provides MAPD values of ~3% for the viscosity when coupled with the HTHP VT PR or HTHP VT SRK EOS. Combined with PC-SAFT-based density predictions, the FV theory had MAPD values of less than 2% for hydrocarbon viscosity values.

The modeling of fluid flow in petroleum reservoirs, production wells, and surface operations requires reliable predictions of the density and viscosity of the brine phase and the hydrocarbon phase(s). This study focuses on the accurate prediction of hydrocarbon density and viscosity at the extreme conditions associated with ultra-deep formations, a new and prolific source of petroleum of great interest to the energy industry with temperatures ranging from ambient to 533 K and pressures from 6.9 to 276 MPa. Because reservoir compositional simulators typically model crude oil, condensate, and natural gas as a mixture of components, this study will focus primarily on hydrocarbons ranging between methane and tetracontane, with CO_2 also included.

There are two general classes of models with which researchers have demonstrated success in modeling the density of multi-component hydrocarbon mixtures. The first such class is the relatively compact cubic equation of state (EoS), which is the main type of EoS presently used in reservoir simulators. Because of their simplicity and predictive capacity, cubic EoSs are widely used to predict the phase behavior and volumetric properties of pure fluids and mixtures. This is especially true in computationally intensive software such as compositional simulators of oil reservoirs. Two cubic EoSs frequently used by the petroleum engineering industry are the Soave-Redlich-Kwong (SRK) EoS [1] and the Peng-Robinson (PR) EoS [2]. The general form of these two–parameter cubic EoSs is given in Eq. 1 [3].

$$P = \frac{RT}{v-b} - \frac{a_c\,\alpha}{v^2 + ubv + wb^2} \tag{1}$$

In Eq. 1, P is the pressure, T is absolute temperature in Kelvins, and v represents molar volume. For the SRK EoS, $u = 1$ and $w = 0$; for the PR EoS, $u = 2$ and $w = -1$. In the attractive term, the α function is a correlation of reduced temperature and acentric factor. The parameter b represents the effective molecular volume in the hard sphere repulsive term. The a_c and b parameters, which depend only upon the gas constant and critical constants, are obtained by applying the constraints shown in Eq. 2 at the critical point.

$$\left(\frac{\partial P}{\partial v}\right)_{T=T_c} = 0, \qquad \left(\frac{\partial^2 P}{\partial v^2}\right)_{T=T_c} = 0 \tag{2}$$

The SRK EoS predicts a critical compressibility factor, $Z_c = \frac{P_c v_c}{RT_c}$, of 0.3333 while the PR EoS predicts a slightly lower value of 0.3074. However, the experimental values of Z_c are generally smaller than those predicted by either cubic equation of state. As a result, the liquid molar volume predictions by cubic EoSs in the near-critical region may deviate significantly from experimental values. However, the performance of cubic EoSs can be considerably improved by utilizing the concept of volume translation.

The other class of predictive models for density is the set of EoS which are based upon the Statistical Associated Fluid Theory (SAFT) EoS, put forth by Chapman, Radosz, and coworkers in the 1990s [4,5,6,7]. Even though SAFT is generally acknowledged as superior to cubic equations of state with respect to describing the physics of the system, it is less frequently used in reservoir simulations because of its complexity. The SAFT theory represents the pure component or mixture Helmholtz free energy as a sum of terms accounting for hard sphere repulsion, dispersion interactions, chain connectivity, and association between segments. Many modifications to the original SAFT equation have been put forth, but the one which is most important and effective for modeling hydrocarbon density is the perturbed-chain SAFT (PC-SAFT) equation [8]. This equation is identical to the original SAFT equation except for the dispersion interaction term, which is calculated for a mixture of hard chains rather than hard spheres. PC-SAFT has been successfully utilized to predict phase behavior for organic compounds of interest to the oil industry including normal and branched alkanes, benzene derivatives, and cycloalkanes. It has also been successfully used for alkenes, gases, ethers, esters, halogenated hydrocarbons, polymer-hydrocarbon mixtures and poorly defined materials such as polymers [8]. The PC-SAFT equation has recently been used to accurately predict, to better than ±1%, the pure component densities of toluene, *n*-pentane, *n*-octane, isooctane, cyclooctane, *n*-decane, *n*-hexadecane, *n*-octadecane, and *n*-eicosane [9,10] at pressures up to ~ 55 MPa. However, PC-SAFT has the tendency to over-predict hydrocarbon density values at pressures greater than ~ 55 MPa [9,10]: this over-prediction can be as much as 5% more than the experimental value at pressures greater than 240 MPa. Such extreme temperatures and pressures are found throughout ultra-deep reservoirs. Therefore, one of the objectives of this study was to modify the PC-SAFT pure component parameters to improve density predictions at temperatures from ambient to 533 K and pressures from 6.9 to 276 MPa.

For most of the n-alkanes and other compounds studied by Gross and Sadowski [8] the three pure-component PC-SAFT parameters (*m*, the number of hard spheres in a molecule; σ, the temperature-independent hard core diameter of a sphere; ε/k_B, the attractive energy of dispersion divided by the Boltzmann constant) were obtained by fitting sub-critical $P\rho T$ data although occasionally both sub-critical and supercritical density data were used. Recently, we derived a new set of pure-component PC-SAFT parameters [11] by fitting the equation to high-temperature, high-pressure (HTHP) density data present in the literature [9,10,12,13,14] that cover almost the entire pressure and temperature ranges of interest for deep reservoirs. Sub-critical density data are not used to obtain these new parameters. For each set of density data, the pressure ranged between ~ 4.2 and ~ 300 MPa, while the temperature was usually ~ 423 K. A downhill simplex method is used to determine the best values for the three pure-component PC-SAFT parameters by minimizing the value of the mean absolute percent deviation (MAPD) between the experimental density and the PC-SAFT prediction.

Recently, the two theories that have shown the greatest amount of promise with regard to the accurate modeling of viscosity at both low pressures and HTHP conditions are the frictional theory (f-theory) of viscosity [15,16] and the free volume theory (FV theory) of viscosity [17,18]. In f-theory, the viscosity is assumed to be a function of a number of temperature-dependent coefficients, and the attractive and repulsive pressures. One-parameter f-theory models, for which necessary inputs include the attractive and repulsive pressure values that can be provided by cubic EoSs such as the Soave-Redlich-Kwong (SRK) and Peng-Robinson (PR) EoSs, yield accurate viscosity predictions (typically within ±4% of experimental value) for n-alkanes at pressures of up to 15000 psi [15]. The f-theory has also been coupled with more recently developed EoSs which better describe the system thermodynamics at a molecular level. For example, in 2006, Quiñones-Cisneros et al. [16] presented an altered, two-parameter f-theory model that, when coupled with the PC-SAFT equation [19], yielded viscosity predictions for n-alkanes of carbon number up to 18 at pressures and temperatures of up to 100 MPa and at least

423 K that were typically accurate to ±1%. However, f-theory was extended to only two non-normal alkanes, carbon dioxide and nitrogen, for which it was accurate to about ±4%.

In FV theory, viscosity is assumed to be dependent on the molecular free-volume fraction, which is expressed as a function of density. This three-parameter model is applicable to both the vapor and liquid phases and has been successfully fit to viscosity data for a number of aromatics, and both normal and branched alkanes at pressures up to ~ 500 MPa [17]; experimental values were used to supply the required density inputs. However, only the parameters for methane and propane were reported. A subsequent effort [18] established that the FV theory could be coupled with the density predictions provided by either SAFT or PC-SAFT EoSs, thereby allowing for the accurate prediction of viscosity values at pressures up to 60 MPa. However, the validity of these parameters over the full range of HTHP conditions was not established. Therefore, another objective of this study will be the evaluation of f-theory and FV theory viscosity predictions at pressures from ambient to 276 MPa and temperatures from ambient to 533 K.

Theory

Volume-Translated Cubics

One previously investigated method of improving the performance of cubic equations of state has been to correct the predicted liquid volume by translating it. In fact, a systematic deviation, c, is observed between the predicted liquid molar volume (v_{EoS}) and the corresponding experimental value ($v_{\exp}$).

$$c = v_{EoS} - v_{\exp} \tag{3}$$

Peneloux et al. [20] proposed an equation for the volume correction constant c which is a function of critical properties. It is equal to the difference between the predicted and the reference saturated liquid volume at $T_r = 0.7$

$$c = v_{EoS} - v_{\exp}\Big|_{T_r=0.7} \qquad (4)$$

$$c(SRK) = 0.40768\,\frac{RT_c}{P_c}(0.29441 - Z_{RA}) \qquad (5)$$

where Z_{RA} is the Rackett compressibility factor.

Other authors have proposed temperature-dependent volume translation correlations. For example, Ji and Lempe [21], and Wang and Gmehling [22] defined such volume correction expressions applied to the SRK EoS as

$$c = c_C\, f(\omega, Z_C, T_r) \qquad (6)$$

where c_c is the volume correction at the critical temperature. The parameters in the function $f(\omega, Z_c, T_r)$ were optimized by fitting experimental saturated liquid densities. The T-dependent volume translated cubic equations of state defined by Ji and Lempe and Wang and Gmehling are designed to fit saturated liquid density data and may provide good results within the low pressure range. However, preliminary calculations by our group indicate that they usually exhibit thermodynamic inconsistencies at high pressures in that the isotherms cross over each other.

Other investigators have incorporated density dependency into the translation function as well as temperature dependency. When fit to saturated density data, this type of T-ρ volume translation expression may substantially improve the predictive ability of a cubic EoS in the near-critical and low pressure regions. However, a T-ρ-translated SRK EoS proposed by Frey et al. [23] provides only a modest improvement in the SRK EoS predictions at HTHP conditions [9].

PC-SAFT equation

Because the PC-SAFT equation is described in detail elsewhere, only a synopsis is given here. The residual Helmholtz free energy $\tilde{a}^{res}$ for a given system is written as a sum of Helmholtz free energies.

$$\tilde{a}^{res} = m\tilde{a}^{hs} + \tilde{a}^{chain} + \tilde{a}^{disp} + \tilde{a}^{assoc} \tag{7}$$

Except for the reduced dispersion Helmholtz energy $\tilde{a}^{disp}$, all the terms on the right hand side of the PC-SAFT equation are equivalent to the analogous terms of the same name in the original SAFT equation [6,7]. The $\tilde{a}^{assoc}$ term, to account for association, is set to zero since there are no such interactions between the nonpolar hydrocarbon molecules considered here. Therefore, Eq. 7 becomes

$$\tilde{a}^{res} = m\tilde{a}^{hs} + \tilde{a}^{chain} + \tilde{a}^{disp} = \tilde{a}^{hc} + \tilde{a}^{disp} \tag{8}$$

where the reduced Helmholtz energy of a hard chain $\tilde{a}^{hc}$ is equal to the sum of the total reduced hard sphere fluid and chain Helmholtz energies, $m\tilde{a}^{hs}$ and $\tilde{a}^{chain}$, respectively. The hard chain Helmholtz energy term can be calculated from the equations given in the original SAFT theory [6,7], as shown in Eqs. 3-8.

$$\tilde{a}^{hc} = m\tilde{a}^{hs} + \tilde{a}^{chain} = m\tilde{a}^{hs} + \sum_i x_i(1-m_i)\ln\left(g_{ii}^{hs}\right) \tag{9}$$

$$m = \sum_i x_i m_i \tag{10}$$

$$\tilde{a}^{hs} = \frac{1}{\xi_0}\left[\frac{(\xi_2)^3 + 3\xi_1\xi_2\xi_3 - 3\xi_1\xi_2(\xi_3)^2}{\xi_3(1-\xi_3)^2} - \left[\xi_0 - \frac{(\xi_2)^3}{(\xi_3)^2}\right]\ln(1-\xi_3)\right] \tag{11}$$

$$\xi_k = \left(\frac{\pi\rho_{num}}{6}\right)\sum_i x_i m_i (d_i)^k ; k = 0,1,2,3 \quad (12)$$

$$d_i = \sigma_i\left[1 - 0.12\exp(-3\varepsilon_i / k_B T)\right] \quad (13)$$

$$g_{ii}^{hs} = \frac{1}{1-\xi_3} + \frac{3d_i}{2}\frac{\xi_2}{(1-\xi_3)^2} + 2\left[\frac{d_i}{2}\right]^2 \frac{\xi_2^2}{(1-\xi_3)^3} \quad (14)$$

In Eqs. 3-8:

m = mixture average segment number

T = absolute temperature, K

x_i = the mole fraction of component i

g_{ii}^{hs} represents the hard-sphere fluid radial distribution function

d_i = hard-sphere diameter at a given temperature, Å

ρ_{num} = total number density of molecules per cubic Ångstrom

The terms m_i, σ_i, and ε_i/k_B represent, for component i, the pure-component PC-SAFT parameters, while ξ_0, ξ_1, ξ_2, and ξ_3 are functions of the average molecular size.

PC-SAFT theory differs from SAFT theory in that the dispersion energy is calculated for a reference fluid comprised of hard chains rather than hard spheres. The PC-SAFT dispersion term, given in Eq. 15, is determined by applying the second-order perturbation theory of Barker and Henderson [24,25] to a hard chain reference fluid.

$$\tilde{a}^{disp} = -\pi\rho_{num}\left[2I_1(\eta,m)\sum_i\sum_j x_i x_j m_i m_j \sigma_{ij}^3 \frac{\varepsilon_{ij}}{k_B T} + mI_2(\eta,m)C\sum_i\sum_j x_i x_j m_i m_j \sigma_{ij}^3 \left(\frac{\varepsilon_{ij}}{k_B T}\right)^2\right] \quad (15)$$

$$\sigma_{ij} = 0.5(\sigma_i + \sigma_j) \quad (16)$$

$$\varepsilon_{ij} = (1 - k_{ij})\sqrt{\varepsilon_i \varepsilon_j} \quad (17)$$

The terms I_1 and I_2 in Eq. 15 represent complicated integrals that are approximated as power series in density [8]. The combining rules given in Eqs. 10 and 11 are used to calculate the size and energy parameters σ_{ij} and ε_{ij} for a pair of unlike segments, with the binary interaction parameter k_{ij} used to correct the value of the segment-segment energy parameter ε_{ij} for unlike segments. C is given as

$$C = \left(1 + m\frac{8\eta - 2\eta^2}{(1-\eta)^4} + (1-m)\frac{20\eta - 27\eta^2 + 12\eta^3 - 2\eta^4}{[(1-\eta)(2-\eta)]^2}\right)^{-1} \tag{18}$$

$$\eta = \frac{\pi}{6}\rho_{num}\sum_i x_i m_i d_i^3 \tag{19}$$

Using the method of Gross and Sadowski [8], the density at a given system pressure is determined by adjusting the reduced density, η, until the calculated pressure and the system pressure are equal. The reader is cautioned not to confuse the η used here with the viscosity reported later in this paper.

Frictional theory of viscosity (f-theory)

A common approach to viscosity modeling is to express viscosity as the sum of two terms as shown below in Eq. 20.

$$\eta = \eta_0 + \Delta\eta \tag{20}$$

In Eq. 20, η_0, the "dilute-gas" term [26] is typically only relevant when gas-like densities are encountered. The second term, $\Delta\eta$, dominates at high densities. In the case of f-theory, this term is determined by envisioning shear-induced fluid flow as two sliding surfaces under shear stress. As the friction laws indicate that the "shear force" of friction is proportional to the normal forces on the surfaces, it follows that the shear stress of friction is proportional to the normal stresses on the surfaces. It is further assumed that the normal stress is equivalent to the system pressure, which is separated into the attractive and repulsive pressure terms P_a and P_r (which must be calculated using an equation of state). From Newton's law of viscosity, shear stress is also proportional to viscosity. After adding second-order terms to improve the quality of the model predictions, the high-pressure viscosity term $\Delta\eta$ is derived [16,27]:

$$\Delta\eta = \kappa_a P_a + \kappa_{aa} P_a^2 + \kappa_r P_r + \kappa_{rr} P_r^2 \tag{21}$$

Previous work [15,16,27] has indicated that accurate viscosity predictions are made using f-theory at pressures up to 100 MPa when the friction coefficients κ_a, κ_{aa}, κ_r, and κ_{rr} are assumed to have an exponential relationship with respect to reduced temperature. F-theory has been successfully used when the PR, SRK, and PC-SAFT equations are used to obtain values for P_a and P_r at a given set of $P\rho T$ values. However, it is necessary to define separate sets of equations for κ_a, κ_{aa}, κ_r, and κ_{rr} each time a different EoS is used to calculate the P_a and P_r inputs to Eq. 21. Previous work [16] indicates that the agreement of f-theory viscosity predictions with experimental data up to 100 MPa is best when the PC-SAFT equation is used to calculated P_a and P_r. In this case, the frictional coefficients κ_a, κ_{aa}, κ_r, and κ_{rr}, are defined as indicated in Eqs. 22-25.

$$\kappa_a = \eta_a \frac{\tilde{\kappa}_a}{P_c} \tag{22}$$

$$\kappa_{aa} = \eta_a \frac{\tilde{\kappa}_{aa}}{P_c^2} \tag{23}$$

$$\kappa_r = \eta_r \frac{\tilde{\kappa}_r}{P_c} \tag{24}$$

$$\kappa_{rr} = \eta_r \frac{\tilde{\kappa}_{rr}}{P_c^2} \tag{25}$$

In Eqs. 4-7, the terms η_a and η_r are adjustable parameters that are specific to the component in question and which determine the strength of the attractive and negative components of the viscosity. They are determined by fitting to experimental viscosity data. In this work, P_c represents the critical pressure of a component as calculated using the PC-SAFT equation (not the literature value). The temperature-dependent friction parameters $\tilde{\kappa}_a$, $\tilde{\kappa}_r$, $\tilde{\kappa}_{aa}$ and $\tilde{\kappa}_{rr}$ represent also depend on the segment number m. Collectively, they are a function of 38 model constants, which are given elsewhere [16].

FV Theory of Viscosity

The derivation of the FV theory viscosity begins in the same manner as the derivation of f-theory, as Eq. 20 serves as a starting point and η_0 is again equal to the Chapman-Enskog dilute-gas viscosity. However, Doolittle [28] proposed that $\Delta\eta$ is proportional to the fraction of free volume f_v (i. e. a measure of the space within a system that is not occupied by molecules) present in a fluid, thereby correlating viscosity to molecular structure via the empirical relationship

$$\Delta\eta = e^{B/f_v} \tag{26}$$

In lattice theory, free volume represents the space in a "cage" in which a molecule is free to wander while its nearest neighbors are frozen in their lattice positions. Therefore f_v is characteristic of the empty space between molecules. Allal and coworkers [17] assumed that f_v is ultimately dependent on PM_w/ρ, the energy required to form vacant vacuums required for diffusion, and $\alpha\rho$, a barrier energy which a molecule must cross to diffuse. Their relatively simple final equation, Eq. 27, requires as inputs only the molecular weight M_w, $P\rho T$ values, and the three pure-component parameters L, α, and B.

$$\Delta\eta = \frac{10^7 \rho L\left(\alpha\rho + \frac{PM_w}{\rho}\right)}{\sqrt{3RTM_w}} e^{\left[B\left(\frac{\alpha\rho + \frac{PM_w}{\rho}}{RT}\right)^{1.5}\right]} \quad (27)$$

In Eq. 27:

$\Delta\eta$ = high-density viscosity term, cP

M_w = molecular weight, kg/mol

ρ = mass density, kg/m^3

T = absolute temperature, Kelvins

R = gas constant, 8.3145 J/mol*K

P = pressure, MPa

L = length parameter, Å

α = energy parameter, m^5/mol*s^2

B = unitless constant parameter

Results and Discussion

Developing a new HTHP volume translated EoS

Recently, our research group proposed a high temperature, high pressure (HTHP) volume translation model to improve the predictive capability of the PR and SRK cubic equations of hydrocarbon densities at temperatures from ambient to 533 K and pressures from 6.9 to 276 MPa [29]. These conditions are representative of those imposed on petroleum in ultra-deep reservoirs and the production wells. Unlike prior volume translation methods, in which the volume correction is correlated to saturated liquid densities typically collected at pressures less than 6.9 MPa, the volume translation term in the HTHP VT-SRK EoS and HTHP VT-PR EoS is correlated to pure component, single-phase density data for short- and long-chain alkanes, cycloalkanes and aromatics over the pressure range ~(7-276) MPa and temperature range ~(278-533) K of interest. The volume correction term, c, is expressed as a linear function of the reduced temperature, T_r,

$$c = \Delta v = v_{EoS} - v_{\exp} = A + B \cdot T_r \tag{28}$$

In Eq. 29, the pure fluid parameters *A* and *B* have been regressed for 17 pure components ranging in size from C_1 to C_{40}. A generalized method for determining *A* and *B* for compounds not included in the limited database is obtained by expressing A and B as functions of molecular weight (*M*) and acentric factor (*ω*).

$$A, B = f(M, \omega) = k_0 + k_1 \exp\left(\frac{-1}{k_2\, M\, \omega}\right) + k_3 \exp\left(\frac{-1}{k_4\, M\, \omega}\right) + k_5 \exp\left(\frac{-1}{k_6\, M\, \omega}\right) \tag{29}$$

Values for the parameters of Eq. 29 are listed in Table 1 for both the PR and SRK EoSs.

Table 1. Parameters of Eq. 29 [29].

Constants		HTHP VT-SRK	HTHP VT-PR
A (cm^3/mol)	k_0	0.2300	-4.1034
	k_1	46.843	31.723
	k_2	0.0571	0.0531
	k_3	23161	188.68
	k_4	0.0003	0.0057
	k_5	267.40	20196
	k_6	0.0053	0.0003
B (cm^3/mol)	k_0	-0.3471	-0.3489
	k_1	-29.748	-28.547
	k_2	0.0644	0.0687
	k_3	-347.04	-817.73
	k_4	0.0010	0.0007
	k_5	-88.547	-65.067
	k_6	0.0048	0.0076

Fig. 1 shows the density predictions for n-decane obtained with the HTHP VT-SRK EoS and HTHP VT-PR EoS models while using the values in Table 1 to calculate A and B. The results of both HTHP models are in very good agreement with the experimental data over the entire pressure and temperature ranges of interest. Furthermore, Fig. 1 demonstrates that these new proposed volume-translated models do not yield isotherm crossovers over a very broad range of temperature and pressure.

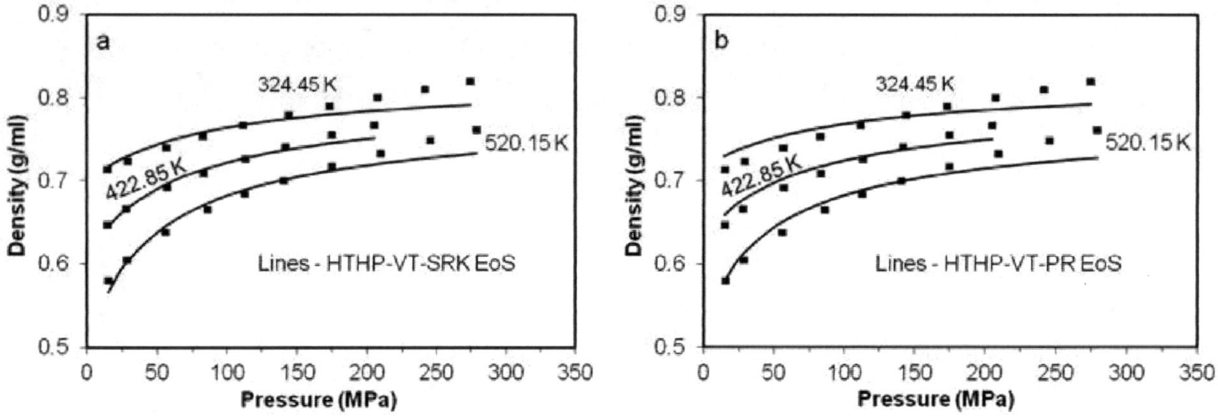

Fig. 1. (a) Comparison of the HTHP VT-SRK EoS with n-decane density data, (b) Comparison of the HTHP VT-PR EoS with n-decane density data. [9,29]

Obtaining a new set of PC-SAFT parameters for use at HTHP conditions

In order to obtain new values for the pure-component PC-SAFT parameters σ, ε/k_B, and m, PC-SAFT density predictions were fit to pure-component experimental density data [11]. These data were obtained at pressures from 4.2 to 300 MPa for compounds for which literature data were available. A downhill simplex method was employed to optimize the values of the pure-component PC-SAFT parameters σ, ε/k_B, and m.

Initially, the PC-SAFT equation was fit to a set of experimental density values spanning both the sub-critical and supercritical regions for the compounds of interest. However, preliminary results indicated that at high pressures PC-SAFT density predictions were not greatly improved over those made with the PC-SAFT parameters derived by Gross and Sadowski [8]. Thus, the PC-SAFT equation was only fit to density values found in a range of pressures between approximately 6.9 and 300 MPa. Typically, the PC-SAFT density values were fit to HTHP density data obtained at a single temperature (typically 423 K) and covering the full range of pressures, as this method yields virtually identical $P\rho T$ values as would be obtained when fitting to experimental data points obtained over the stated pressure range and the full range of temperatures from ambient to 533 K. By plotting the respective parameters for n-alkanes as functions of molecular weight [11], the following predictive expressions (Eqs. 23-25) can be used to determine the HTHP PC-SAFT parameters for normal and branched alkanes for which insufficient experimental data points are available to make the fit. These correlations were used in making all calculations involving HTHP PC-SAFT discussed in the paper.

$$m = 0.03614MW + 1.33810 \tag{30}$$

$$m\sigma = 0.13217MW + 3.51050 \tag{31}$$

$$\frac{\varepsilon}{k_B} = 35.03 \ln MW + 55.60 \tag{32}$$

HTHP PC-SAFT density predictions, illustrated with grey lines in Fig. 2, exhibit a marked improvement over the corresponding G-S PC-SAFT density predictions at pressures above 55 MPa, and are reliable over the whole range of HTHP conditions. The HTHP PC-SAFT parameters also provide superior density predictions for a mixture of 80 mol % n-hexane and 20 mol % n-hexadecane (see Fig. 3). Especially encouraging is that the mixture density predictions were made with a k_{ij} value of 0. Thus, for this n-alkane mixture, the HTHP PC-SAFT parameters are fully predictive. This observation is in line with previous observations on PC-SAFT predictions for multicomponent mixtures [30].

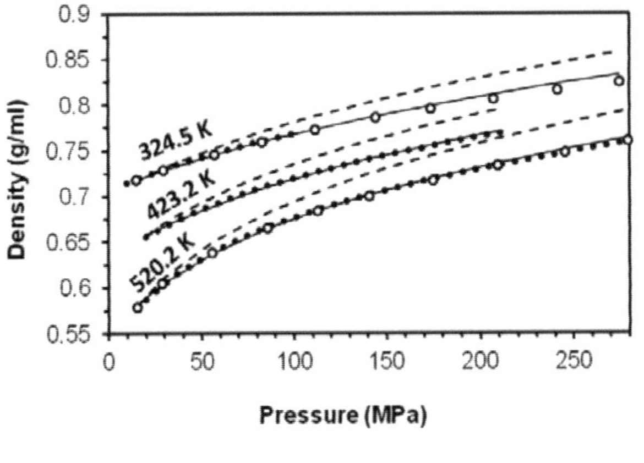

Fig. 2. HTHP PC-SAFT corrects over-prediction of n-decane density by G-S PC-SAFT in the HTHP region, agreeing with reference data to better than ±1%. Adapted from Burgess et al. [11]
----- G-S PC-SAFT —— HTHP PC-SAFT o Liu et al. [9]
••••• Density values given by smoothed NIST correlation [13]

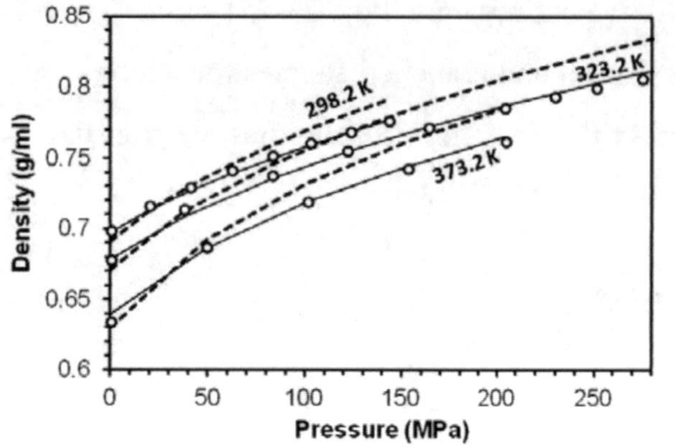

Fig. 3. HTHP PC-SAFT predictions for density of a 0.8:0.2 n-hexane/n-hexadecane mixture agree with reference data to within better than ±1%. Adapted from Burgess et al. [11]
----- G-S PC-SAFT —— HTHP PC-SAFT o Dymond et al. [31]

Viscosity Predictions from f-theory and FV Theory

Next, literature viscosity models were evaluated for their predictive ability in the HTHP region. From Fig. 4, f-theory gives good viscosity predictions for n-decane at pressures up to ~ 150 MPa when coupled to the PC-SAFT EoS for prediction of the attractive and repulsive pressure inputs P_a and P_r.. However, at pressures approaching 276 MPa, viscosity under-prediction can often be in excess of 10% and even as high as 20%. The viscosity under-prediction is particularly noticeable at temperatures near ambient.

Therefore, predictions made using the free volume theory were examined. Very few literature values are present for the pure-component FV theory parameters L, α, and B. However, free volume theory parameters have been published for n-alkanes ranging in size from methane to n-dodecane by Tan et al. [18], who used the G-S PC-SAFT equation to predict the density input to the FV theory. For this study, the experimental viscosity data used to make the parameter fit were obtained at pressures no higher than 60 MPa, and temperatures no higher than 490 K. The n-decane viscosity predictions are good to within ~ 3 %) at pressures up to ~ 100 MPa (see Fig. 5). However, n-decane density predictions can deviate from the experimental value by > 15% at pressures near 276 MPa.

Next, we turned our attention to the work of Allal et al. [17], who obtained a set of pure component FV parameters for methane by fitting to experimental viscosity data collected over a larger interval of pressures and temperatures – that is, pressures from 0.01 to 200 MPa and temperatures from 90.7 to 600 K. Methane viscosity predictions obtained using these parameters, however, are only good at pressures under ~ 100 MPa, as indicated in Fig. 6.

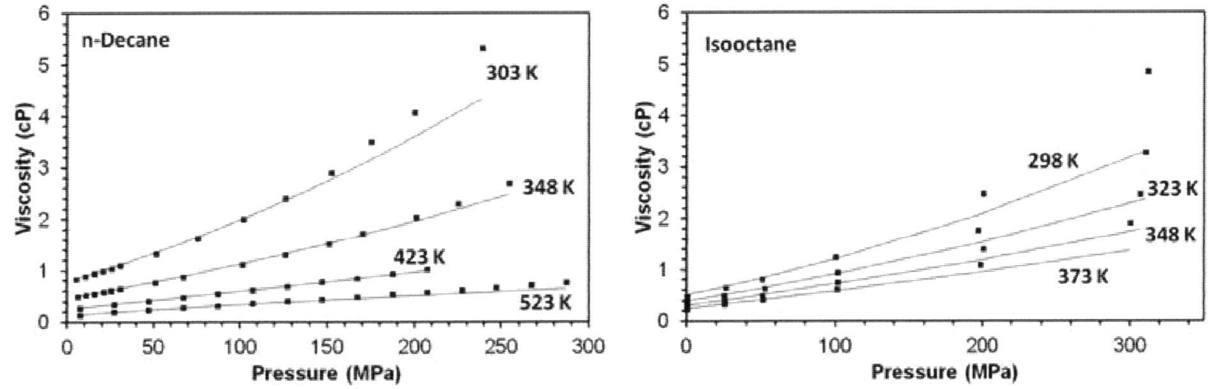

Fig. 4. When coupled to the PC-SAFT EoS, the f-theory gives accurate n-decane viscosity predictions (left panel) at pressures to 150 MPa but under-predicts viscosity at higher pressures. The f-theory gives adequate viscosity predictions for isooctane (right panel) at pressures to ~ 100 MPa but completely fails at higher pressures.

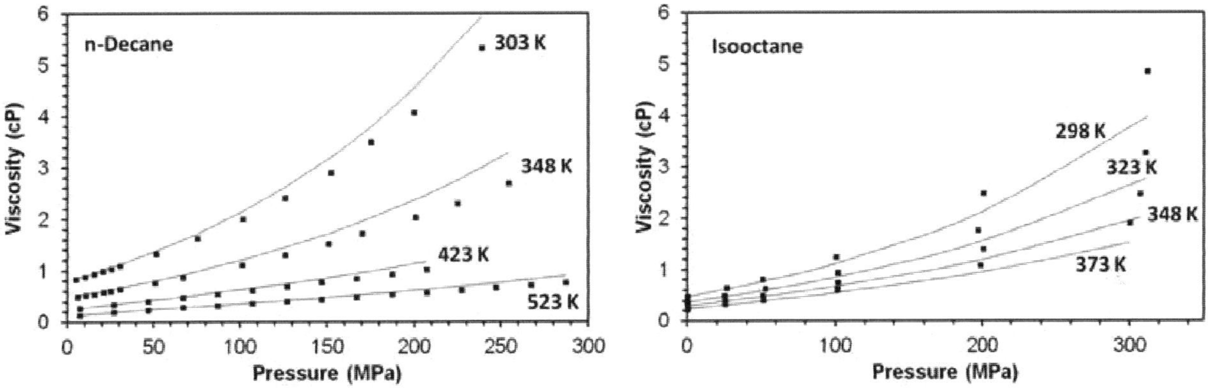

Fig. 5. The FV theory parameters put forward by Tan et al. [18] give adequate n-decane and isooctane viscosity predictions at pressures below ~ 100 MPa but are unreliable at higher pressures. PC-SAFT was used to calculate the density inputs to FV theory.

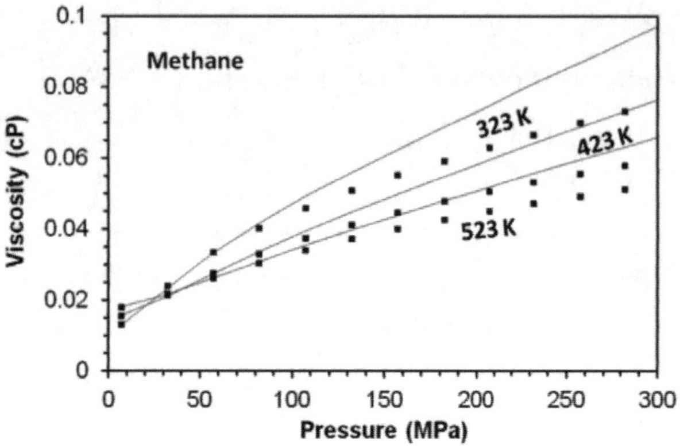

Fig. 6. Methane viscosity predictions obtained using the FV theory parameters of Allal et al. [17]. Experimental data were used as the density inputs to FV theory.

Conclusions

Existing literature models were evaluated for their ability to accurately reproduce experimental density and viscosity values. Results in this arena were mixed. Methods based on both cubic and non-cubic EoS are capable of giving accurate density predictions in the HTHP range, while existing viscosity models do not perform adequately above at most 150 MPa.

Two-parameter, temperature-dependent volume-translation terms were developed by Baled et al. [29] by fitting both the PR and SRK EoSs to experimental HTHP density values for normal and branched alkanes, cycloalkanes, and aromatics. These models yield density predictions that typically deviate < 2% from experimental values, although near 276 MPa density predictions may deviate from the experimental value by up to 5%.

By correlating the PC-SAFT equation of state to HTHP density data typically collected at 423 K and pressures from 6.9 to 276 MPa, we have produced a new set of pure-component parameters that, when utilized with the PC-SAFT equation, typically give absolute percent deviations of less than 1% over the this range of conditions. In so doing, we have eliminated the over-prediction of pure-component density in the HTHP region that is observed when using the parameters put forth by Gross and Sadowski. In addition, correlations were developed for predicting the HTHP PC-SAFT parameters for n-alkanes and branched alkanes for which insufficient experimental data are available to obtain them by best fit to reference data. Initial results for alkane-alkane mixture densities are encouraging, but are limited by the relative scarcity of mixture density data in the HTHP region and especially at temperatures near 533 K.

Hence, it will first be necessary to obtain mixture $P\rho T$ data over the full temperature and pressure range of interest.

On the other hand, the f- and FV theories of viscosity give good (within ~ ±2% of experimental value) predictions up to ~ 100 MPa over the full temperature range of interest. However, they over-predict viscosity by as much as 25% at pressures near 276 MPa, indicating that the actual **predictive** power of the pure-component f- and FV theory model parameters is significantly weaker than that of the HTHP-VT-EoS and HTHP PC-SAFT parameters. In order to obtain improved models for viscosity at HTHP conditions, it will be necessary to fit the f-theory and FV theory parameters to experimental viscosity data covering the entire temperature and pressure range of interest.

References

[1] G. Soave, Chemical Engineering Science, 27 (1972) 1197-1203.

[2] D.Y. Peng, D.B. Robinson, Industrial & Engineering Chemistry Fundamentals, 15 (1976) 59-64.

[3] G. Schmidt, H. Wenzel, Chemical Engineering Science, 35 (1980) 1503-1512.

[4] W.G. Chapman, K.E. Gubbins, G. Jackson, M. Radosz, Ind. Eng. Chem. Res., 29 (1990) 1709-1721.

[5] W.G. Chapman, G. Jackson, K.E. Gubbins, Mol. Phys., 65 (1988) 1057-1079.

[6] S.H. Huang, M. Radosz, Ind. Eng. Chem. Res., 29 (1990) 2284-2294.

[7] S.H. Huang, M. Radosz, Ind. Eng. Chem. Res., 30 (1991) 1994-2005.

[8] J. Gross, G. Sadowski, Ind. Eng. Chem. Res., 40 (2001) 1244-1260.

[9] K. Liu, Y. Wu, M.A. McHugh, H. Baled, R.M. Enick, B.D. Morreale, J. of Supercrit. Fluids, 55 (2010) 701-711.

[10] Y. Wu, B. Bamgbade, K. Liu, M. A. McHugh, H. Baled, R. M. Enick, W. A. Burgess, D. Tapriyal, B. D. Morreale, Fluid Phase Equilib. 311 (2011) 17-24.

[11] W. A. Burgess, D. Tapriyal, B. D. Morreale, Y. Wu, M. A. McHugh, H. Baled, R. M. Enick, Fluid Phase Equilib. (2012), doi:10.1016/j.fluid.2012.01.032.

[12] D.R. Caudwell, J.P.M. Trusler, V. Vesovic, W.A. Wakeham, Int. J. Thermophys., 25 (2004) 1339-1352.

[13] P. J. Linstrom, W. G. Mallard, Eds., NIST Chemistry WebBook, NIST Standard Reference Database Number 69, June 2005, National Institute of Standards and Technology, Gaithersburg MD, 20899 (http://webbook.nist.gov/chemistry/fluid/)

[14] A. K. Doolittle, J. Chem. Eng. Data, 9 (1964) 275-279.

[15] S. E. Quiñones-Cisneros, C. K. Zéberg-Mikkelsen, E. H. Stenby, Fluid Phase Equilib. 178 (2001), 1-16.

[16] S. E. Quiñones-Cisneros, C. K. Zéberg-Mikkelsen, J. Fernández, J. García, AIChE Journal 52 (2006) 1600-1610.

[17] A. Allal, C. Boned, Q. Baylaucq, Phys. Rev. E 64 (2001), 011203.

[18] S. P. Tan, H. Adidharma, B. F. Towler, M. Radosz. Ind. Eng. Chem. Res. 44 (2005), 8409-8418.

[19] J. Gross, G. Sadowski, Ind. Eng. Chem. Res., 40 (2001) 1244-1260.

[20] A. Peneloux, E. Rauzy, Fluid Phase Equilib., 8 (1982) 7-23.

[21] W.R. Ji, D.A. Lempe, Fluid Phase Equilib., 130 (1997) 49-63.

[22] L.S. Wang, J. Gmehling, Chemical Engineering Science, 54 (1999) 3885-3892.

[23] K. Frey, J. Tester, M. Modell, Fluid Phase Equilib., 279 (2009) 56-63.

[24] J. A. Barker, D. Henderson, J. Chem. Phys., 47 (1967) 2856.

[25] J. A. Barker, D. Henderson, J. Chem. Phys., 47 (1967) 4714.

[26] T.-H. Chung, M. Ajlan, L. L. Lee, K. E. Starling, Ind. Eng. Chem. Res., 27 (1988) 671-679.

[27] S. E. Quiñones-Cisneros, C. K. Zéberg-Mikkelsen, E. H. Stenby, Fluid Phase Equilib., 169 (2000) 249-276.

[28] A. K. Doolittle, J. Appl. Phys. 22 (1951), 1471.

[29] H. Baled, R. M. Enick, Y. Wu, M. A. McHugh, W. Burgess, D. Tapriyal, B. D. Morreale, Fluid Phase Equilib., 317 (2012) 65-76.

[30] P. D. Ting, P. C. Joyce, P. K. Jog, W. G. Chapman, M. C. Thies, Fluid Phase Equilib., 206 (2003) 267-286.

[31] J. H. Dymond, K. J. Young, J. D. Isdale, J. Chem. Thermodyn. 11 (1979) 887-895.

SAFT Model for Upstream Asphaltene Applications

Sai R. Panuganti[1], *Francisco M. Vargas*[2], *Doris L. Gonzalez*[3] *and Walter G. Chapman*[1]*

1.Department of Chemical and Biomolecular Engineering, Rice University, Houston, USA

2. Department of Chemical Engineering, Petroleum Institute, Abu Dhabi, UAE

3. Data Quality Group, Schlumberger, Houston, USA

Authors Email Address: sai@rice.edu, fvargas@pi.ac.ae, DGonzalez21@slb.com and wgchap@rice.edu

Corresponding Author: email- wgchap@rice.edu, Phone- 001(713) 348-4900

Abstract

The increasing incidence of flow assurance problems caused by asphaltene deposition during oil production has motivated the development of numerous theoretical models and experimental methods to analyze this complex phenomenon. Even more challenging are the prediction of the occurrence and the magnitude of asphaltene deposition. It is normally well accepted that precipitation of asphaltenes is a necessary but not a sufficient condition for deposition. Hence, a significant amount of work has been devoted to the understanding of the conditions at which asphaltenes precipitate from the crude oil. Although, several models seem to work well for correlating available data of onsets of asphaltene precipitation, we have found that they usually lack good prediction capabilities.

In this article, we review the properties of asphaltene and present a thermodynamic model based on the Perturbed Chain version of the Statistical Associating Fluid Theory (PC-SAFT) equation of state, which has been proven to provide excellent prediction capabilities for both the asphaltene onset pressure and the bubble point of the crude oil, in a wide range of temperatures, pressures and compositions. Furthermore, several case studies are presented where we analyze the effect of gas injection, commingling of oils, and contamination with oil based mud, as well as the effect of asphaltene polydispersity. Finally, the asphaltene compositional grading that can lead in some cases to the formation of tar-mats is studied using the same thermodynamic model. In all the cases, very good agreement with experimental data and field observations is obtained. From the results presented in this work, we are confident that the PC-SAFT equation of state is an excellent tool to understand the phase behavior of asphaltenes in crude oil.

Keywords: flow assurance, asphaltene, PC-SAFT, phase behavior

1. Introduction

Asphaltenes are of particular interest to the petroleum industry because of their deposition tendencies in production equipment that cause considerable production costs. In addition, precipitated asphaltenes impart high viscosity to crude oils, negatively impacting production. Unlike wax and gas hydrates, asphaltenes pose a special challenge because asphaltenes are not well characterized and can deposition even at high temperature. Paradoxes like light oils (very low in asphaltene content) reporting more asphaltene deposition problems than heavy oils (high in asphaltene content), makes the problem even more interesting. Thus, ability to predict the thermodynamic and transport properties of crude oil systems containing asphaltenes are important in planning for and possibly eliminating asphaltene related flow assurance problems.

In the second section of this paper, the properties of the asphaltene fraction of crude oils are reviewed in detail. Based on this background and with the recent advances in asphaltene science, the solubility theory with liquid-liquid equilibrium using a SAFT based equation of state (EoS) is chosen as the most appropriate to model the thermodynamics of asphaltenes. The third section of this paper explains SAFT and in particular the PC-SAFT EoS with all the parameters required to predict the phase behavior of high molecular weight fluids such as asphaltenes. The fourth section of this paper reviews the limitations of previous characterization procedures and hence the subsequent fifth section introduces the new characterization methodology with asphaltene as one of the components. Results section includes the successful PC-SAFT thermodynamic modeling and understanding of asphaltene-crude oil system along with applications.

2. Asphaltene Background

Asphaltenes are the polydisperse, heaviest and most polarizable fraction of crude oil. The term asphaltene was first introduced by Boussignault in 1837 to designate the material that precipitates out of petroleum on the addition of petroleum ether [1]. Today, it is operationally defined in terms of its solubility as the component of crude oil which at ambient conditions is completely miscible in aromatic solvents, such as benzene, toluene or xylenes, but insoluble in light paraffinic solvents, such as n-pentane or n-heptane. The following aspects, discuss asphaltene properties in detail.

2.1 Chemical Composition

Asphaltene is a solubility class of molecules found in crude oil, along with resins, aromatic hydrocarbons, and alkanes; primarily consists of carbon, hydrogen, nitrogen, oxygen and sulfur, as well as trace amounts of iron, vanadium and nickel. An asphaltene macro molecule (asphaltene aggregate entity) has a carbon number in the range of 40 to 80. The carbon to hydrogen ratio is approximately 1:1.2, depending on the asphaltene source [2].

2.2 Molecular Structure

Asphaltenes are polydisperse in nature and hence does not have a specific chemical identity. In fairly early stages (1970's) of asphaltene research, Yen proposed a condensed aromatic cluster model that assumes extensive condensation of aromatic rings into large sheets with substitute side chains (Figure 1a) [3]. This type of structure was developed based on spectroscopic studies. Later on, a very different structural organization, the bride aromatic model (Figure 1b) was proposed by Murgich and Abaner based on structural mechanics [4]. A number of other investigators have also attempted to postulate model structures for asphaltene [5, 6]. But the Modified Yen Model is the current accepted micro and macro structure for asphaltene [7].

The first macrostructure of an asphaltene was proposed by Dickie and Yen in 1961 (Figure 2) and is more popularly known as the Yen model [8]. Over the past decade, asphaltene science has progressed dramatically and the Modified Yen Model (Figure 3) proposed by Mullins in 2010 explained many of the observed phenomena for asphaltene. According to the Modified Yen model, the predominant part of the asphaltene structure has a single, moderately large polycyclic aromatic hydrocarbon with peripheral alkanes. Asphaltene molecules form asphaltene nano-aggregates with aggregation numbers of ~6 and these aggregates can further cluster with aggregation numbers estimated to be ~8.

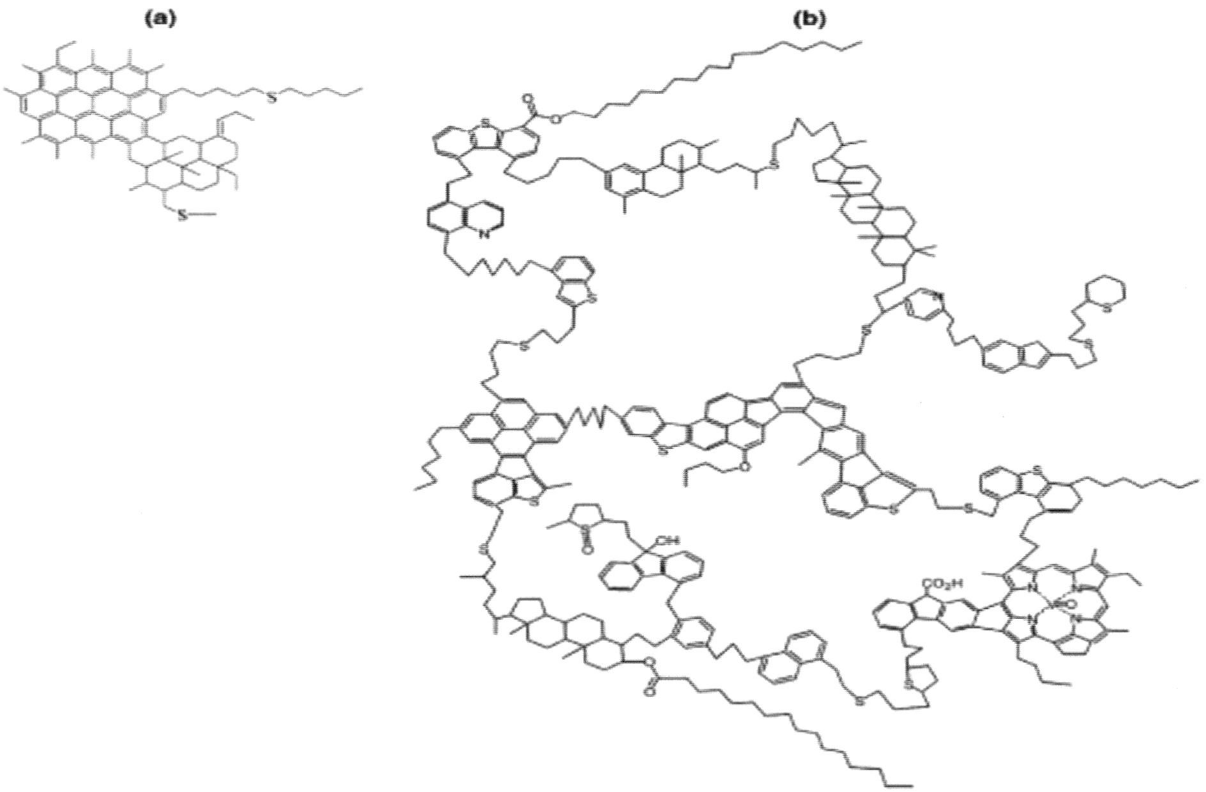

Figure 1. Asphaltene molecular structural models

(a) Condensed aromatic cluster model [3]

(b) Bridge aromatic model [4]

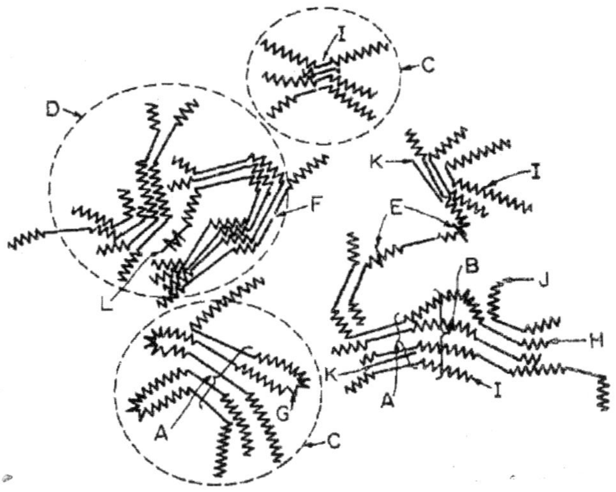

A. Crystallite	B. Chain Bundle
C. Particle	D. Micelle
E. Weak Link	F. Gap and hole
G. Intra-cluster	H. Inter-cluster
I. Resin	J. Single Layer
K. Petroporphyrin	L. Metal

Figure 2. Yen Model [8]

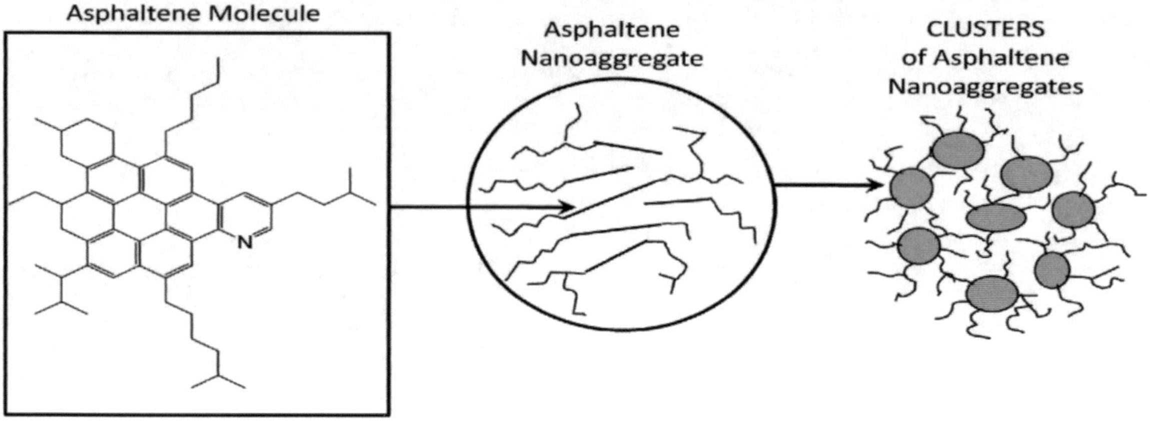

Figure 3. Modified Yen Model [7]

2.3 Molecular Weight

Depending on the solvent, concentration and the measurement technique employed, several molecular weight ranges were reported for asphaltene. Different techniques to estimate molecular weight include vapor-pressure osmometry, viscometry, gel permeation chromatography, florescence depolarization, ultracentrifuge, electron microscope studies. The vapor-pressure osmometry technique results in an asphaltene molecular weight in the range of 800 to 3000 g/mol in good solvents [9]. Alboudwarej et al, estimated an average molecular weight of asphaltene as 1800 g/mol for heavy oil and bitumen [10]. Adopting the Modified Yen Model, Zuo et al. concluded the molecular weight of asphaltene nano-aggregate as ~1600 gm/mole [11]. Long before Mullins work, in 1984 Hirschberg and Hermans estimated the average asphaltene molecular weight to be between 1300 and 1800 g/mol based on phase behavior studies [12].

2.4 Density

The density of asphaltene varies between 1.1 and 1.2 g/cc at ambient conditions [13]. Asphaltene density is calculated indirectly based on density of asphaltene and toluene mixture.

Diallo et al. estimated an average asphaltene density as 1.12 g/cc based on (NPT) isothermal isobaric ensemble dynamic molecular simulations followed by energy minimization [14]. This estimated density compares favorably with the measured density, 1.16 g/cc of petroleum asphaltene reported by Yen et al. using helium displacement [15].

2.5 Diffusion

Asphaltene diffusion coefficient is a function of concentration dependent size of asphaltene aggregates. At low concentrations, the degree of association of asphaltene is small, but increases with increase in concentration until a critical size is reached. The higher the state of association, the larger the entity size and lower the diffusion coefficient. The diffusion coefficient of asphaltene in toluene estimated by low field nuclear magnetic resonance and florescence spectroscopy measurements was in the order of 10^{-10} m^2/s for a particle size of 1-2 nm [16]. The diffusion coefficient in toluene at infinite asphaltene dilution is $2.2x10^{-10}$ m^2/s [17].

2.6 Interfacial Characteristics

The oil-water interfacial properties in the presence of asphaltene are peculiar. The initial rapid diffusion of asphaltene towards the interface is followed by a long reorganization and progressive building of layers. The overall adsorption process as observed from Figure 4 is slow, but for different oils, has similar time scales (Figure 5) [18]. When asphaltene concentration is varied, Figure 5 shows that interfacial equilibrium is reached faster at higher asphaltene concentrations for good solvents and slower in the presence of poor solvents. Asphaltene molecules also adsorb at oil-air interface, forming skins which impart stability to foamy oils in which bubbles persist for a long duration of time [19, 20]. Figure 5 shows the adsorption time versus asphaltene concentration for oil-water and oil-air interfaces with different solvents:

toluene (star), toluene-heptane (50:50) (circles) and toluene-heptane-maltene (45:45:10) (triangles) [18]. The black squares are for oil-air interface where the oil used was toluene.

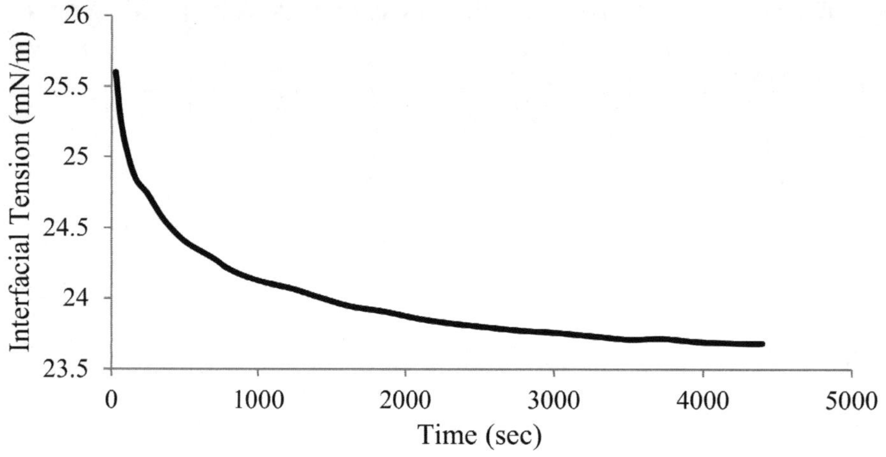

Figure 4. Interfacial tension between water and toluene with 3 wt% asphaltene solutions, versus time

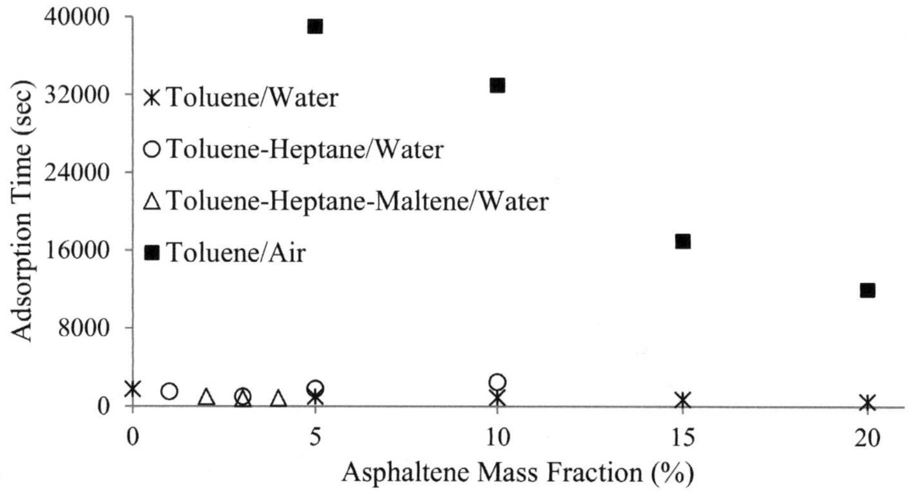

Figure 5. Adsorption time for different interfaces under a varying asphaltene concentration

In the presence of surfactant there is a co-adsorption of both surfactant and asphaltene at the interface. Surfactants being smaller molecules than asphaltene arrive at the interface before the arrival of asphaltene. At intermediate times, surface population is over compressed by surfactants adsorbed at the interface and incoming asphaltene [21]. This over-population of the surface gives rise to a minimum in the dynamic surface tension [22]. At long enough times, the interfacial tension increases with time, a phenomenon not seen with asphaltene or surfactant alone. Such a phenomenon indicates the presence of transfer of species across the interface, at which location other species settle [23]. Figure 6 shows the dynamic surface tension experiment results of an oil-water interface in the presence of asphaltene and surfactant [18].

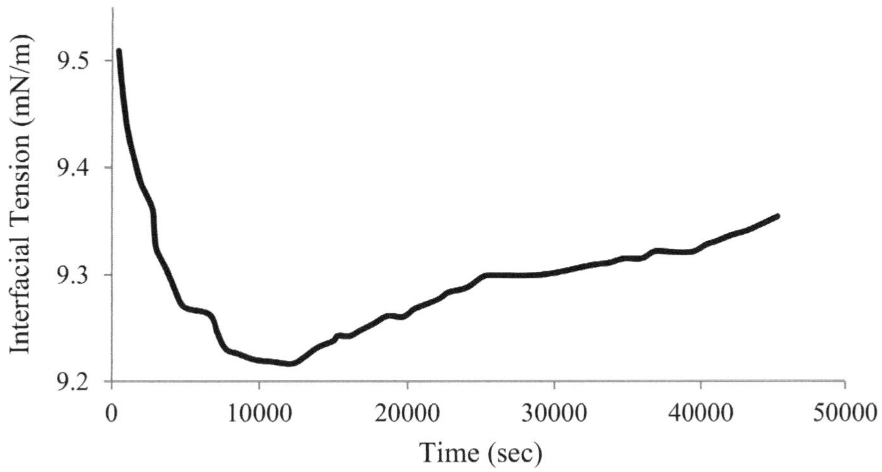

Figure 6. Interfacial tension versus time between a toluene solution of asphaltene (3 wt%) and an aqueous solution of Triton X405 (1 wt%)

2.7 *Viscosity*

The presence of asphaltene means additional difficulties related to transport and processing due to the increased crude oil viscosity caused by the asphaltene [24]. The experimental results of Peng and Yongan on asphaltene content and temperature affecting the crude oil viscosity are plotted in Figure 7 [25]. Sirota et al. working independently also reported the same evidence of increased asphaltene content and low temperatures significantly increasing the oil viscosity [26].

Presence of precipitated asphaltene particles results in an increase in the viscosity of the oil. This increase becomes remarkable at and after the onset of flocculation and hence can be used to detect the asphaltene precipitation onset conditions [27]. Compositional grading is a well known phenomena and with asphaltene compositional variation, the crude oil viscosity changes significantly with depth. Modeling the crude oil viscosity variation with depth due to asphaltene compositional grading, can help in predicting the occurrence of tar-mat; characterized as the highly viscous oil zone enriched in asphaltene [28].

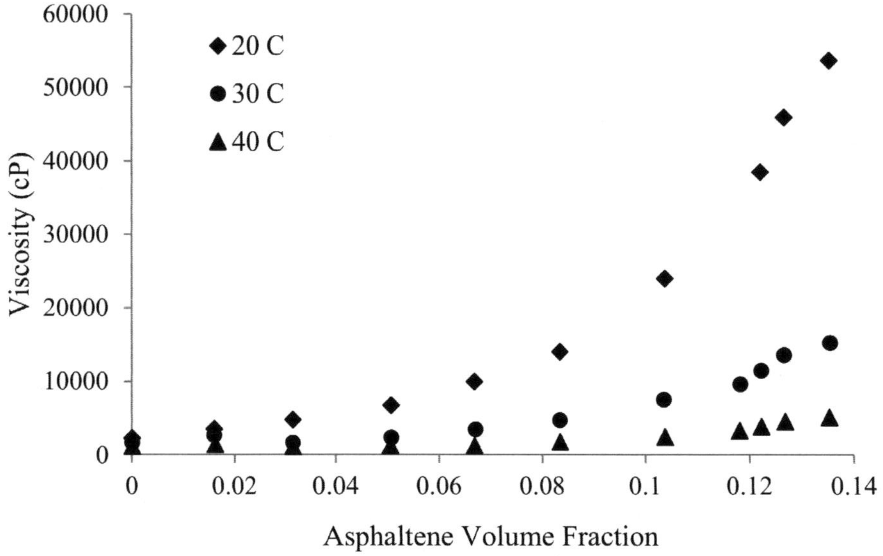

Figure 7. Measured viscosity of a heavy oil sample versus the asphaltene volume fraction at different temperatures

2.8 *Solubility Parameter*

Solubility parameter *(δ)* is an important tool based on which asphaltene phase behavior can be explained. The solubility parameter for a non-polar fluid is given by Hildebrand as,

$$\delta = \sqrt{CED} = \sqrt{-\frac{U^{rsm}}{\upsilon}} \qquad (1)$$

Where, *CED* is the crude oil cohesive energy density, U^{rsm} is the residual internal energy obtained by subtracting the ideal gas contribution from that of the real fluid and υ is the pure liquid volume.

Asphaltene has the highest solubility parameter among the crude oil components and is between 19 to 24 $MPa^{0.5}$ [29]. If asphaltene is assumed to be dissolved in the crude oil, the equilibrium can be altered by a change in temperature, pressure or composition of the crude. For example, if significant amounts of low-molecular weight hydrocarbons are dissolved into the liquid phase, the solubility parameter is reduced and asphaltenes precipitate may form. Changes in temperature and pressure will likewise alter the solubility parameter, thus causing conditions that induced precipitation.

2.9 Refractive Index

The refractive index (*RI*) of an asphaltene solution can be measured in place of the solubility parameter based on the observation that London dispersion interactions dominate the asphaltene phase behavior [30]. The London dispersion properties of a material can be characterized by the wavelength dependence of the refractive index or the "dispersion" of visible light. Refractive index and solubility parameter can be inter-converted using a linear correlation shown in Figure 8 [31].

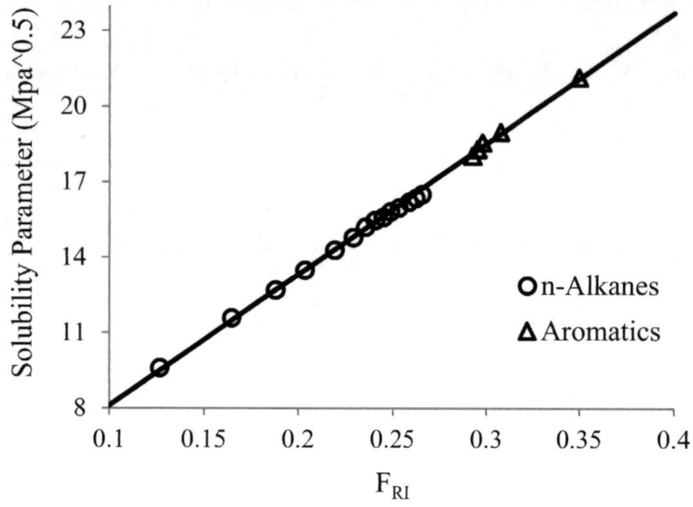

$$\delta = 52.042 F_{RI} + 2.904 \quad (2)$$
$$R^2 = 0.998$$

$$\text{(Where, } F_{RI} = \frac{RI^2 - 1}{RI^2 + 2}\text{)}$$

Figure 8. Comparison of solubility parameter and F_{RI}

Using the above correlation, the refractive index of pure asphaltene is observed to range from 1.54 to 1.74.

2.10 Precipitation

Asphaltene precipitation is treated as the growth of primary particles from the solution, due to super-saturation. Thus, asphaltene precipitation consumes super-saturation until the system reaches an equilibrium asphaltene concentration. Very limited data is available in the literature about the rate of asphaltene precipitation and the reported spectrophotometry data on 0.1 g of asphaltene/L of toluene and mixed with a 50 volume % n-heptane solution, can be approximately modeled using a first order precipitation rate constant ~ 0.01 min^{-1} [32].

2.11 Aggregation

Aggregation does not consume super-saturation and is considered as the growth of primary particles forming larger, secondary particles (asphaltene aggregate). Concentration of asphaltene aggregates less than the critical micelle concentration of 3-4 g of asphaltene/L toluene is controlled by diffusion while at higher concentrations of asphaltene the aggregation will be controlled by impact [33]. Most of the literature available does not consider precipitation and aggregation separately. Researchers using microscopy have defined the onset of asphaltene precipitation based on the resolution limit of microscopes. The redissolution kinetics of precipitated asphaltene can be considerable in the time scale of well-bore travel while that of aggregated asphaltene is slow and hence negligible [34]. It is experimentally observed that asphaltene aggregate particle size gets saturated with time and assuming exponential approach to equilibrium, Anisimov et al. obtained [34]: $nr^3 = n_\infty R_\infty^3 (1 - e^{-t/\tau})$ *(3)*

2.12 Deposition/Adsorption

Photothermal surface deformation studies of asphaltene adsorption onto surfaces revealed that the adsorption resulted in multilayer formation [35]. Recent studies performed using quartz crystal microbalance showed saturation adsorption isotherms with rapid initial response followed by a slower approach to equilibrium [36]. The initial response was indicative of a diffusion controlled process. At longer timescales, the data was found to follow first-order kinetics. The overall first order rate constant was estimated as ~ $2x10^{-3}$ min^{-1} [37]. But, all the above observations were reported for asphaltene solubilized in toluene and the adsorption kinetics of precipitated asphaltene are expected to be different. Xie et al. and Goual et al. reported one case each where deposition of precipitated asphaltene never reached equilibrium even at very long times [38, 39].

2.13 Asphaltene - Crude Oil System

Understanding of the asphaltene stability in crude oil is based mainly on two different thermodynamic models: colloidal and solubility models. The colloidal approach describes asphaltene phase behavior assuming that the crude oil can be divided into polar and non-polar subfractions or solvent in which resins stabilize asphaltene as in micelle formation [40]. The solubility model uses the molecular solubility approach to describe an asphaltene containing fluid as a mixture of solute (asphaltene) and solvent (bulk oil) in a homogeneous liquid state [41]. The asphaltene precipitation can be treated as solid–liquid or liquid–liquid equilibrium and is reversible [42, 43]. But, as discussed in the *Aggregation* section, the kinetics of re-dissolution varies significantly depending on the physical state of the system.

The colloidal model assumes that asphaltenes exist in the oil as solid particles; and, describes the stability of asphaltene in terms of micelle formation where asphaltene self-associate into an aggregate to form the core and resins adsorb onto the core (based on polar-polar interactions) to form a steric shell shown in Figure 9. These points of view are based on the observation that the increase in resins content improves the stability of asphaltene. According to these models, asphaltene precipitates when a sufficient amount of diluent is added to the system because the concentration of resins is diluted, with resins becoming increasingly soluble in the liquid phase that separate them from the asphaltene micelles. However, the addition of solvents like toluene actually takes asphaltene back into solution. Using impedance analysis, Goual has demonstrated that the diffusion coefficient of asphaltene is same in the presence and absence of resin [44]. Thus, resins are unlikely to coat asphaltene nanoaggregates and do not provide the steric stabilizing layer that the colloidal model proposes. Also, the polar nature of asphaltene was never confirmed in terms of charge carried by the asphaltene particles through hydrophilic-lipophilic balance [45]. While colloidal models of asphaltene were beneficial in the early stages of understanding asphaltene behavior, these analogies are not justified.

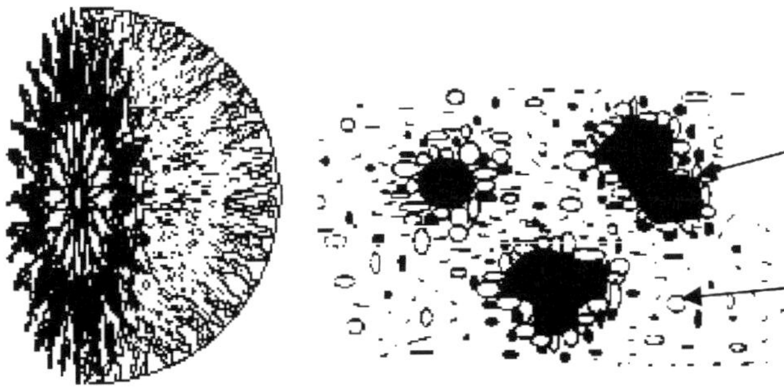

Figure 9. Asphaltene colloidal model [46]

Solubility models assume that asphaltene is dissolved in the crude oil. This equilibrium can be solid-liquid or liquid-liquid type. In the solid-liquid type by Nghiem and Coombe [42], precipitated asphaltene is treated as a single component residing in the solid phase, while oil and gas phases are modeled with a cubic equation of state. However, pure asphaltene is never precipitated [47]. Also asphaltene is a substance with no defined melting point, high solubility parameter and moderately high molecular weight. Having a high aromatic content, it forms true solution in aromatic solvents. However, its molecular weight being moderately high, the entropy of mixing is not sufficient to keep it in solution if the interaction with the remainder of the solution becomes unfavorable, such as upon the addition of an aliphatic fluid. The phase separation which results is thus a thermodynamically driven liquid-liquid phase separation.

The liquid-liquid treatment was initially proposed by Hirschberg, et al. [48] starting with a detailed compositional model based on the Soave equation of state to calculate liquid phase compositions and molar volumes before asphaltene precipitation. Then, using a Flory–Huggins-regular solution theory, the amount of precipitated asphaltene was estimated by calculating the solubility parameter from Hildebrand's definition. This initial approach was difficult to extend to model asphaltene solubility under reservoir conditions because the solubility parameters must be corrected with correlations or an equation of state for high pressure and temperatures. Also cubic

EoS models cannot describe the phase behavior of systems with large size disparities and they cannot accurately describe their fluid densities.

A more modern equation of state is the Statistical Associating Fluid Theory family. This EoS based on statistical mechanics can accurately model mixtures of different molecular sizes. Thus, solubility model with liquid-liquid equilibrium using the SAFT based EoS, is the most appropriate theory for modeling asphaltene phase behavior. SAFT based equations of state have demonstrated excellent performance and predictive capabilities applied to a wide range of mixtures, including petroleum systems [49].

3. SAFT Theory

The statistical association fluid theory is a molecular-based equation of state derived by Chapman et al. by applying and extending Wertheim's first-order perturbation theory to chain molecules [50-52]. In this theory, molecules are modeled as chains of bonded spherical segments and the properties of a fluid are obtained by expanding about the same properties of a reference fluid. SAFT has a similar form to group contribution theories in that the fluid of interest is initially considered to be a mixture of unconnected groups or segments. As shown in Figure 10, SAFT includes a chain connectivity term to account for the bonding of various groups to form polymers and an explicit association term to account for intermolecular attractions like hydrogen bonding. SAFT describes the residual Helmholtz free energy (A^{res}) of a fluid as

$$A^{res} = A - A_{ideal} = A^{segment} + A^{chain} + A^{assoc} = m\left(A_o^{hs} + A_o^{disp}\right) + A^{chain} + A^{assoc} \qquad (4)$$

Where $A^{segment}$, A^{chain}, A^{assoc}, A_o^{hs} and A_o^{disp} are the segment, chain, association, hard-sphere, and dispersion contributions respectively to the mixture's residual Helmholtz free energy. The average segment number of the mixture, m, is an average of the pure species' segment number, m_i, weighted by the species' compositions, x_i. $m = \sum_i x_i m_i$ *(5)*

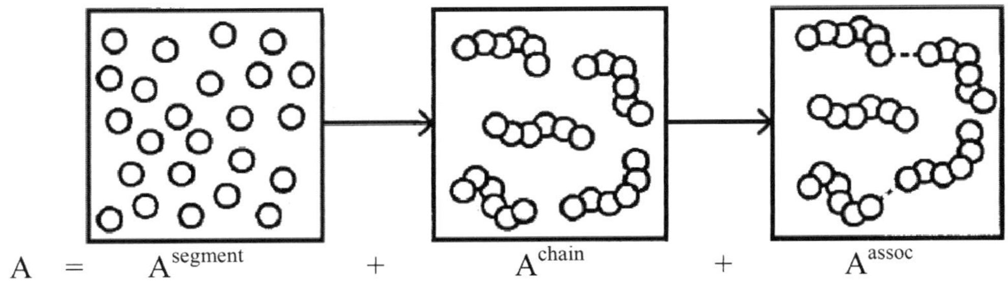

Figure 10. SAFT treats molecules as chains of spherical segments. The fluid's free energy is then the sum of the independent segment free energy and the change in free energy due to chain formation and association

Statistical association fluid theory and its various modifications and extensions have demonstrated the ability to accurately predict the phase behavior of simple, hydrogen bonding, polar, and polymeric mixtures [53]. It is particularly useful for modeling systems with significant chain contribution effects (as in polymer mixtures). When considering asphaltene phase behavior dominated by nonpolar interactions, PC-SAFT is chosen among different forms of SAFT [54]. Gross and Sadowski proposed the perturbed chain modification (PC-SAFT) to account for the effects of chain length on the segment dispersion energy, by applying the Barker-Henderson perturbation theory truncated at the second-order term to a hard chain reference fluid [55]. PC-SAFT employs a hard sphere reference fluid described by the Mansoori-Carnahan-Starling-Leland equation of state. This version of SAFT properly predicts the phase behavior of mixtures containing high molecular weight components similar to the large asphaltene molecules.

The free energy contribution of the hard-sphere mixtures given by Mansoori-Carnahan-Starling-Leland equation of state is [56]

$$\frac{A_o^{hs}}{RT} = \frac{6}{\pi\rho}\left[\frac{\zeta_2^3 + 3\zeta_1\zeta_2\zeta_3 - 3\zeta_1\zeta_2\zeta_3^2}{\zeta_3(1-\zeta_3)^2} - \zeta_0\left(1 - \frac{\zeta_2^3}{\zeta_3^2}\right)\ln(1-\zeta_3)\right] \qquad (6)$$

Where $\zeta_k = \frac{\pi\rho}{6}\sum_i x_i m_i d_{ii}^k$ *(7)*

And $d_{ii}^k = \sigma_i\left[1 - 0.12\exp\left(\frac{-3u_i}{kT}\right)\right]$ *(8)*

Here, ρ is the number density (molecules/A^3), d_{ii} the temperature-dependent segment diameter of species i in units of Angstroms and σ_i is the adjustable PC-SAFT temperature-independent segment diameter (A). A temperature dependent hard sphere diameter accounts for the soft repulsion between molecules. Calculated for Barker-Henderson theory, the temperature dependence was fit by Chen and Kerglewski [57].

The contribution to free energy due to chain formation from Wertheim theory is given by [52]

$$\frac{A^{chain}}{RT} = \sum_i x_i(1-m_i)\ln g_{ii}^{hs}(d_{ii}) \qquad (9)$$

Where $g_{ii}^{hs}(d_{ii}) = \left[\frac{1}{1-\zeta_3} + \frac{3d_{ii}\zeta_2}{2(1-\zeta_3)^2} + \frac{d_{ii}^2\zeta_2^2}{2(1-\zeta_3)^3}\right]$ *(10)*

Perturbed chain version of SAFT was initially developed for alkanes by incorporating the effects of chain length on the segment dispersion energy. With PC-SAFT, the dispersion contribution becomes [55]

$$\frac{A_o^{disp}}{RT} = \frac{A_1}{RT} + \frac{A_2}{RT} \tag{11}$$

$$\frac{A_1}{RT} = -2\pi\rho I_1(\eta, m)\sum_i\sum_j \frac{x_i x_j m_i m_j \sigma_{ij}^3 \varepsilon_{ij}}{kT} \tag{12}$$

$$\frac{A_2}{RT} = -2\pi m I_2(\eta, m)W^{-1}\sum_i\sum_j \frac{x_i x_j m_i m_j \sigma_{ij}^3 \varepsilon_{ij}^2}{(kT)^2} \tag{13}$$

$$W = 1 + m\frac{8\eta - 2\eta^2}{(1-\eta)^4} + (1-m)\frac{20\eta - 27\eta^2 + 12\eta^3 - 2\eta^4}{(1-\eta)^2(2-\eta)^2} \tag{14}$$

$$\varepsilon_{ij} = (1-k_{ij})\sqrt{\varepsilon_{ii}\varepsilon_{jj}} \tag{15}$$

$$\sigma_{ij} = \frac{\sigma_i + \sigma_j}{2} \tag{16}$$

Where η is the package fraction, I_1 and I_2 are functions of the system packing fraction, ε is the segment-segment dispersion energy and k_{ij} is the binary interaction parameter between species *i* and *j*. Equation 11 was derived from Barker–Henderson perturbation theory, and A_1 and A_2 correspond to the first and second order contributions respectively [58]. Reexamining the equations from 6 to 16, it is observable that in PC-SAFT for pure components only three parameters are required to be estimated, which are the number of segments per molecule (*m*), the temperature-independent diameter of each molecular segment (σ) and the segment-segment dispersion energy (ε). Although PC-SAFT requires only three parameters, the equation of state produces accurate vapor pressures and liquid densities. Cubic equations of state with volume translation typically require four or more pure component parameters to fit the same data.

The main presumption in the PC-SAFT approach for the thermodynamic modeling of asphaltene is that asphaltene associates to form preaggregates and further association is not considered during precipitation. Therefore, asphaltene phase behavior can be qualitatively explained in terms of London dispersion interactions, and polar interactions are assumed to have an insignificant contribution. Aromatic ring compounds making the core of asphaltene are highly polarizable; therefore, the polarizability determines the ability of hydrocarbons to serve as a precipitant or as a solvent for asphaltene. The evidence obtained through the works of Buckley and Hirasaki validates the presumption [30, 59] that polarity does not play a significant role in asphaltene phase behavior. Thus the association term in SAFT can safely be neglected for asphaltene thermodynamic modeling.

4. Characterization of Reservoir Fluid

Crude oil has numerous components and characterizing the oil as a mixture of well defined fractions that represent blends of similar components instead of handling the components individually, will aid in reducing the computational cost significantly. The earliest study on crude oil characterization dates back to 1978 by Katz & Firoozabadi [60] where boiling point temperatures of n-paraffins were used for separating the carbon number fraction. The resulting densities are for paraffinic oils and therefore very low [61].

A later method proposed by Whitson, as a correction of Katz and Firoozabadi, is widely applied in the industry to characterize the stock tank oil (STO) [62]. Whitson's method is based on the average boiling point of each single carbon number cut and correlations from Riazi and Daubert [63]. Whitson's method presents a set of physical properties for the petroleum fractions C_6 through C_{45}. The calculated properties include the average boiling point, the specific gravity, and the molecular weight based on the analysis of the physical properties of liquid hydrocarbons and condensates. However, this characterization method leads to significant errors when applied to heavier components [64].

Whitson's method was followed by the paraffins-naphthenes-aromatics method to characterize crude liquid phase. A new set of correlations reported by Riazi on the properties of single carbon number from C_6 to C_{50} is used to estimate the paraffins-naphthenes-aromatics composition [65]. Leelavanichkul et al. used the paraffins-naphthenes-aromatics technique to characterize different hydrocarbon fluids in a solid-liquid model to determine the asphaltene precipitation onsets [66]. However, the solubility parameter for C_{50} fraction is low to represent the heaviest fractions in a crude oil. Also, the maximum refractive index does not reach the expected 1.7 value that has been estimated for asphaltene in different investigations [59].

Even with the availability of the SAFT models in commercial simulators such as PVT Sim, Multiflash and VLXE, the lack of a standard characterization procedure incorporating the heavy components such as asphaltene, hinders the industrial use of SAFT based models [67].

5. The New Characterization Methodology

The characterization procedure described below is based on saturates-aromatics-resins and-asphaltene (SARA) analysis of stock tank oil. A reservoir fluid which is usually monophasic, when flashed from reservoir to ambient conditions yields residual liquid/stock tank oil and evolved gas phase/flashed gas. Such a flash of reservoir fluid is carried out under controlled environment in a PVT cell, and the liberated gas and residual liquid are analyzed for composition using gas chromatography. Table 1 report the compositional analysis of crude oil A and Table 2 shows the properties of crude oil A which will be used for the characterization.

Table 1: Compositional Analysis of a Middle Eastern crude oil (crude oil A)

Component	MW	Density	Flashed gas		STO		Reservoir fluid (GOR-787 scf/stb)	
	g/mol	g/cc	wt%	mol%	wt%	mol%	wt%	mol %
N_2	28.04	0.809	0.270	0.280	0	0	0.047	0.163
CO_2	44.01	0.817	5.058	3.340	0	0	0.874	1.944
H_2S	34.08	0.786	0	0	0	0	0	0

C_1	16.04	0.300	31.858	57.716	0	0	5.503	33.600
C_2	30.07	0.356	13.431	12.981	0.044	0.279	2.356	7.557
C_3	44.10	0.508	17.571	11.581	0.296	1.294	3.280	6.742
iC_4	58.12	0.567	5.280	2.640	0.251	0.835	1.120	1.884
nC_4	58.12	0.586	11.74	5.871	0.923	3.066	2.792	4.695
iC_5	72.15	0.625	4.593	1.850	0.999	2.673	1.620	2.195
nC_5	72.15	0.631	5.139	2.070	1.589	4.250	2.202	2.984
C_6	84.00	0.690	3.497	1.210	3.593	8.254	3.576	4.162
Mcyclo-C_5	84.16	0.749	0	0	0.447	1.024	0.369	0.429
Benzene	78.11	0.876	0	0	0.143	0.354	0.119	0.148
Cyclo-C_6	84.16	0.779	0	0	0.322	0.739	0.267	0.310
C_7	96.00	0.727	1.222	0.370	3.604	7.245	3.193	3.251
Mcyclo-C_6	98.19	0.770	0	0	0.619	1.217	0.512	0.510
Toluene	92.14	0.867	0	0	0.702	1.471	0.581	0.616
C_8	107.00	0.749	0.258	0.070	3.805	6.862	3.192	2.916
C_2-Benzene	106.17	0.866	0	0	0.224	0.407	0.185	0.171
m&p Xylene	106.17	0.860	0	0	0.644	1.171	0.533	0.491
o Xylene	106.17	0.860	0	0	0.038	0.069	0.032	0.029
C_9	121	0.768	0.083	0.020	3.936	6.277	3.270	2.642
C_{10}	134	0.782	0	0	4.605	6.632	3.809	2.779
C_{11}	147	0.793	0	0	3.787	4.971	3.132	2.083
C_{12}	161	0.804	0	0	3.241	3.885	2.682	1.628
C_{13}	175	0.815	0	0	3.096	3.414	2.561	1.431
C_{14}	190	0.826	0	0	2.929	2.975	2.423	1.247
C_{15}	206	0.836	0	0	2.83	2.651	2.341	1.111
C_{16}	222	0.843	0	0	2.437	2.150	2.046	0.901
C_{17}	237	0.851	0	0	2.356	1.918	1.949	0.804
C_{18}	251	0.856	0	0	2.128	1.636	1.761	0.686
C_{19}	263	0.861	0	0	2.231	1.637	1.845	0.686
C_{20}	275	0.866	0	0	2.193	1.539	1.814	0.645
C_{21}	291	0.871	0	0	1.900	1.260	1.572	0.528
C_{22}	300	0.876	0	0	1.805	1.161	1.493	0.486
C_{23}	312	0.881	0	0	1.628	1.007	1.346	0.422
C_{24}	324	0.885	0	0	1.512	0.900	1.250	0.377
C_{25}	337	0.888	0	0	1.417	0.811	1.172	0.340
C_{26}	349	0.892	0	0	1.377	0.761	1.139	0.319
C_{27}	360	0.896	0	0	1.269	0.680	1.050	0.285
C_{28}	372	0.899	0	0	1.280	0.664	1.059	0.278
C_{29}	382	0.902	0	0	1.079	0.545	0.893	0.228
C_{30}	394	0.903	0	0	1.031	0.505	0.853	0.212
C_{31}	404	0.907	0	0	0.937	0.448	0.775	0.188

C_{32}	415	0.910	0	0	0.883	0.411	0.731	0.172
C_{33}	426	0.913	0	0	0.803	0.364	0.664	0.152
C_{34}	437	0.916	0	0	0.694	0.307	0.574	0.129
C_{35}	445	0.919	0	0	0.666	0.289	0.551	0.121
C_{36}+	594	0.941	0	0	27.673	8.991	22.893	3.767

Table 2. Properties of crude oil A

Origin	S field
GOR (scf/stb)	787
MW of reservoir fluid (g/mol)	97.750
MW of flashed gas (g/mol)	29.06
MW of STO (g/mol)	192.99
STO Density (g/cc)	0.82
Saturates (wt %)	66.26
Aromatics (wt %)	25.59
Resins (wt %)	5.35
Asphaltene (wt %)	2.80

It has been reported in the literature that light components in oil affect both the bubble pressure and asphaltene onset pressure (AOP) significantly [68]. Hence considering the lightest fractions of oil individually, will result in better prediction of asphaltene onset pressures. Thus gas phase is characterized to consist of seven components: N_2, CO_2, methane, ethane, propane and heavy gas pseudo-component that represents a mixture of hydrocarbons heavier than propane. The PC-SAFT EoS parameters for the pure components, N_2, CO_2 and C_1-C_3, are available in the literature [55]. The average molecular weight of the heavy gas pseudo-component (mostly consisting of saturates) is used to estimate the corresponding PC-SAFT parameters through correlations shown in Table 3 [69]. Table 4 represents the characterized gas phase of crude oil A. If H_2S is also present in the flashed gas, the number of components in the characterized gas phase will become eight with H_2S being considered individually.

Table 3. PC-SAFT correlations for saturates as a function of molecular weight

Parameter	Function
m	$(0.0257 * MW) + 0.8444$
$\sigma(A)$	$4.047 - \frac{4.8013 * Ln(MW)}{MW}$
$\varepsilon(K)$	$\exp\left(5.5769 - \frac{9.523}{MW}\right)$

Table 4. Characterized gas phase for crude oil A

			PC-SAFT Parameters		
Component	MW (g/mol)	Mole %	m	σ (A)	ε (K)
N_2	28.04	0.28	1.21	3.31	90.96
CO_2	44.01	3.34	2.07	2.78	169.21
C_1	16.04	57.72	1.00	3.70	150.03
C_2	30.07	12.98	1.61	3.52	191.42
C_3	44.10	11.58	2.00	3.62	208.11
Heavy gas	65.49	14.10	2.53	3.74	228.51

The liquid fraction characterization into saturates, aromatics plus resins and asphaltene is based on the stock tank oil composition and the SARA analysis. Because of their similar effect on the asphaltene phase behavior, the aromatics and resin fractions are combined into a single lumped pseudo-component defined in terms of the degree of aromaticity (γ). Table 5 shows the PC-SAFT parameters for the aromatics + resins pseudo fraction. The aromaticity parameter determines the aromatics plus resins tendency to behave as poly-nuclear-aromatic ($\gamma = 1$) or benzene derivative ($\gamma = 0$) and is tuned to meet the values of density and bubble point simultaneously [70]. Asphaltenes exist as pre-aggregated molecules even in good solvents such as toluene [7, 9] and the average molecular weight for such a pre-aggregated asphaltene is considered as 1700 g/mol [71-73]. The asphaltene PC-SAFT parameters in the recombined oil are tuned to meet the onset of asphaltene precipitation. Table 6 shows the characterized stock tank oil before recombining. The gas phase and stock tank oil are then recombined as per the gas-to-oil ratio (GOR) at reservoir conditions and Table 7 represents the characterized live oil after tuning the parameters. The constant set of temperature independent PC-SAFT binary interaction parameters are well established through one of our recent works, by fitting to binary vapor-liquid equilibrium data for the combination of pure components [74]. With all the pseudo components and parameters set, PC-SAFT can be used like any other equation of state and Figure 11 plots the resultant asphaltene phase behavior.

Table 5. PC-SAFT correlations for aromatics+resins pseudo component as a function of molecular weight. The equations are of the form, Parameter = (1- γ)(Benzene derivatives correlation) + γ(Poly-nuclear-aromatic correlation)

Parameter	Function
m	$(1-\gamma)(0.0223 * MW + 0.751) + \gamma(0.0101 * MW + 1.7296)$
$\sigma(A)$	$(1-\gamma)\left(4.1377 - \frac{38.1483}{MW}\right) + \gamma\left(4.6169 - \frac{93.98}{MW}\right)$
$\varepsilon(K)$	$(1-\gamma)(0.00436 * MW + 283.93) + \gamma\left(508 - \frac{234100}{(MW)^{1.5}}\right)$

Table 6. Characterized stock tank oil for crude oil A

			PC-SAFT Parameters		
Component	MW (g/mol)	Mole %	m	σ (A)	ε (K)
Saturates	167.68	75.97	5.15	3.90	249.69
Aromatics + Resins (γ = To be tuned)	253.79	23.71	6.41	3.99	285.00
Asphaltene	1700	00.32	To be tuned	To be tuned	To be tuned

Table 7. Characterized crude oil A

Component	MW	Contribution from gas	Contribution from STO	Moles in live oil	PC-SAFT Parameters		
	g/mol	moles	moles	Basis 100	m	σ (A)	ε (K)
N_2	28.04	0.163	0	0.163	1.21	3.31	90.96
CO_2	44.01	1.944	0	1.944	2.07	2.78	169.21
C_1	16.04	33.600	0	33.600	1.00	3.70	150.03
C_2	30.07	7.557	0	7.557	1.61	3.52	191.42
C_3	44.10	6.742	0	6.742	2.00	3.62	208.11
Heavy gas	65.49	8.198	0	8.198	2.53	3.74	228.51
Saturates	167.68	0	31.743	31.743	5.15	3.90	249.69
Aromatics + Resins ($\gamma = 0.0$)	253.79	0	9.907	9.907	6.41	3.99	285.00
Asphaltene	1700	0	0.133	0.133	33.00	4.20	353.50

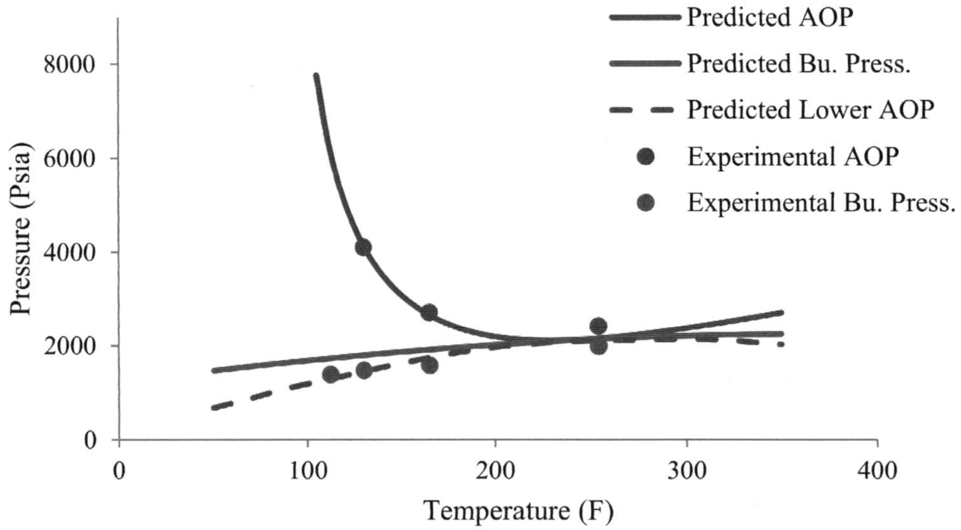

Figure 11. Asphaltene phase envelope for crude oil A generated by PC-SAFT

The asphaltene onset pressure is the cloud point at a fixed temperature for which the crude oil will split up into two liquid phases of asphaltene rich and lean phases. Such measurements can involve depressurization of live oil or titration of dead oil with a precipitant. The lower asphaltene onset represents the pressure below which asphaltene and oil coexist in a single phase.

As observed from the characterization procedure, one of the important inputs on which the crude oil is characterized is the SARA. Unfortunately, a disadvantage of the SARA analysis is that fraction measurements by different techniques can show large differences [75, 76]. Table 8 shows the SARA reported by thin layer chromatography with flame ionization detection (TLC-FID) and high pressure liquid chromatography (HPLC) for the same light crude oil B. In the process of quantifying SARA, TLC-FID lost significant amount of light ends in the form of saturates, and hence reported a higher amount of aromatics and asphaltene than actually present. When the SARA for another light crude oil C from a different field in the same region was made available through TLC-FID, the corrected weight fractions are obtained by multiplying and renormalizing the crude oil C data with the corresponding ratio of HPLC/TLC-FID obtained from crude oil B. From Table 9, the corrected SARA is very close to the actual SARA measured by HPLC for crude oil C. Thus, under data constraints, such SARA estimations can be used for crude oils of similar nature.

Table 8. SARA analysis as reported by TLC-FID and HPLC for the same crude oil B

	TLC-FID	HPLC (IP 143)	HPLC/TLC-FID (P)
Saturates	49.5	73.0	1.47
Aromatics	40.2	19.1	0.48
Resins	7.2	7.6	1.06
Asphaltene	3.1	0.2	0.06

Table 9. SARA analysis corrected for crude oil C

	TLC-FID	Corrected (= TLC-FID*P)	HPLC (IP 143)
Saturates	44.9	69.39	67.09
Aromatics	46.3	23.05	25.84
Resins	6.7	7.41	6.92
Asphaltene	2.1	0.14	0.15

6. Results and Discussion

6.1 Pressure

The effect of pressure on asphaltene phase behavior can be explained by analyzing the depressurization of a model crude oil, where the toluene solution before combining with methane contains 0.93 gm of asphaltene/100 ml toluene at 20 ^{o}C [54]. During pressure depletion at constant temperature, asphaltene precipitate formation is observed within a range above and below the bubble point. In Figure 12, as pressure drops during production from the reservoir pressure, the oil expands, reducing the oil solubility parameter, and becomes a poor solvent for asphaltene. At low enough pressure, the asphaltene precipitation onset is reached and asphaltene begin to precipitate. Upon further depressurization, the system arrives at its bubble point, where the light components, which are asphaltene precipitants, escape from the liquid phase. As this happens, the solubility parameter of the oil increases until the oil becomes a better asphaltene solvent and the oil stabilizes again.

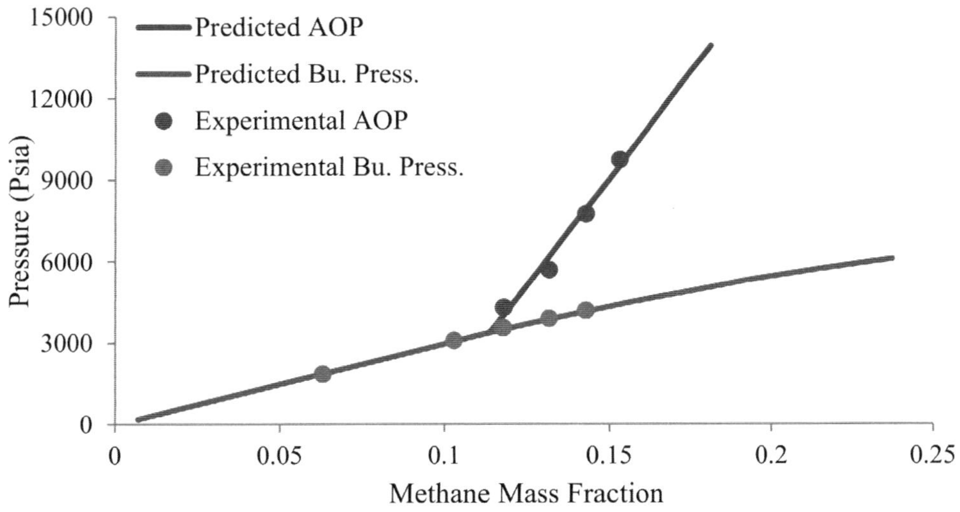

Figure 12. PC-SAFT generated asphaltene phase diagram for a model oil composing of methane, toluene and asphaltene at 20°C

We can follow the same depressurization on a plot of solubility parameter shown in Figure 13 where asphaltenes are unstable below the asphaltene instability line. We see that the solubility parameter of the oil decreases as the pressure is decreased to the bubble point. Upon further depressurization, the solubility parameter of the oil increases. The maximum driving force for asphaltene precipitation is at the bubble point as observed from Figure 13.

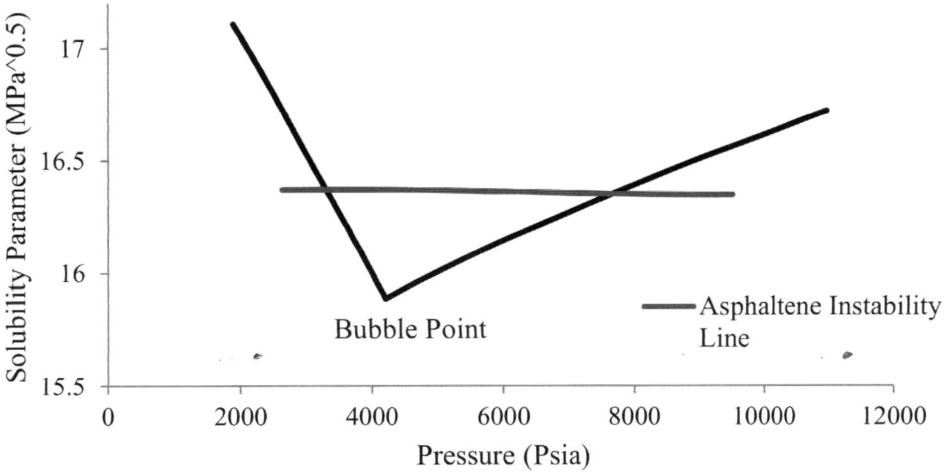

Figure 13. PC-SAFT generated solubility parameter during pressure depletion for the model oil (methane, toluene and asphaltene) at 20°C with 0.143 mass fraction methane

6.2 Temperature

The effect of temperature on asphaltene phase behavior can be explained by analyzing the phase diagram of crude oil A presented in Figures 11. The nearly vertical phase boundary at about 130 °F is an upper critical solution temperature type boundary to the left of which asphaltene precipitate out of the crude oil. As the temperature rises above this boundary, the entropy gain from mixing just overcomes the enthalpically favored phase splitting and the system moves from a region of two liquid phases to a single liquid phase stable with asphaltene.

There is another phase boundary observed from Figure 11 as the temperature is increased at constant pressure (Eg., at P=2500 Psia and T=350 °F in Figure 11). This is the phase boundary corresponding to lower critical solution temperature. As the temperature is increased at constant pressure, the crude oil expands lowering the solubility parameter of the solvent and becomes a poor solvent for the asphaltene. Prediction of this type of phase boundary requires an equation of state because it is the compressibility of the system that causes the phase splitting. The lower critical solution temperature phase boundary is more pronounced as the critical temperature of the crude oil is approached, because the system is most compressible in the critical region.

6.3 Gas Injection

Changes in composition occur during gas injection processes employed in enhanced oil recovery. Gas injection includes processes such as miscible flooding with CO_2, N_2 or Natural Gas or artificial gas lifting. The effect is studied by the addition of increasing amounts of natural gas at different temperatures for crude oil B as shown in Figure 14. The PC-SAFT predictions match the experimental observations of both the asphaltene onset and bubble pressures for varying injection rates of the natural gas. The dissolved gas decreases asphaltene solubility and the asphaltene becomes more unstable. Thus along with an increasing trend of bubble pressure (due to more volatiles) asphaltene onset pressure also increases. Specific effects on asphaltene precipitation onset due to CO_2, N_2 and methane injection are described by Gonzalez who also showed increasing asphaltene instability with increasing injection of the gases [77].

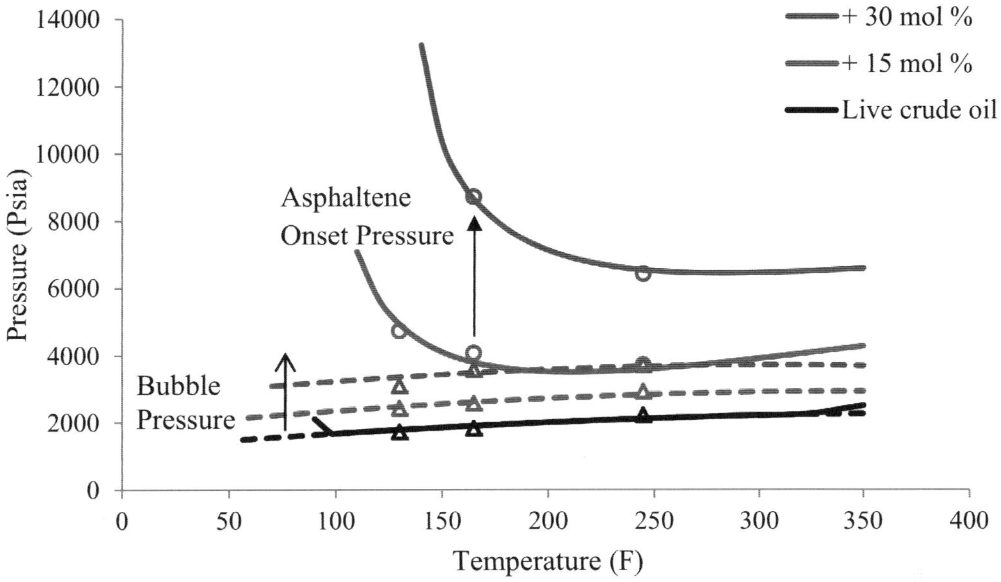

Figure 14. Asphaltene phase behavior for crude oil B after the addition of natural gas. Injected gas composition (mole %): N_2-0.5%, CO_2-4.5%, C_1-87.4%, C_2-7.2% and C_3-0.4%

6.4 *Comingling of Crude Oils*

Changes in composition also arise from the relative proportions of mixing fluids. Reducing the number of flow lines by commingling streams from several wells or manifolds is a common subsea system design consideration. This decision can reduce capital and operational costs but might cause problems if the fluids in the streams are not compatible. The mixture of different oil streams can generate asphaltene precipitation. This problem might occur when fluids of very different API gravities are mixed. For example, a problem may occur when a condensate is mixed with black oil. The density of the entire liquid phase decreases, reducing the crude oil solubility parameter and the asphaltenes precipitate.

The effect of mixing crude oils on asphaltene phase behavior can be analyzed by considering comingling of two crudes oils D and E as shown in Figure 15. None of these fluids showed asphaltene instability at their corresponding reservoir condition. Asphaltene precipitation due to pressure depletion is not expected to occur for any of the fluids at their respective reservoir temperature and gas-to-oil ratio. Asphaltenes would precipitate only above their reservoir gas-to-oil ratio. However, crude oil E is unstable in asphaltene with respect to the reservoir conditions of crude oil D and the resultant commingled oil is a good case to study the effect of crude oils commingling on asphaltene instability. The crude oils when combined in the volume ratio of D:E = 50:50, Figure 15 shows the PC-SAFT predicted asphaltene phase behavior of the commingled system [78]. The asphaltene precipitation is not expected to occur at the mixture gas-to-oil ratio of 1453 scf/stb which is consistent with the field observations.

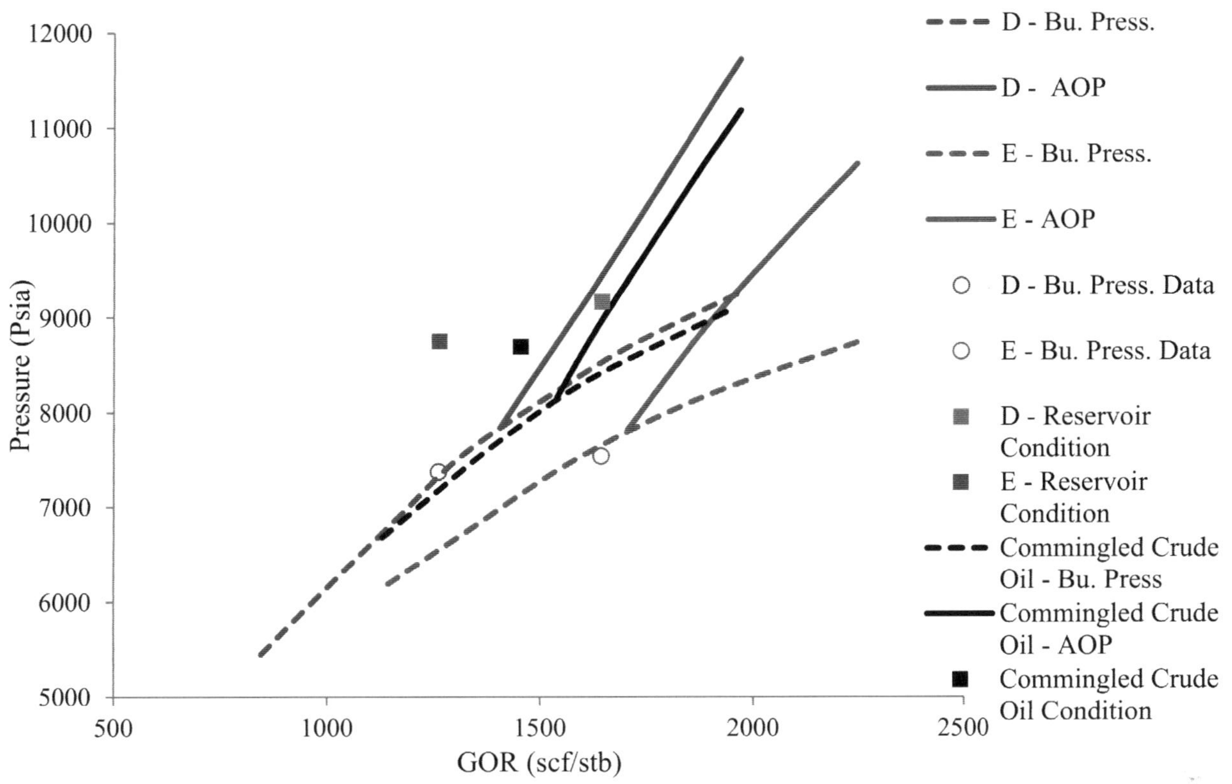

Figure 15. PC-SAFT predicted asphaltene stability of commingled crude oils D and E in the volume ratio 50 and 50 respectively at 150 °F

6.5 *Oil Based Mud*

Oil based mud (OBM) that is used to increase borehole stability during drilling can contaminate near wellbore reservoir fluids. Significant overbalance pressure during the drilling process can result in the mud invasion into the formation and mix with the reservoir fluid. An oil based mud can significantly modify the composition and predicted phase behavior of the asphaltene in the formation fluid causing wrong data interpretation.

OBM being a precipitating agent for asphaltene is expected to increase the pressure at which asphaltene start precipitating. Simulations using the PC-SAFT equation of state performed for the clean and OBM contaminated oil show that this is not necessarily the case. According to the results presented in Figure 16, for a crude oil F, both the asphaltene precipitation onset and bubble pressure decrease when successive amounts of OBM are added [79]. The AOP and the bubble point curves estimated by PC-SAFT closely follow the experimental findings. A decrease in the gas-to-oil ratio due to oil based mud addition is also noted. Although the oil based mud is a precipitant for asphaltene, the OBM contamination dilutes the gaseous components of the oil that are stronger asphaltene precipitants. As the gas-to-oil ratio decreases, the asphaltene precipitation onset pressure and bubble point pressure decrease. The correction for oil based mud contamination can be significant in the temperature range of the production as observed from Figure 17 for the same crude oil F [79].

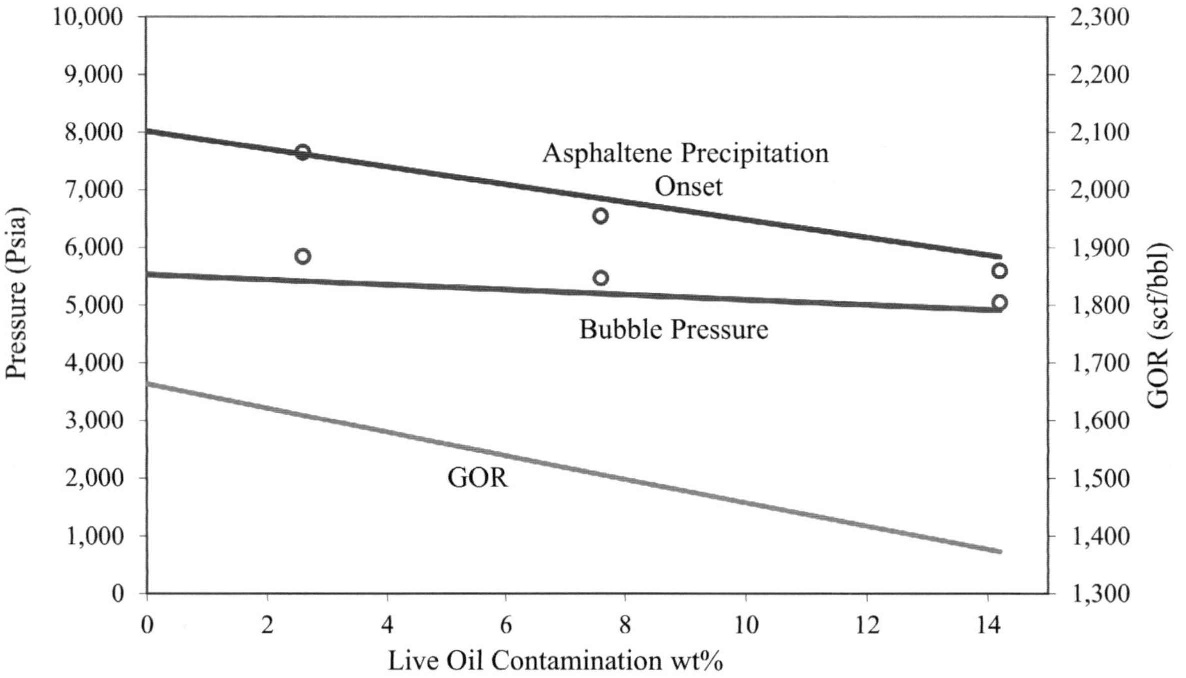

Figure 16. PC-SAFT simulation of OBM contamination effect on the asphaltene phase behavior of crude oil F at 178°F.

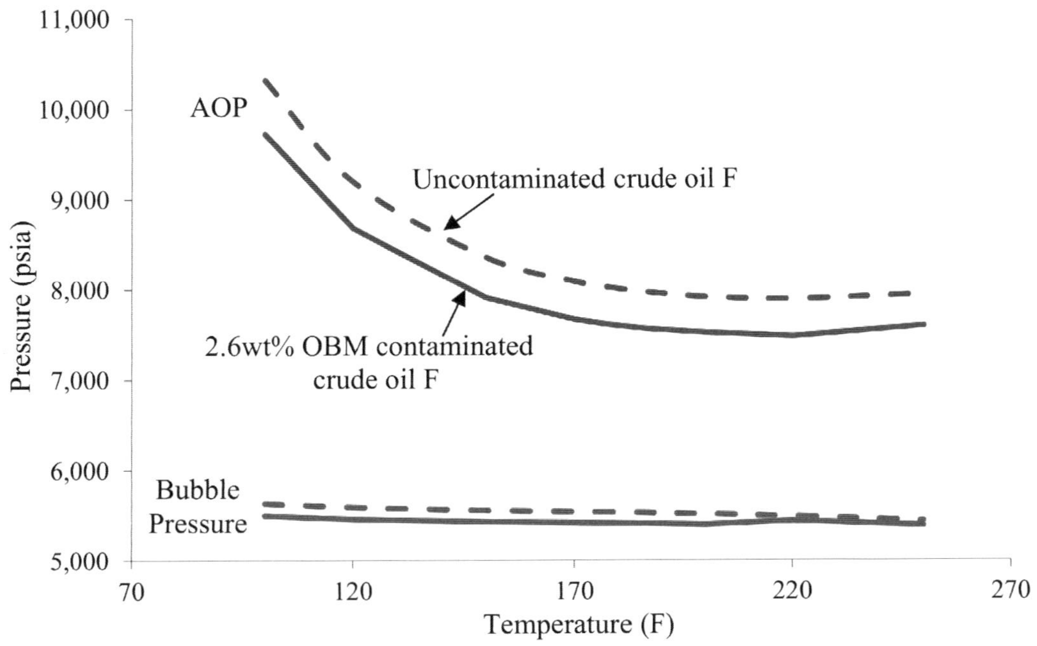

Figure 17. Asphaltene phase behavior of crude oil F calculated with the PC-SAFT

6.6 *Asphaltene Polydispersity*

The effect of asphaltene's polydispersity on its phase behavior can be analyzed by comparing the monodispese and polydisperse asphaltenes modeled using PC-SAFT. Figure 18 shows the different precipitation profiles obtained for mono- versus polydisperse asphaltenes upon titration with an n-alkane precipitant [80]. The polydisperse asphaltene component is modeled as a mixture of three pseudo-components that represent the incremental amount of asphaltene precipitated on a 20:1 dilution of the oil with heptane, undecane, and pentadecane, respectively. From Figure 18, we can see that polydispersity affects the amount of asphaltene precipitated. For monodisperse asphaltene, the amount of asphaltene that precipitates increases quickly upon the addition of precipitant beyond the precipitation onset condition. For polydisperse asphaltene, the amount of asphaltene precipitated increases more slowly upon adding the precipitant. As expected, any precipitate can redissolve at very high dilution in a precipitant. The amount of dilution that is "high enough" for asphaltene depends upon the highest molecular-weight components of the asphaltenes fraction.

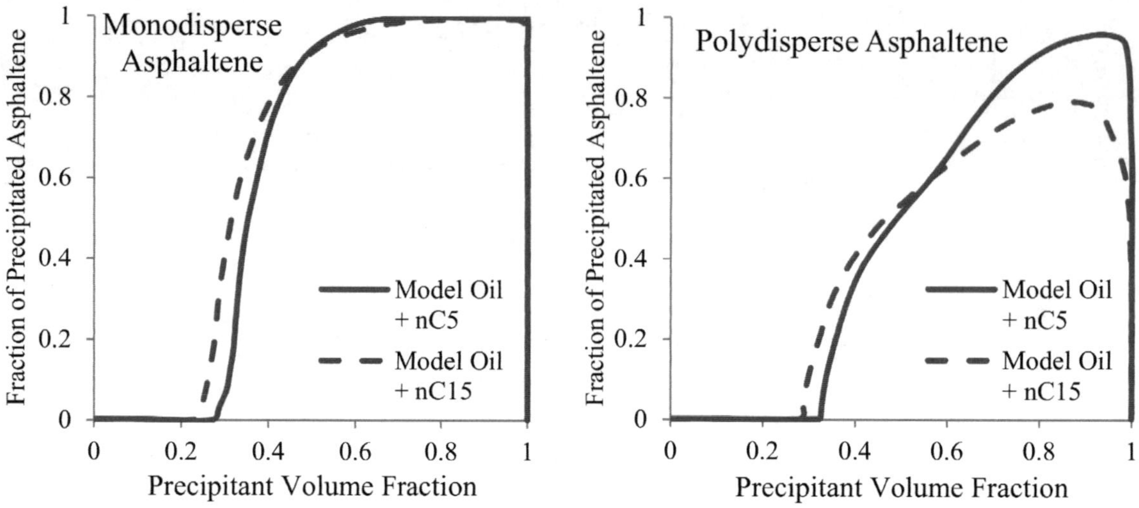

Figure 18. Phase behavior of monodisperse and polydisperse asphaltene in model oil (7.5 g asphaltene/100 ml toluene) mixed with n-alkanes at 20 °C and 1 bar.

6.7 Comparison of cubic and PC-SAFT EoS

Despite their poor prediction of asphaltene properties, cubic equations of state are widely used in the oil industry due to the simplicity of models. However, it is seen that the parameters fit using a cubic equation of state for a particular data set fails to predict another situation for the same well [81]. This is demonstrated in Figure 19 where the parameters for both SRK-P and PC-SAFT are estimated to the saturation pressures and asphaltene onset pressures for various temperatures of crude oil G with 5 mol% gas injection. The same parameters are then used to predict the saturation pressure and temperature dependence of asphaltene onset pressure for different amounts of gas injected. The predictions made by the PC-SAFT and cubic EoS are compared in Figures 19A, 19C and 19D. We can see that only PC-SAFT does a good job in predicting the phase behavior of asphaltene for various gas injection amounts.

The major limitation of any cubic equation of state is that they cannot describe adequately the phase behavior of mixtures of molecules with large size differences and they are unable to accurately calculate liquid densities of the precipitated phase. Accurate modeling of liquid density is essential for an equation of state to predict liquid-liquid equilibrium and their corresponding parameters, such as the solubility parameter, over a range of conditions.

6.8 Asphaltene Composition Gradients

Fluid properties vary along the depth of a continuous reservoir due to the gravitational field, an effect called compositional grading. A considerable elevation difference produces the thermodynamic drive creating the composition gradient [82]. A resultant of the compositional grading is greater gas-to-oil ratio from the upper parts of a formation. Whitson's analysis accounts for the gas-to-oil ratio and composition of equilibrium fluids as measured by the down-hole fluid analysis [83]. Down-hole fluid analysis also measures crude oil optical density using infrared light. This optical density is directly proportional to asphaltene content [84]. But, Whitson or others did not analyze compositional grading related to asphaltene. Here, the PC-SAFT equation of state is extended to model asphaltene distribution in oil columns.

For the Tahiti field in the Gulf of Mexico, Figure 20 shows the changes in asphaltene composition with depth which could result from the attainment of thermodynamic equilibrium in a uniform gravitational field as predicted by PC-SAFT and compared with field data. A close match is observed between the predicted asphaltene concentration gradient and the field data procured by down-hole fluid analysis. This asphaltene compositional gradient can be further used to evaluate the connectivity of wells and reservoir compartmentalization by following the analysis of Mullins [85].

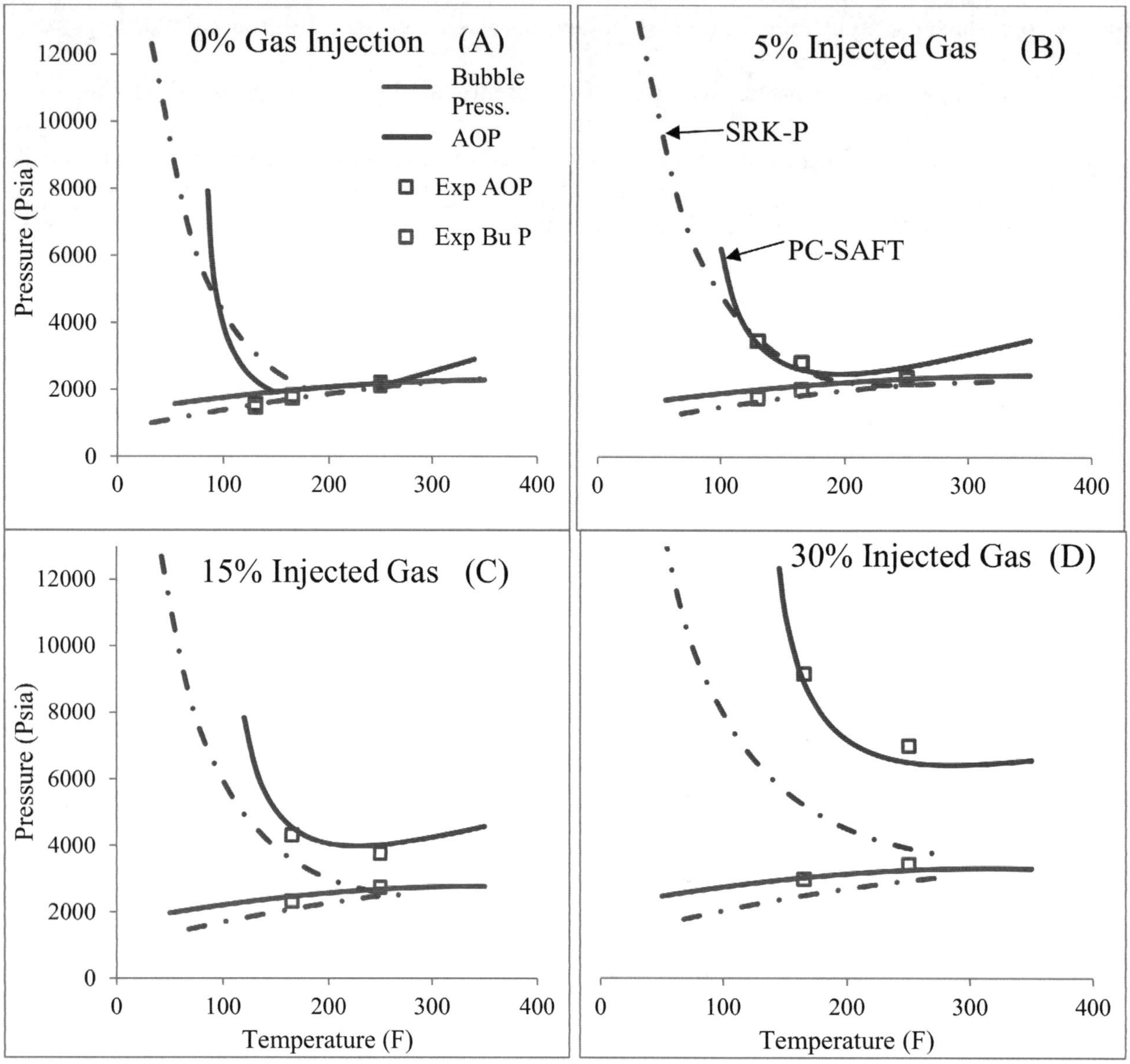

Figure 19. PC-SAFT and SRK-P phase behavior predictions for crude oil G after estimating the parameters for 5 mol% of gas injection data. Injected gas composition (mole %): N_2-0.4%, CO_2-3.9%, C_1-71.4%, C_2-12%, C_3-7.2%, heavy gas-5.1%

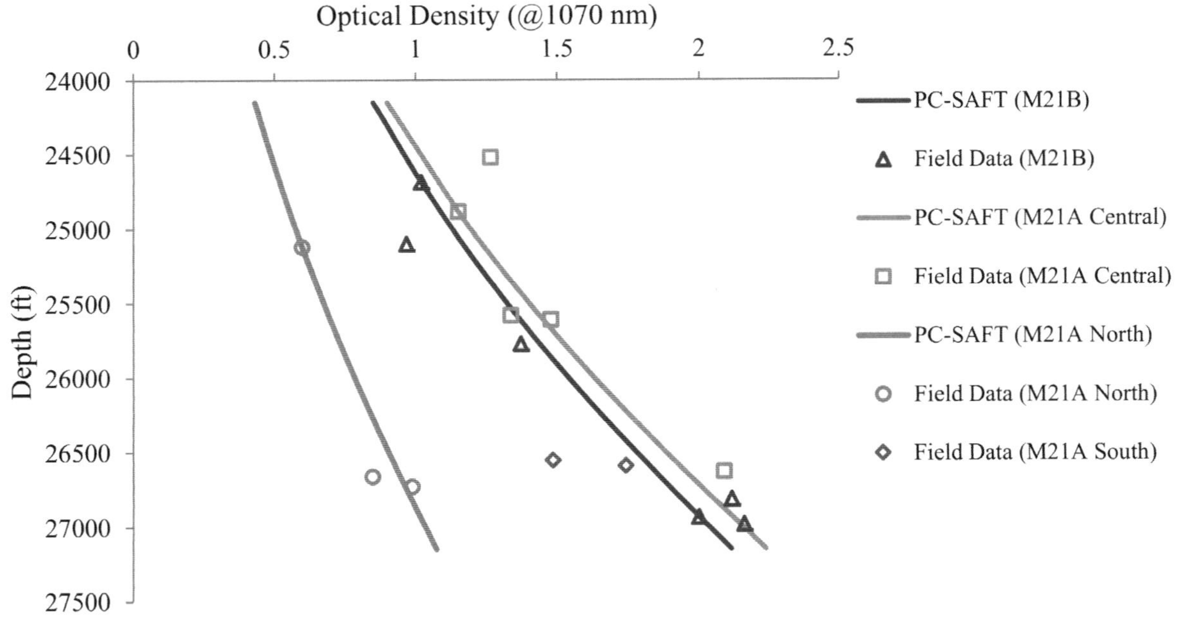

Figure 20. PC-SAFT generated optical density profile of Tahiti crude oil with depth in the Tahiti field at the reservoir temperature of 200 °F

6.9 Tar-Mat

Asphaltene compositional gradient can lead to significant variations in crude oil viscosity and extreme cases of asphaltene compositional grading can lead to tar-mat formation [86]. A Tar-mat represents a reservoir zone of highly viscous extra heavy oil enriched with asphaltene (20-60 wt. %) relative to the oil column [87]. A tar-mat is differentiated from heavy oil in that the tar-mat is characterized by high oil saturation associated with high residual oil saturation during logging [88]. Tar-mat is an important organic barrier for an oil reservoir and must be modeled as accurately as possible in order to obtain correct estimate of producible reserves, effective placement of injection wells and ultimately, predictive production profiles.

The reservoir considered for tar-mat study is the S field consisting of crude oil A. The asphaltene compositional grading was the most feasible tar-mat formation mechanism for the S field tar-mat [89]. The PC-SAFT asphaltene compositional grading is extended to model the possibility of this tar-mat formation due to gravitational segregation of asphaltene, resulting in Figure 21. After a depth of ~ 9000 ft, the asphaltene content suddenly increased from 15 wt % to 48 wt % while measurements of the tar-mat samples report between 26 to 80 weight percent [90]. Above 9000 ft asphaltene was stable in oil. Between 9000 to 9045 ft asphaltene concentration increased to an extent of phase separation. Thus, the tar-mat formed is because of transport of asphaltene in oil along chemical and gravitational potential gradients in the reservoir to the zone of asphaltene enrichment at the site of tar-mat.

The PC-SAFT observed tar-mat formation is in accordance with the field observation of tar-mat depth and the asphaltene content of the tar-mat. Hence, the PC-SAFT model successfully predicted the tar-mat occurrence depth from just knowing the pressure, temperature and reservoir oil composition at a reference depth in the upper parts of the formation. Such a prediction is possible only with an equation of state because along with asphaltene phase splitting the compressibility of oil needs to be taken into consideration.

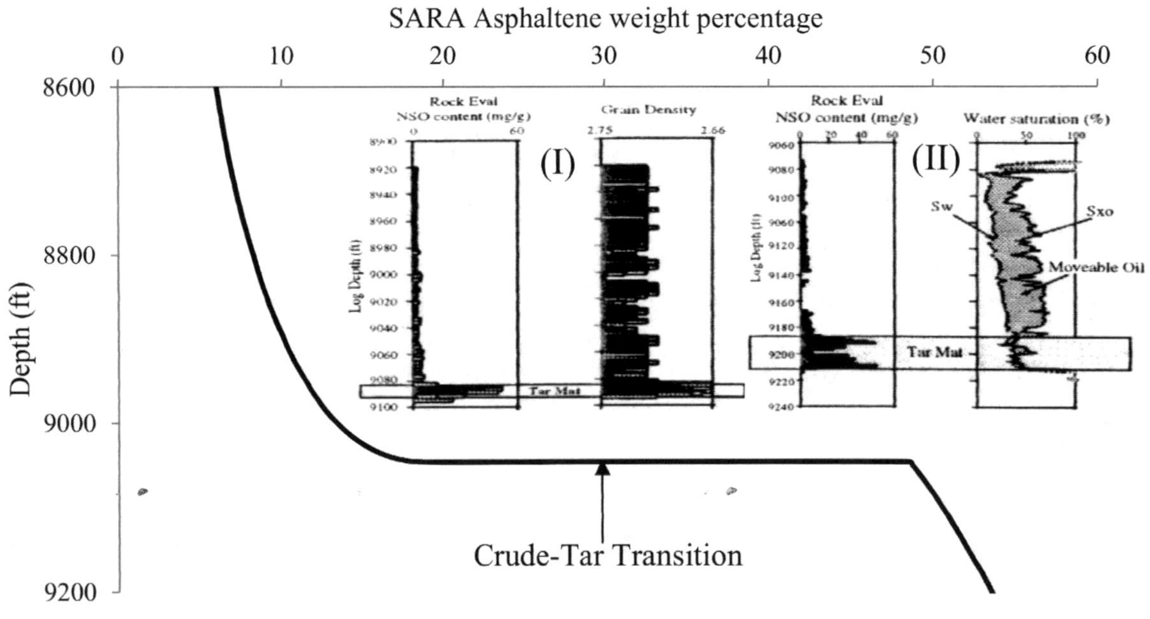

Figure 21. PC-SAFT prediction of tar-mat in the S field at the reservoir temperature of 250 F. The inserted figures (I and II) are well logs from two different wells identifying the tar-mat in the S field [91]

6.9.1 *Tar-mat analysis*

The Tahiti field asphaltene compositional grading curve, when extended to deeper depths using PC-SAFT, does not show any kink in the curve; indicating no tar-mat (Figure 22). But the S field showed a sharp asphaltene compositional variation correlating to a tar-mat. Recently Zuo et al. have reported observing large asphaltene gradients even without asphaltene instabilities [92]. Thus a general analysis on when asphaltene show large compositional contrasts is needed.

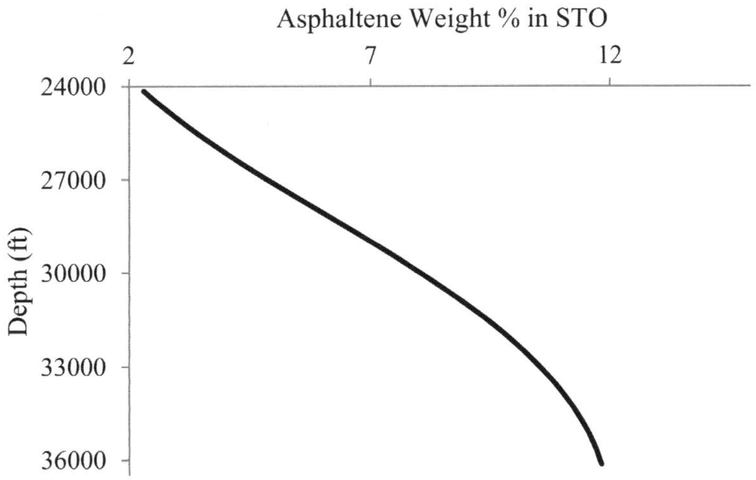

Figure 22. Asphaltene compositional variation with depth in Tahiti field as generated by PC-SAFT

In the S field, PC-SAFT asphaltene compositional grading plots generated with different starting pressure at the reference depth, 8000 ft are plotted together in Figure 23. The sharp compositional contract represents the tar-mat formation depth where the crude oil is phase splitting into 2 liquid phases (asphaltene lean and rich phases) existing at equilibrium. The boundaries of these liquid-liquid phase separations when joined using the dashed black line represents the phase boundary and is analogues to PV isotherms. Observed from Figure 23, as one approaches the critical point region on this composition-depth phase diagram, there exists sharp asphaltene compositional gradients even without phase transitions. Thus it is concluded that, there need not be phase separation to have large compositional gradients and, based on the system's instability to asphaltene the tar-mat formation depth and the asphaltene content in the tar-mat vary.

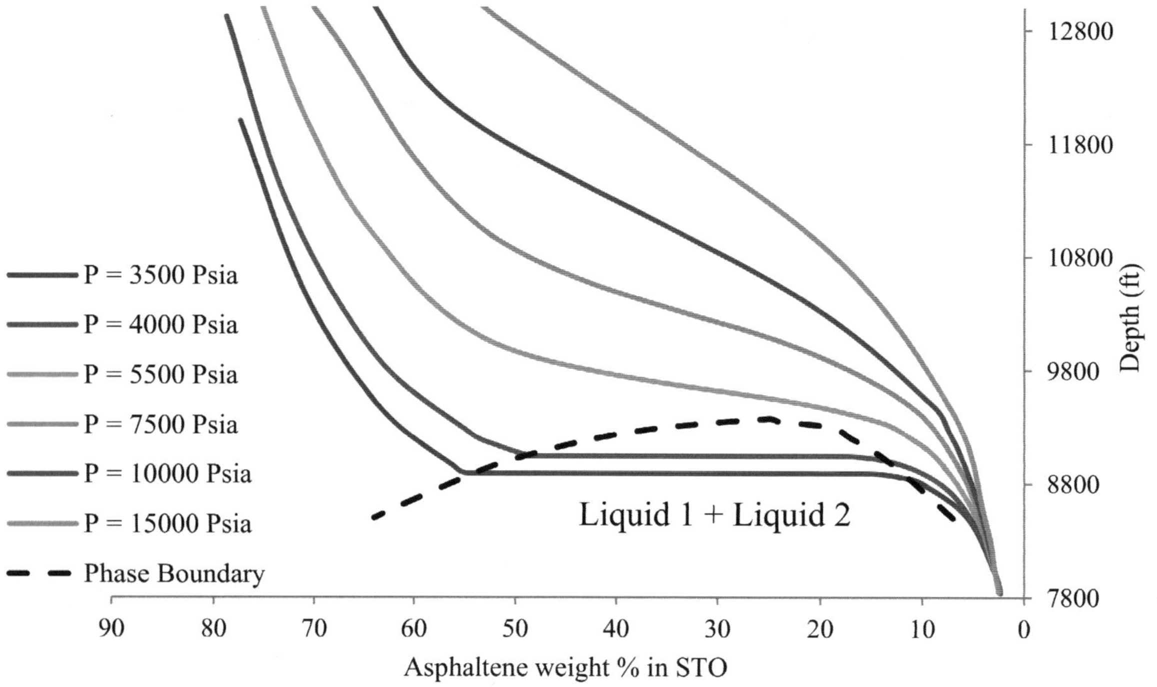

Figure 23. Asphaltene compositional grading isotherms with different starting pressures at the reference depth of 8000 ft. (Liquid 1 : ashaltene lean phase, Liquid 2 : asphaltene rich phase)

Conclusions

Asphaltenes constitute a potential problem in oil production because of the tendency of this petroleum fraction to precipitate and deposit. Predicting asphaltene flow assurance issues requires the ability to model phase behavior of asphaltene as a function of the operating conditions. In this work we have presented a modeling method based on the PC-SAFT equation of state, in which, for a given crude oil, a single set of simulation parameters was sufficient to predict the phase behavior with various gas injection amounts. Among the experimental information necessary to fit the simulation parameters, the SARA requires special attention because, unfortunately, measurements performed using different techniques lead to results that are significantly different. A methodology to correct the SARA values obtained with TLC-FID technique was presented, which seems to provide satisfactory results for the case of light crude oils of similar nature.

The PC-SAFT equation of state has been successful in modeling the phase stability of asphaltene in crude oil over a wide range of conditions and for a variety of cases, including:

- Effect of pressure: In reservoir depressurization, we explained that there is a pressure range where asphaltene become unstable. As the pressure decreases along the wellbore, oil expands making it a poor solvent for asphaltene. Eventually, asphaltene undergo a phase separation, at a pressure that is known as the asphaltene onset pressure. As the pressure keeps decreasing the bubble point is reached, and the light fraction of hydrocarbons, which are strong asphaltene precipitants, escape from the liquid phase. As this happen, asphaltene become stable again.

- Effect of temperature: For a given pressure, asphaltene can become stable or unstable by increasing the temperature depending on the location of the system on the phase diagram. At low temperatures, asphaltene can get re-dissolved by increasing the temperature because the entropy gain overcomes the enthalpically favored phase splitting. However, at much higher temperatures, the volume of the oil increases making it a poor solvent for asphaltene and causing their phase separation. The latter case might explain the fouling of heat exchangers produced during the refining process.

- Gas addition: An increase of the asphaltene onset pressure and bubble point is observed upon addition of miscible gas, which decreases the solubility of asphaltene. The prediction of asphaltene precipitation onset and bubble pressure is satisfactory for all the gas percentages tested.

- Commingling of crude oils: A mixture of incompatible oils might induce the precipitation of asphaltene. For the case of nearly compatible crude oils D and E, the asphaltene phase behavior of the commingled system did not show instability towards asphaltene. An extreme mixing condition for asphaltene instability might be when a gas condensate is mixed with black oil.

- Oil based mud contamination: Although at first one might think that the contamination with oil based mud could increase the instability of asphaltene, it has been confirmed that this is not the case. The addition of oil based mud dilutes the light fractions present in the live oil and asphaltene are stabilized. The PC-SAFT equation of state correctly predicts this phenomenon, and the simulation results obtained match well with the available data.

Furthermore, preliminary results on the effect of asphaltene polydispersity have been presented and discussed. Finally, isothermal asphaltene compositional grading was analyzed from knowing just the pressure, temperature and composition in the upper farts of the formation. The PC-SAFT generated asphaltene compositional grading showed a close agreement with the field data. Compositional grading can lead to significant variations in crude oil properties and in some cases to the formation of tar-mats. A novel representation of the tar-mat formation produced by asphaltene compositional grading is presented in the form of a composition – depth phase diagram. The sharp compositional contract represents the tar-mat formation depth, where the crude oil is phase splitting into two liquid phases (asphaltene lean and rich phases), which co-exist at equilibrium. The tar-mat formation depth and the asphaltene content in the tar-mat vary based on the extent of asphaltene's instability in a crude oil.

In overall, the modeling method for asphaltene phase behavior presented in this article is clearly very versatile and offers great potential for the design of improved experimental procedures and for a better understanding of the asphaltene behavior observed in the daily operations of oil fields across the world.

Acknowledgement

This work was undertaken with the funding from ADNOC's Oil R&D Subcommittee. The authors are thankful to Shahin Negahban, Dalia Abdullah, Sanjay Misra and the Enhanced Oil Recovery Technical Committee for valuable support. The authors also thank Oliver Mullins, Jefferson Creek and George Hirasaki for helpful discussions. The authors thank the technical committee of "The 1st International Conference on Upstream Engineering and Flow Assurance" for inviting the presentation of this work.

Acronyms

SAFT: Statistical associating fluid theory

PC-SAFT: Perturbed chain form of the statistical associating fluid theory

EoS: Equation of state

STO: Stock tank oil

SARA: Saturates, aromatics, resins and asphaltene

GOR: Gas-to-oil ratio

AOP: Asphaltene onset pressure

TLC-FID: Thin layer chromatography with flame ionization detection

HPLC: High pressure liquid chromatography

OBM: Oil based mud

Nomenclature

δ : Solubility parameter

CED : Cohesive energy density

U^{rsm} : Residual internal energy

υ : Pure liquid volume

RI : Refractive index

n : Concentration of asphaltene flocs

r : Mean radius of aggregates

n_{∞} : Final concentration of asphaltene flocs

R_{∞} : Final mean radius of aggregates

t : Time of aggragation

τ : Characteristic time

A^{res} : Residual Helmholtz free energy

$A^{segment}$: Segment contribution to residual Helmholtz free energy

A^{chain} : Chain contribution to residual Helmholtz free energy

A^{assoc} : Association contribution to residual Helmholtz free energy

A_o^{hs} : Hard-sphere contribution to residual Helmholtz free energy

A_o^{disp} : dispersion contribution to residual Helmholtz free energy

m : Average segment number of the mixture

x_i : Composition of species i

m_i : Segment number of species i

R : Universal gas constant

T : Temperature

ρ : Molecular number density

d_{ii} : Temperature-dependent segment diameter of species i

σ_i : Temperature-independent segment diameter of species i

I : System packing fraction function

η : Package fraction

ε : Segment-segment dispersion energy

k_{ij} : Binary interaction parameter between species i and j

γ : Aromaticity

Appendix

Details of crude oils used in the current study are listed below. Crude oil C is not used for asphaltene modeling work, and hence its oil properties are not reported. Properties of the crude oils D and E cannot be reported due to proprietary data.

	Crude Oil A	Crude Oil B	Crude Oil F	Crude Oil G	Tahiti crude oil
GOR (scf/stb)	787	852	1664	798	510
MW of reservoir fluid (g/mol)	97.750	92.78	70.27	96.15	131.5
MW of flashed gas (g/mol)	29.06	30.24	22.59	28.54	25.83
MW of STO (g/mol)	192.99	182.02	227.58	191.02	243.26
STO Density (g/cc)	0.82	0.817	0.8623	0.823	0.88
Saturates (wt %)	66.26	73.42	52.3	75.56	52.90
Aromatics (wt %)	25.59	19.32	28	20.08	29.70
Resins (wt %)	5.35	7.05	11.3	4.14	13.20
Asphaltene (wt %)	2.80	0.17	8.4	0.22	4.00

References

1. Boussignault, M., *Mémoire sur la composition du bitume.* Ann. Chim. Phys., 1837. **2**: p. 64–141.

2. Carnahan, N.F., *Precipitation of Asphaltenes in Heavy Oil and Tar Sands*, in *Asphaltenes and asphalts*, Chilingar, G.V., Yen, T.F., Editor. 2000, Elsevier B.V.

3. Yen, T.F., *Present Status of Structure of Petroleum Heavy Ends and its Significance to Various Technical Applications.* Prepr .Am. Chem. Soc. Div. Pet. Chem., 1972. **17**(4): p. F102.

4. Murgich, J., Abaner, J.A., *Molecular Recognition in Aggregates Formed by Asphaltene and Resin Molecules from the Athabasca Oil Sand.* Energy & Fuels, 1999. **13**(2): p. 278-288.

5. Altamirano et al., IMP Bulletin, 1986.

6. Carbognani, *INTEVEP.* S.A. Technical Report, 1992.

7. Mullins, O.C., *The Modified Yen Model.* Energy & Fuels, 2010. **24**: p. 2179-2207.

8. Dickie, J.P., Yen, T.F., *Macrostructure of Asphaltic Fractions by Various Instrumental Methods.* Analytical Chemistry, 1967. **39**(14): p. 1847-1852.

9. Spiecker, P.M., Gawrys, K.L., Kilpatrick, P.K., *Aggregation and solubility behavior of asphaltenes and their subfractions.* Journal of Colloid and Interface Science, 2003. **267**(1): p. 178–193.

10. Alboudwarej, H., Akbarzedeh, K., Beck. J., Svrcek, W.Y., Yarranton, H.W, *Regular Solution Model of Asphaltene Precipitation from Bitumen.* AICHE Journal, 2003. **49**(11): p. 2948-2956.

11. Zuo J.Y., Mullins, O.C., Chengli, D., Zhang D., *Modeling of Asphaltene Grading in Oil Reservoirs.* Natural Resources, 2010. **1**: p. 19-27.

12. Hirschberg, A., Hermans, L.J.P.C.M., *Asphaltene Phase Behavior: A Molecular Thermodynamic Model.* in *International Sysmposium on Characterization of Heavy Crude Oils and Petroleum Residues*. June 25-27, 1984. Lyon, France.

13. Yarranton, H.W., Masliyah J.H., *Molar Mass Distribution and Solubility Modeling of Asphaltenes.* AICHE Journal, 1996. **42**(12): p. 3533-3543.

14. Diallo, M.S., Cagin, T., Faulon, J.L., Goddard, W.A., *Thermodynamic Properties of Asphaltenes: A Predictive Approach Based on Computer Assisted Structure Elucidation and Atomistic Simulations.* in Asphaltenes and Asphalts, Chilingar, G.V., Yen, T.F., 2000, Elsevier B.V.

15. Yen, T.F., Erdman J.G., Hanson W.E., *Reinvestigation of Densimetric Methods of Ring Analysis.* Journal of Chemical & Engineering Data, 1961. **6**(3): p. 443–448.

16. Ostlund, J.A., Lofroth, J.E., Holmberg, K., Nyden, M., *Flocculation Behavior of Asphaltenes in Solvent/Nonsolvent Systems.* Journal of Colloid and Interface Science, 2002. **253**: p. 150–158.

17. Ostlund, J.A., Andersson S.I., Nydén, M., *Studies of asphaltenes by the use of pulsed-field gradient spin echo NMR.* Fuel, 2001. **80**(11): p. 1529-1533

18. Jeribi, M., Almir A.B., Langevin, D., Henaut, I., Argillier, J.F., *Adsorption Kinetics of Asphaltenes at Liquid Interfaces.* Journal of Colloid and Interface Science, 2002. **256**(2): p. 268-272.

19. Bauget, F., Langevin, D., Lenormand, R., *Dynamic Surface Properties of Asphaltenes and Resins at the Oil-Air Interface.* Journal of Colloid and Interface Science, 2001. **239**(2): p. 501-508.

20. Maini, B.B., Sarma, H.K., George, A.E., *Significance of Foamy-Oil Behavior in Primary Production of Heavy Oils.* Journal of Canadian Petroleum Technology, 1993. **32**(9): p. 50-54.

21. Gorevski, N., Reinhard, M., Ferri, J.K., *Non-equilibrium exchange kinetics in sequential non-ionic surfactant adsorption: Theory and experiment.* Colloids and Surfaces A-Physicochemical and Engineering Aspects, 2008. **323**(1-3): p. 12-18.

22. Ferraria, M., Liggierib, L., Raverab, F., Amodiob, C., Miller, R., *Adsorption Kinetics of Alkylphosphine Oxides at Water/Hexane Interface: 1. Pendant Drop Experiments.* Journal of Colloid and Interface Science, 1997. **186**(1): p. 40-45.

23. Liggieria, L., Raverab, F., Ferraria, M., Passeronea, A., Miller, R., *Adsorption kinetics of alkylphosphine oxides at water/hexane interface .2. Theory of the adsorption with transport across the interface in finite systems.* Journal of Colloid and Interface Science, 1997. **186**(1): p. 46–52.

24. Werner, A., Behar, F., Hemptinne, J.C., Beha, E., *Viscosity and phase behaviour of petroleum fluids with high asphaltene content.* Fluid Phase Equilibria, 1998. **147**(1-2): p. 343–356.

25. Peng, L., Yongan, G., *Effects of asphaltene content on the heavy oil viscosity at different temperatures.* Fuel 2007. **86**(7-8): p. 1069–1078.

26. Sirota, E.B., Lin, M.Y., *Physical Behavior of Asphaltenes.* Energy & Fuels, 2007. **25**(5): p. 2809-2815.

27. Escobedo, J., Manaoori, G.A., *Theory of Viscosity as a Criterion for Determination of Onset of Asphaltene Flocculation.* SPE, 1996. SPE 28729.

28. Hirschberg, A., *Role of Asphaltenes in Compositional Grading of a Reservoir's Fluid Column.* Journal of Petroleum Technology, 1988. **40**(1): p. 89-94.

29. Andersen, S.I., Speight, J.G., *Petroleum Resins: Seperation, Character and Role in Petroleum.* Petroleum Science and Technology, 2001. **19**(1-2): p. 1-34.

30. Buckley, J.S., *Predicting the Onset of Asphaltene Precipitation from Refractive Index Measurements.* Energy & Fuels, 1999. **13**(2): p. 328-332.

31. Wang, J.X., Buckley, J.S., *A Two-Component Solubility Model of the Onset of Asphaltene Flocculation in Crude Oils.* Energy & Fuels, 2001. **2**(5): p. 1004-1012.

32. Khoshandam, A., Alamdari, A., *Kinetics of Asphaltene Precipitation in a Heptane-Toluene Mixture.* Energy & Fuels, 2010. **24**(3): p. 1917-1924.

33. Yudin, I.K., Nikolaenko, G.L., Gorodetskii, E.E., Markhashov, E.L., Agayan, V.A., Anisimov, M.A., Sengers, J.V., *Crossover kinetics of asphaltene aggregation in hydrocarbon solutions.* Physica A, 1998. **251**(1-2): p. 235-244

34. Anisimov, M.A., Yudin, I.K., Nikitin, V., Nikolaenko, G., Chernoutsan, A., Toulhoat, H., Frot, D., Briolant, Y., *Asphaltene Aggregation in Hydrocarbon Solutions Studied by Photon Correlation Spectroscopy.* Journal Physical Chemistry, 1995. **99**(23): p. 9576-9580.

35. Acevedo, S., Castillo, J., Fernández, A., Goncalves, S., Ranaudo, M.A., *A Study of Multilayer Adsorption of Asphaltenes on Glass Surfaces by Photothermal Surface Deformation. Relation of This Adsorption to Aggregate Formation in Solution.* Energy & Fuels, 1998. **12**(2): p. 386–390.

36. Rudrake, A., Karan, K., Horton, J.H., *A combined QCM and XPS investigation of asphaltene adsorption on metal surfaces.* Journal of Colloid and Interface Science, 2009. **332**(1): p. 22–31.

37. Acevedo, S., Ranaudo, M.A., García, C., Castillo, J., Fernández, A., Caetano, M., Goncalves, S., *Importance of asphaltene aggregation in solution in determining the adsorption of this sample on mineral surfaces.* Colloids and Surfaces A: Physicochemical and Engineering Aspects, 2000. **166**(1-3): p. 145-152

38. Xie, K., Karan, K., *Kinetics and thermodynamics of asphaltene adsorption on metal surfaces: A preliminary study.* Energy & Fuels, 2005. **19**(4): p. 1252–1260.

39. Goual, L., Abudu A., *Predicting the Adsorption of Asphaltenes from Their Electrical Conductivity.* Energy & Fuels, 2010. **24**(1): p. 469–474.

40. Nabzar, L., Aguiléra, M.E., *The Colloidal Approach. A Promising Route for Asphaltene Deposition Modelling.* Oil & Gas Science and Technology, 2008. **63**(1): p. 21-35.

41. Vargas, F.M., Panuganti, S., Yap, F., Chai, J., Chapman, W.G., *Advances in Modeling Asphaltene Precipitation and Deposition in Crude Oil Systems.* in *SPE ATW Flow Assurance and Production Chemistry – Integration of Disciplines to Address the Challenges of Tomorrow* May 30 - June 1, 2011. Dubai, UAE.

42. Nghiem, L.X., Coombe, D.A., *SPE 36106.* Soc. Pet. Eng. J., 1997. **2**: p. 170-176.

43. Panuganti, S.R., Punnapala, S., Gonzalez, D.L., Kurup, A.S., Vargas, F.M., Chapman, W.G., *Prediction of Asphaltene Stability and Compositional Gradients in Crude Oil Systems using the PC-SAFT Equation of State*, in *SAFT Discussion Meeting.* October, 24-25, 2011: Pau, France.

44. Goual, L., in *10th International Conference on Petroleum Phase Behavior and Fouling.* June 14-18, 2009. Rio de Janeiro, Brazil.

45. Czarnecki, J., in *10th International Conference on Petroleum Phase Behavior and Fouling*. June 14-18, 2009. Rio de Janeiro, Brazil.

46. Nellensteyn, F.J., Journal of the Institute of Petroleum Technologist, 1928. **14**: p. 134-138.

47. Abudu, A., Goual, L., *Adsorption of Crude Oil on Surfaces Using Quartz Crystal Microbalance with Dissipation (QCM-D) under Flow Conditions.* Energy & Fuels, 2009. **23**(Sp. Iss. SI): p. 1237-1248

48. Hirschberg, A., Dejong, L.N.J., Schipper, B.A., Meijer, J.G, *Influence of Temperature and Pressure on Asphaltene Flocculation.* SPE 11202, 1984. **24**(3): p. 283-293.

49. Chapman, W.G., Sauer, S.G., Ting, D., Ghosh, A., *Phase behavior applications of SAFT based equations of state—from associating fluids to polydisperse, polar copolymers.* Fluid Phase Equilibria, 2004. **217**(2): p. 137–143.

50. Chapman, W.G., Gubbins, K.E., Jackson, G., Radosz, M., *SAFT: Equation-of-state Solution Model for Associating Fluids.* Fluid Phase Equilibria 1989. **52**: p. 31-38.

51. Chapman, W.G., Gubbins, K.E., Jackson, G., Radosz, M., *New reference equation of state for associating liquids.* Industrial Engineering and Chemistry Research 1990. **29**(8): p. 1709-1721.

52. Chapman, W.G., Jackson, G., Gubbins, K.E., *Phase equilibria of associating fluids Chain molecules with multiple bonding sites.* Molecular physics, 1988. **65**(5): p. 1057-1079.

53. Economou, I.G., *Statistical Associating Fluid Theory: A Successful Model for the Calculation of Thermodynamic and Phase Equilibrium Properties of Complex Fluid Mixtures.* Industrial & Engineering Chemistry Research, 2002. **41**(5): p. 953-962.

54. David, T., Hirasaki, G.J., Chapman, W.G., *Modeling of asphaltene phase behavior with the SAFT equation of state.* Petroleum Science and Technology 2003. **21**(3-4): p. 647–661.

55. Gross, J., Sadowski, G., *Perturbed-chain SAFT: An equation of state based on a perturbation theory for chain molecules.* Industrial Engineering and Chemistry Research 2001. **40**(4): p. 1244-1260.

56. Mansoori, G.A., Carnahan, N.F., Starling, K.E., Leland, T.W., *Equilibrium Thermodynamic Properties of the Mixture of Hard Spheres.* Journal of Chemical Physics, 1971. **54**(4): p. 1523-1525.

57. Chen, S.S., Kreglewski, A., *Applications of Augmented van der Waals Theory of Fluids .1. Pure Fluids.* Berichte der Bunsengesellschaft für Physikalische Chemie, 1977. **81**(10): p. 1048-1052.

58. Barker, J.A., Henderson, D., *Perturbation Theory and Equation of State for Fluids. II. A Successful Theory of Liquids.* The Journal of Chemical Physics 1967. **47**(11): p. 4714-4721.

59. Buckley, J.S., Hirasaki, G.J., Liu, Y., Von Drasek, S., Wang, J.X., Gill, B.S., *Asphaltene precipitation and solvent properties of crude oils.* Petroleum Science and Technology, 1998. **16**(3-4): p. 251-285.

60. Katz, D.L, Firoozabadi A., *Predicting Phase Behavior of Condensate/Crude-Oil Systems Using Methane Interaction Coefficients.* Journal of Petroleum Technology, 1978. **30**(11): p. 1649-1655.

61. Roenningsen, H.P., Skjevrak, I., Osjord, E., *Characterization of North Sea petroleum fractions: hydrocarbon group types, density and molecular weight.* Energy & Fuels, 1989. **3**(6): p. 744–755.

62. Whitson, C.H., *Characterizing Hydrocarbon Plus Fractions.* Society of Petroleum Engineers 1983. **23** (4): p. 683-694.

63. Riazi, M.R, Daubert T.E., *Simplify Property Predictions.* Hydrocarbon Processing 1980. **60**(3): p. 115-116.

64. Riazi, M.R., *Characterization and Properties of Petroleum Fractions.* 2005, Philadelphia, USA: ASTM.

65. Riazi, M.R., Taher, A., *Physical Properties of Heavy Petroleum Fractions and Crude Oils.* Fluid Phase Equilibria, 1996. **117**: p. 217-224.

66. Leelavanichkul, P., Deo, M.D., Hanson, F.V., *Crude Oil Characterization and Regular Solution Approach to Thermodynamic Modeling of Solid Precipitation at Low Pressure.* Petroleum Science and Technology, 2004. **22**: p. 973–990.

67. Pedersen, K.S., Christensen, P.L., *Phase Behavior of Petroleum Reservoir Fluids* 2007: CRC Press. 406.

68. Vargas, F.M., Panuganti S.R., Gonzalez, D.L., Chapman, W.G., *Advances in Predicting Asphaltene Precipitation at High Pressure and Temperature.* in *SPE ATW.* November 29-30 2010. Abu dhabi, UAE: SPE.

69. Panuganti, S.R., Vargas, F.M., Kurup, A.S., Chapman, W.G., *Advanced Crude Oil Characterization for Modeling of Asphaltene Phase Behavior and Compositional Grading*. in *Offshore Technology Conference*. May 2-5, 2011. Houston, USA.

70. Panuganti, S.R., Vargas, F.M., Gonzalez, D.L., Chapman, W.G., *PC-SAFT Characterization of Crude Oils and Modeling of Asphaltene Phase Behavior and Compositional Gradients*. in *PetroPhase*. July, 10-11 2011. London, England.

71. Zuo, J.Y., Mullins, O.C., Dong, C., Zhang, D., *Modeling of Asphaltene Grading in Oil Reservoirs*. Natural Resources, 2010. **1**: p. 19-27.

72. Ting, D., Hirasaki, G.J., Chapman, W.G., *Modeling of Asphaltene Phase Behavior with the SAFT Equation of State*. Petroleum Science and Technology 2003. **21**: p. 647–661.

73. Gonzalez, D.L., Ting, D., Hirasaki, G.J., Chapman, W.G., *Prediction of Asphaltene Instability under Gas Injection with the PC-SAFT Equation of State*. Energy & Fuels, 2005. **19**: p. 1230-1234.

74. Panuganti, S.R., Vargas, F.M., Gonzalez, D.L., Kurup, A.S., Chapman, W.G., *PC-SAFT Characterization of Crude Oils and Modeling of Asphaltene Phase Behavior*. Fuel, 2011. doi:10.1016/j.fuel.2011.09.028.

75. Fan, T., Buckley, J.S., *Rapid and accurate SARA analysis of medium gravity crude oils*. Energy & Fuels, 2002. **16**: p. 1571-1575.

76. Kharrat, A.M., Zacharia, J., Cherian, V.J., Anyatonwu, A., *Issues with Comparing SARA Methodologies*. Energy & Fuels, 2007. **21**: p. 3618-3621.

77. Gonzalez, D.L., Vargas, F.M., Hirasaki, G.J., Chapman, W.G., *Modeling Study of CO2-Induced Asphaltene Precipitation*. Energy & Fuels, 2008. **22**(2): p. 757–762.

78. Gonzalez, D.L., *Modeling of Asphaltene Precipitation and Deposition Tendency using the PC-SAFT Equation of State*, in *Chemical Engineering*. 2008, Rice: Houston.

79. Gonzalez, D.L., Hirasaki, G.J., Chapman, W.G., *Modeling of asphaltene precipitation due to changes in composition using the perturbed chain statistical associating fluid theory equation of state.* Energy & Fuels, 2007. **21**(3): p. 1231-1242.

80. Vargas, F.M., Gonzalez, D.L., Hirasaki, G.J., Chapman, W.G., *Modeling Asphaltene Phase Behavior in Crude Oil Systems Using the Perturbed Chain Form of the Statistical Associating Fluid Theory (PC-SAFT) Equation of State.* Energy & Fuels 2009. **23**(2): p. 1140–1146.

81. Chapman, W.G., Panuganti, S.R., Kurup, A.S., *Issues in Equation of State Modeling for Flow Assurance Applications.* in *AIChE/SPE Joint Workshop on Challenges in Flow Assurance.* September 26-28, 2011. Houston, USA.

82. Panuganti, S.R., Vargas, F.M., Kurup, A.S., Chapman, W.G., *Modeling of Reservoir Connectivity and Tar-Mat using Gravity-Induced Asphaltene Compositional Grading.* in *AICHE Annual Meeting.* October 16-21, 2011. Minneapolis, USA: AICHE.

83. Haier, L., Whitson, C.H., *Compositional Grading-Theory and Practice* SPE, 2001. **4**(6): p. 525-535.

84. Freed, D.E., Mullins, O.C., Zuo, J.Y., *Theoretical Treatment of Asphaltene Gradients in the Presence of GOR Gradients.* Energy & Fuels, 2010. **24**(7): p. 3942-3949.

85. Mullins, O.C., *The Physics of Reservoir Fluids: Discovery through Downhole Fluid Analysis.* 1st edition ed. 2008, Houston: Schlumberger.

86. Carpentiera, B., Arab, H., Pluchery, E., Chautru, J.M., *Tar mats and residual oil distribution in a giant oil field offshore Abu Dhabi.* Journal of Petroleum Science and Engineering, 2007. **58**(3-4): p. 472-490.

87. Wilhelms, A., Carpentiera, B., Huc, A.Y., *New Methods to Detect Tar Mats in Petroleum Reservoirs.* Journal of Petroleum Science and Engineering, 1994. **12**(2): p. 147-155.

88. Wilhelms, A., Larter, S.R., *Origin of tar mats in petroleum reservoirs. Part I: introduction and case studies.* Marine and Petroleum Geology, 1994. **11**(4): p. 418-441

89. Panuganti, S.R., Vargas, F.M., Chapman, W.G., *Modeling Reservoir Connectivity and Tar Mat Using Gravity-Induced Asphaltene Compositional Grading.* Energy & Fuels, 2011. Proceedings of PetroPhase 2011 Conference: doi:10.1021/ef201280d.

90. Kaye, M.N.D., *ADCO Proprietary Report.* 2006.

91. Carpentier, B., Huc, A.Y., Marquis, F., Badr, A.E.R., Al-Aldarous, A.A., Al-Baker, S., *Distribution and Origin of a Tar Mat in the S. Field*, in *Abu Dhabi International Petroleum Exhibition and Conference.* 11-14 November 1998, SPE: Abu Dhabi, United Arab Emirates.

92. Zuo, J.Y., Mullins, O.C., Mishra, V.K., Garcia, G., Dong, C., Zhang, D., *Asphaltene Grading and Tar Mats in Oil Reservoirs.* Energy & Fuels, 2012. DOI: 10.1021/ef201218m.

Prediction of Gas Injection Effect on Asphaltene Precipitation using Different Thermodynamic Models

Doris Gonzalez[1]*, *Frank Lim*[2], *Nikhil Joshi*[3], *Wilson Cañas*[4], *Raul Osorio*[4]

1. Schlumberger, Oilphase-DBR, Houston, Texas, USA
2. Anadarko Petroleum Corporation, The Woodlands, Texas, USA
3. Moulinex Moulinex Business Services LLC, Shenandoah, Texas, USA
4. Ecopetrol S.A., Bogotá D.C., Colombia

Authors Email Address:
DGonzalez21@slb.com, Frank.Lim@anadarko.com, njoshi@moulinexbusiness.com, Raul.Osorio@ecopetrol.com.co, Wilson.CMarin@ecopetrol.com.co
Corresponding Author: email- DGonzalez21@slb.com, Phone- 1(832) 633-7566

Abstract

Gas injection is one of the main types of Enhanced Oil Recovery (EOR) techniques that can recover significant amounts of residual oil. One of the challenges of applying EOR gas injection in reservoirs is the potential of asphaltene precipitation and subsequent deposition. The probability of asphaltene precipitation increases dramatically when light gases are considered for miscible gas injection. Previous studies indicated that the addition of hydrocarbon or nitrogen to live fluids substantially increased the asphaltene onset pressure (AOP) [Lim *et al*, 2008]. Proper predictions of the asphaltene onset at these conditions represent a challenge for reservoir and production modeling.

Anadarko Petroleum Corporation, Ecopetrol S.A. and Schlumberger have jointly investigated the effects of gas addition to the asphaltene phase behavior in a live reservoir oil from deepwater Gulf of Mexico (GoM). In this work, the prediction of the asphaltene phase behavior is investigated in a range of pressures and temperatures in the presence of carbon dioxide, nitrogen and methane using two cubic-plus-association equations of state and the statistical mechanics model developed at Rice University.

The study compares flow assurance simulation results with experimental data recently obtained for a GoM live fluid and for additions of nitrogen, carbon dioxide, methane and n-heptane at high pressure and temperature conditions. Fluids from the field under study presented compositional variation with asphaltene contents from 4 to 15wt%.

Results of the experimental measurements and simulation assessments revealed that the black oil has high propensity for asphaltene precipitation due to addition of gas. The addition of nitrogen, carbon dioxide or methane significantly aggravates the asphaltene precipitation condition of reservoir fluids.

Keywords: flow assurance, asphaltene, equation of state, PC-SAFT, phase behavior

1. Introduction

Gas injection has traditionally played an important role for oil recovery in oil field development. Injection of a gas that is miscible, or having the ability to mix and remain one phase with the oil, allows the recovery of oil that would otherwise be trapped in the tight pores of the rock.

The application of enriched or dry natural gas, CO_2 or N_2 flooding schemes to EOR can induce destabilization and deposition of asphaltenes due to changes in composition. The insolubility of asphaltene in light paraffin liquids and other incompatible fluids is a source of problems during crude oil production operations. This study review and compare some of the most common thermodynamic models simulation results for miscible gas effect on asphaltene precipitation with experimental data obtained at reservoir conditions.

2. Experimental Measurements of Gas Additions to Deepwater GoM Reservoir Oil

Anadarko Petroleum Corporation, Ecopetrol and Schlumberger have jointly investigated the effects of gas addition to the asphaltene precipitation phase behaviour of GoM live fluids. The study focuses on the experimental results from various tests showing the resulting instability of asphaltene in oil due to gas addition of nitrogen, carbon dioxide and methane. N-heptane was also included in this studied for comparison. The study collected PVT and flow assurance data for original live fluid and for fluids prepared after gas additions at reservoir conditions. Fluids from the field presented a compositional variation with a variety of asphaltene contents. Fluids with 4% and 15.5% asphaltene content were selected for this study. Tables 1 and 2 summarize composition and fluid properties:

Table 1. Composition of the Hydrocarbon Fluids

Component	MW	Live Fluid 15.5% Asph	Live Fluid 4.0% Asph
Carbon Dioxide	44.01	0.16	0.13
Nitrogen	28.01	0.34	0.56
Methane	16.04	32.90	46.90
Ethane	30.07	8.15	6.13
Propane	44.10	6.28	4.67
Butane	58.12	1.13	3.05
Pentane	72.15	3.50	2.25
C6	86.20	1.55	1.68
C-Pentane	84.16	2.15	0.40
Benzene	78.11	3.32	0.06
Cyclohexane	84.16	0.38	0.36
C7	100.20	0.06	2.26
C-Hexane	98.19	0.23	0.67
Toluene	92.14	3.10	0.25
C8	107.00	0.46	2.76
E-Benzene	106.17	0.28	0.17
Xylene	106.17	3.07	0.53
C9	121.00	0.12	2.03
C10	134.00	0.29	2.45
C11	147.00	0.20	2.02
C12+	291.40	24.40	20.68
MW		132.94	109.81

Table 2. Fluids Properties Summary

	Live Fluid 15.5% Asph	**Live Fluid 4.0% Asph**
Reservoir Conditions		
Pressure, psia	~8,029	8,342
Temperature, F	208	208
Bubble Point Pressure		
At Tres, psia	2,420	3,617
At 150F, psia	2,128	3,460
At 100F, psia	1,850	3,215
At 40F, psia	1,533	2,820
Gas Oil Ratio (GOR)		
Single-state flash, scf/bbl	565	778
Properties at Reservoir Conditions		
Viscosity, cP	~7.71	1.05
Compressibility, 10^{-6}/psi	~ 9.3	7.5
Density, g/cc	~ 0.786	0.766
Properties at Saturation Conditions		
Viscosity, cP	1.26	0.7
Compressibility, 10^{-6}/psi	13.6	12.8
Density, g/cc	0.754	0.719
Properties at 60F (STO)		
Molar mass	289.99	266.63
API Gravity	22.5	28.4
Density, g/cc	0.919	0.885
Gas Gravity	0.97	0.808
SARA Analysis		
Saturates, wt%	39.23	57.5
Aromatics, wt%	35.9	30.4
Resins, wt%	9.01	8.3
Asphaltenes, wt%	15.52	3.7
Inorganic, wt%	0.34	0.11

Measurements include asphaltene onset pressure (AOP), saturation pressure (Pb) and swelling tests. Other basic measurements of the corresponding dead oil include SARA analysis, Refractive Index, viscosity and density.

Asphaltene Pre-screening based on SARA analysis

The SARA analysis indicates that these reservoir fluids have a high content of aromatics+resins (~ 40%) and the asphaltene content is in the range for a crude with low asphaltene precipitation tendency (>1%) [de Boer, 1995]. Based on de Boer criteria for asphaltene stability, these values indicate a relatively asphaltene stable fluid. However, these concentrations along with the gas compositional information need to be integrated in a single model to determine the potential for asphaltene precipitation. The following pre-screening criteria have been applied to estimate asphaltene stability based on SARA analysis:

- asphaltene/resin ratio,
- Colloidal Instability Index (CII) suggested by Baker-Petrolite and,
- Asphaltene Stability Index (ASI) criterion developed by Schlumberger.

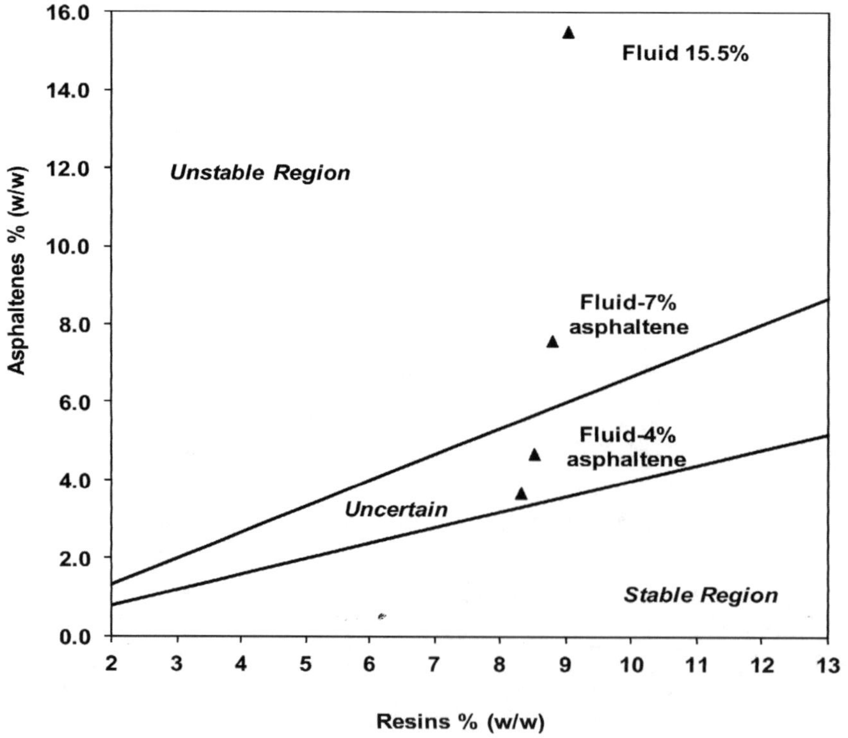

Figure 1. Asphaltene Pre-Screening by SARA Analysis – Asphaltene / Resin ratio

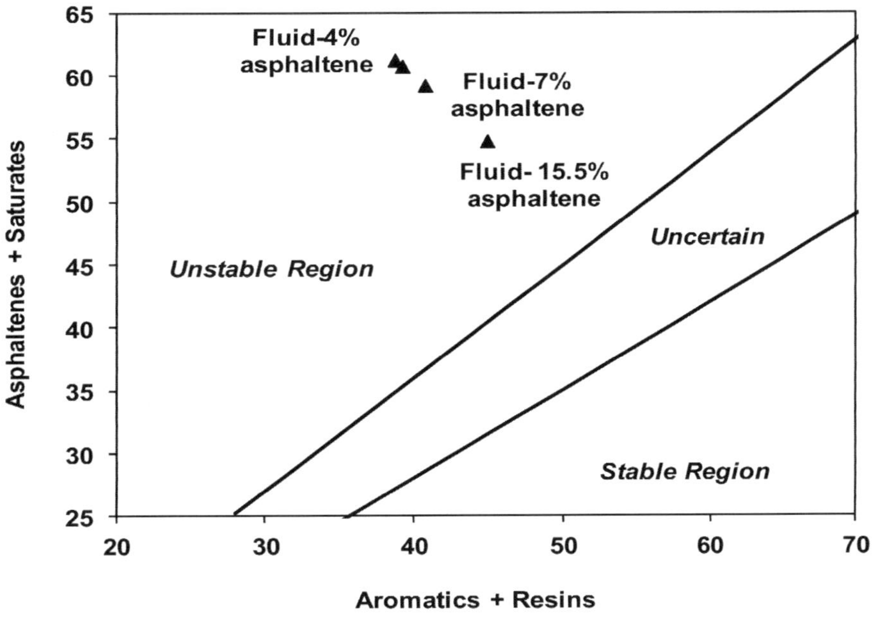

Figure 2. Asphaltene Pre-Screening by SARA Analysis – Colloidal Instability Index

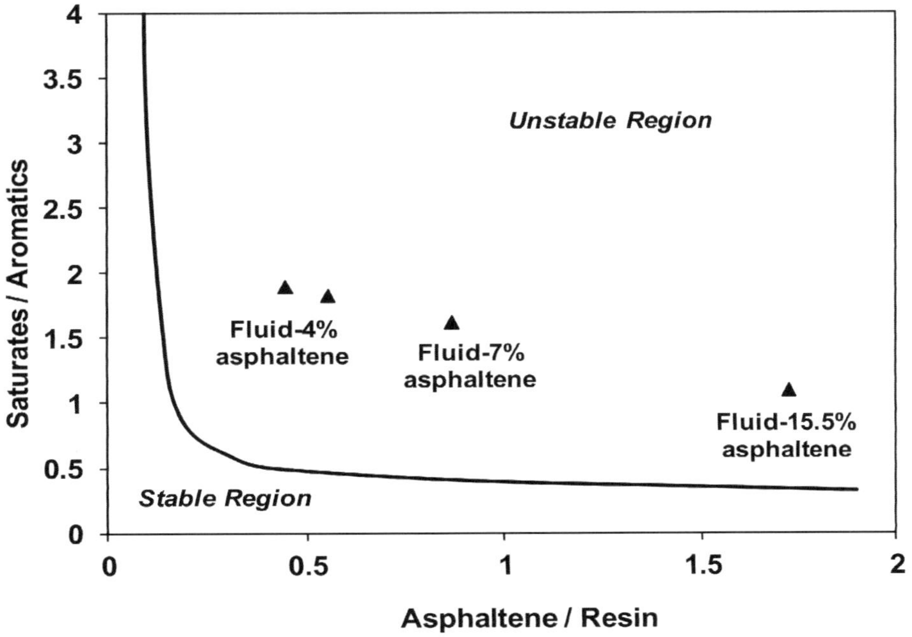

Figure 3. Asphaltene Pre-Screening by SARA Analysis – Asphaltene Stability Index

High and low asphaltene crude oils fall in the Unstable or Uncertain regions. According to this analysis, the crude oil exhibits high tendency to precipitate asphaltenes.

Asphaltene Pre-screening based on deBoer Plot

The deBoer plot is a method to screen crude oils for their tendency to precipitate asphaltenes based on the published work of de Boer and Leerlooyer (1992). It essentially evaluates the loss of asphaltene solubility as a reservoir fluid sample is depressurized. This PVT screen is simply a cross-plot of in situ density and the degree of undersaturation with respect to gas (the difference between reservoir and saturation pressures). Because this screen assumes the fluid is saturated with respect to asphaltene at reservoir conditions, it is extremely conservative (Shell Global Solutions).

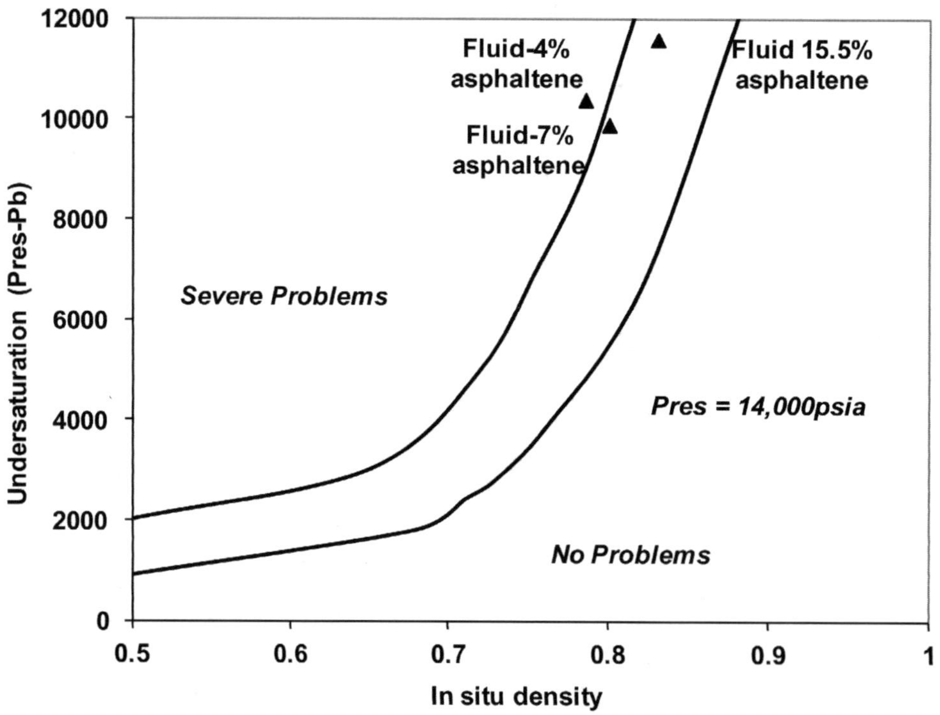

Figure 4. Asphaltene Pre-Screening by deBoer plot

According to this plot, the GoM fluids considered in this project fell in or near the "Severe Problem" region; therefore asphaltene related production problems are expected.

Reservoir Fluid Phase Behavior

The asphaltene onset of precipitation (AOP) for live fluids with 4% and 15.5%-asphaltene content were measured at reservoir temperature by depressurization starting from 15,000psi and using fixed wavelength near infrared (NIR), high pressure microscopy (HPM) and Particle Size Analysis (PSA) techniques. Description of these techniques is found in reference by Jamaluddin et al. [Jamaluddin, 2002]. Figure 5 presents the transmissibility signal behavior of fluid 15.5%-asphaltene at 208°F under depressurization. Asphaltene precipitation onset pressure is detected at 4,600psia.

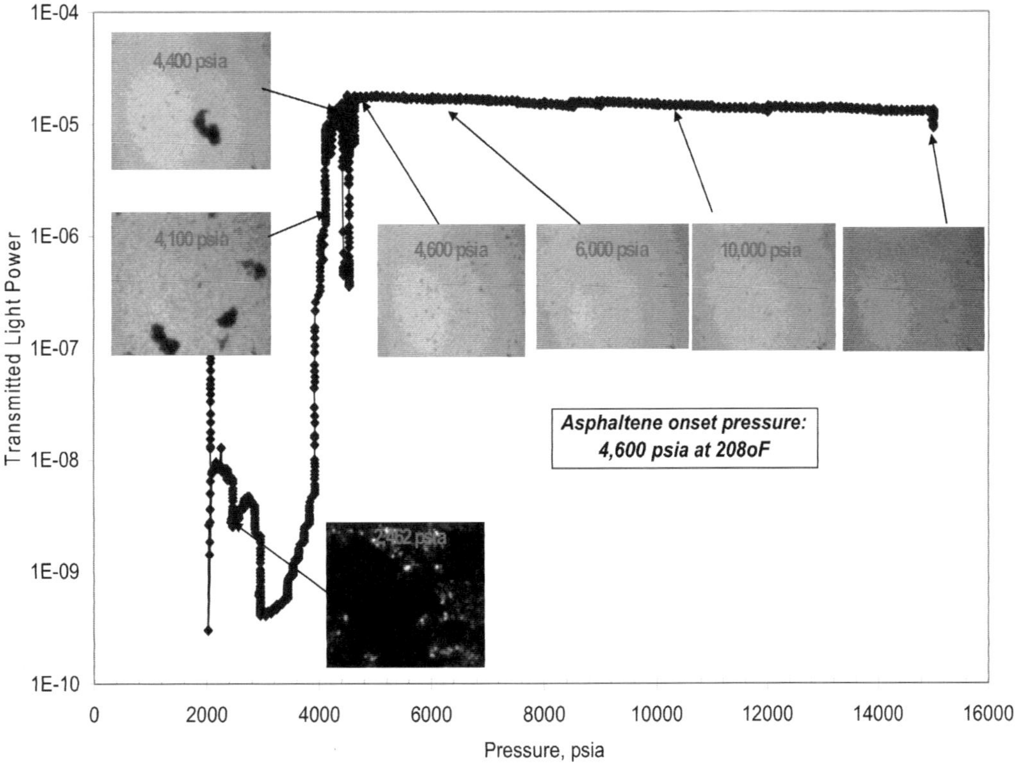

Figure 5. Asphaltene Precipitation Onset (AOP) measurement using NIR and HPM

Asphaltene Onset Pressures (AOP) for the GoM-fluids at different temperatures are listed in the following table.

Table 3. Experimental Saturation Pressures (Pb) and AOPs

	Fluid 15.5% asphaltene		Fluid 7% asphaltene		Fluid 4% asphaltene	
Temp, F	AOP	Pb	AOP	Pb	AOP	Pb
208	4,600	2,420	5,200	4,115	5,000	3,617
150	5,050	2,128	6,100	3,815	5,300	3,460
100	6,700	1,850	8,500	3,565	6,000	3,215
88	8,000	1,750		3,103		
80	8,600	1,700				
70					9,200	3,030
40		1,533				2,820

Live Wax Appearance Temperature (WAT) is measured at constant pressure during a cooling process using the same techniques, fixed wavelength near infrared (NIR), high pressure microscopy (HPM) and Particle Size Analysis (PSA). Figure 6 presents the transmissibility signal behavior of fluid 15.5%-Asph at 14,000psi during cooling process from 208°F down to 50°F. Wax Appearance Temperature (WAT) for live 15.5%-Asph-fluid was detected at 79.8°F at 14,000 psi and 84°F at 20,000 psi. WAT for 15.5%-Asph-STO sample (at ambient pressure) was measured as 85.28°F using Cross Polar Microscope.

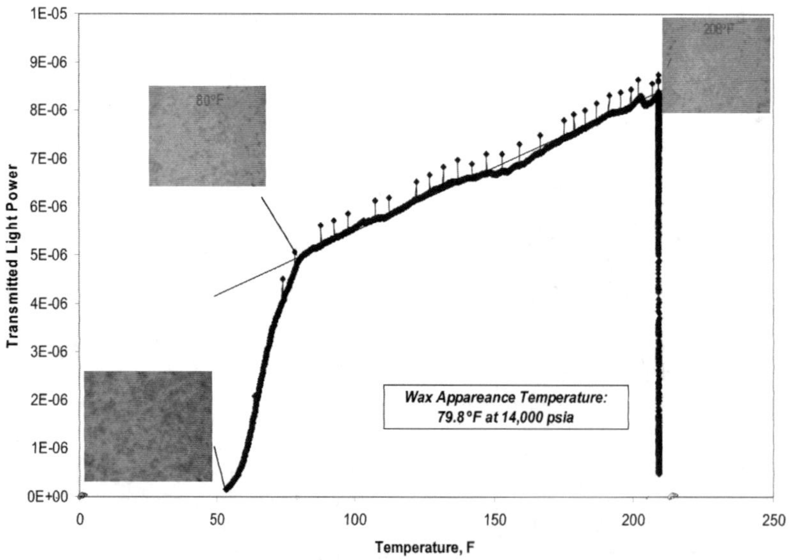

Figure 6. Wax Appearance Temperature (WAT) measurement using NIR and HPM

The following figure summarizes the results of the Flow Assurance analyses conducted on the GoM hydrocarbon samples.

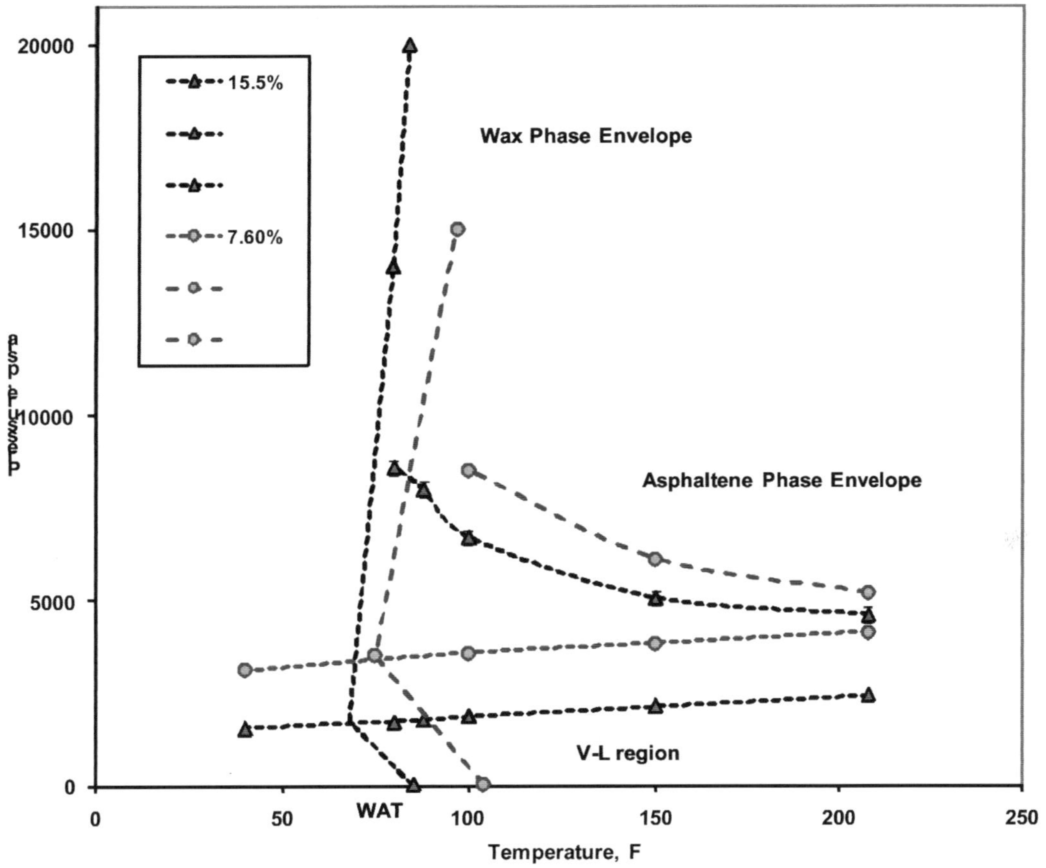

Figure 7. Asphaltene Phase Behavior Comparison

Gas Addition effect on Reservoir Fluid Phase Behavior

Nitrogen, CO2 and methane were individually added to the GoM live fluids to increase their content in 10%mole. The saturation pressure and asphaltene precipitation onset showed an increase at different temperatures as shown in Table 4 and Figure 8.

Table 4. Gas Addition Effect: Pb and AOPs for 15.5wt% asphaltene fluid

	15.5%-asph - Original		15.5%-asph + N2		15.5%-asph + CH4		15.5%-ASPH + CO2	
Temp, F	AOP	Pb	AOP	Pb	AOP	Pb	AOP	Pb
208	4,600	2,420	9,700	4,515	8,200	3,505	6,200	2,722
150	5,050	2,128	12,200	4,516	10,500	3,205	7,100	2,419
125			13,500	4,500				
100	6,700	1,850	19,600	4,416	13,000	2,818	9,800	2,115
88	8,000	1,750*					11,000	2,000*
80	8,600	1,700*						
70						2,585		1,819
40		1,533				2,575		1,716

*Estimated value

The experimental measurements show the increase on saturation pressure and asphaltene precipitation phase envelopes of the fluid with 15.5% asphaltene with 10% increase in nitrogen, methane and carbon dioxide.

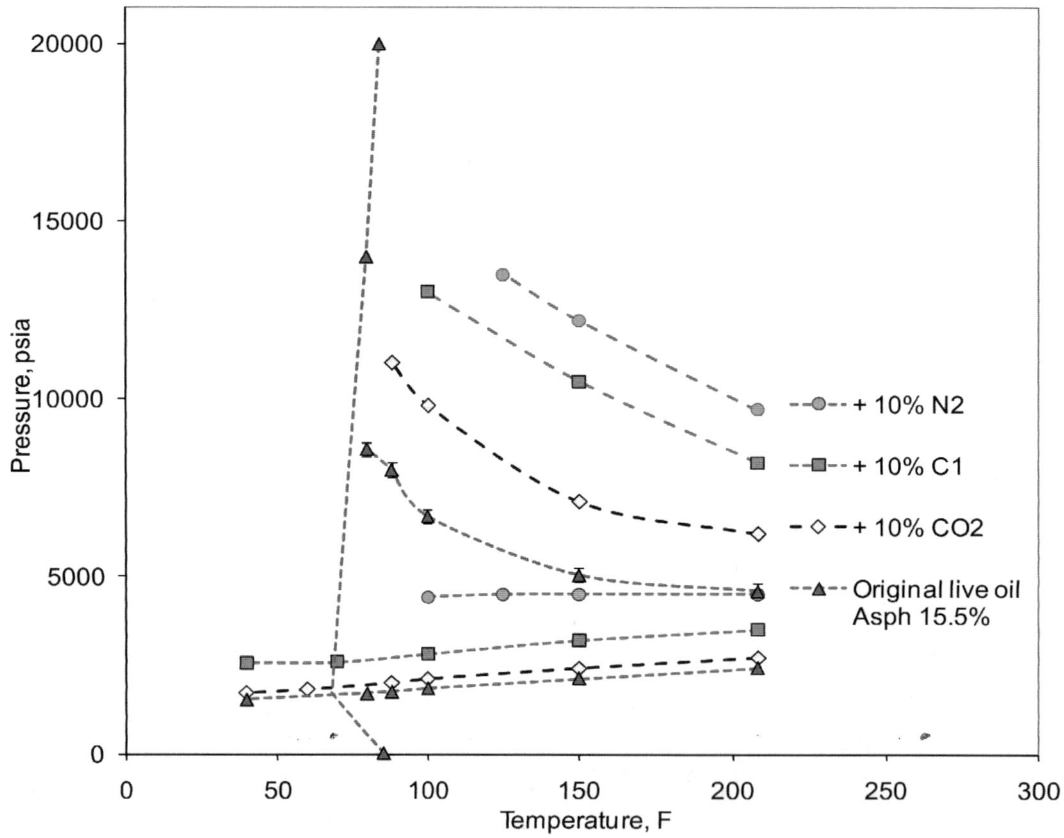

Figure 8. Effect of Gas Addition on the saturation and Asphaltene Onset pressure for the high asphaltene content fluid.

Effect of n-Heptane Addition on phase behavior

High asphaltene content Live fluid was mixed with n-Heptane to increase its content in 10%mole. Figure 9 shows the effect of this addition on the Saturation and Asphaltene onset pressures at different temperatures.

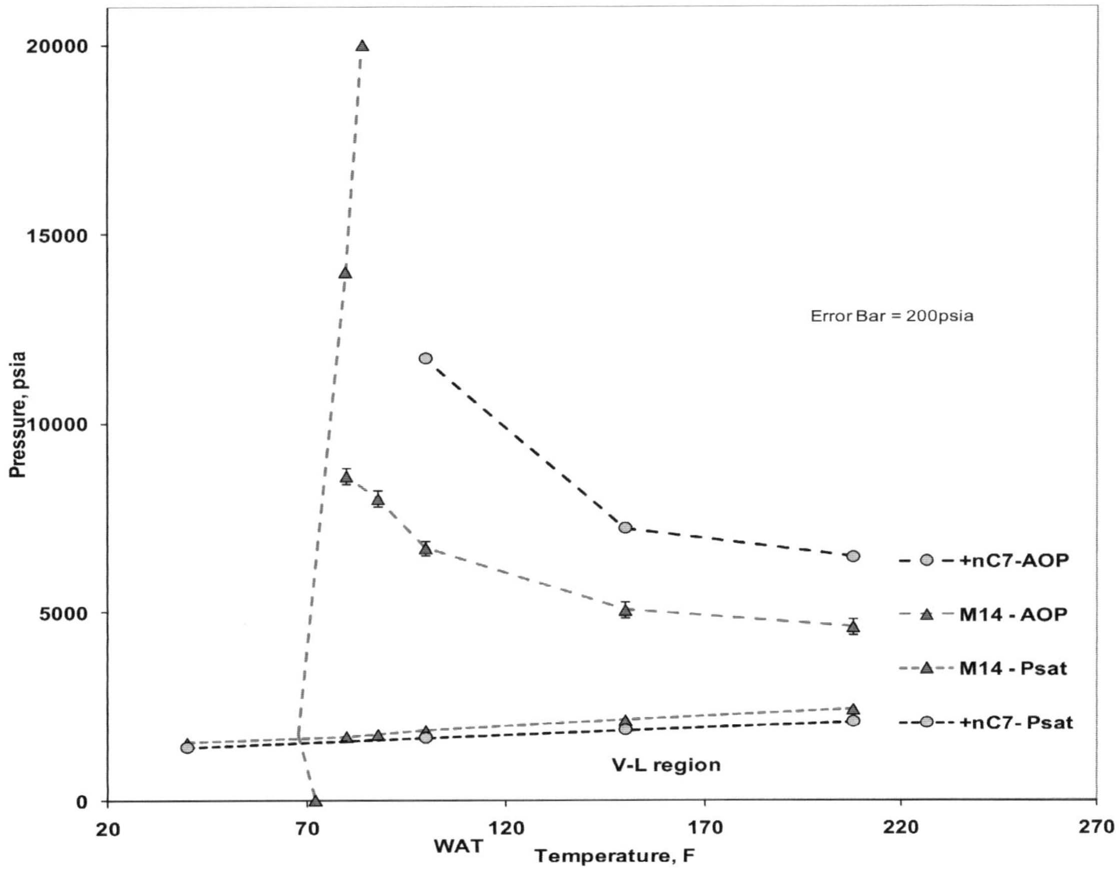

Figure 9. Effect of n-Heptane (n-C7) addition on Saturation and Asphaltene Onset pressures

According to the experimental data, the addition of n-Heptane to the high asphaltene content live fluid slightly decreases its saturation pressure but strongly increases its asphaltene onset pressure. The effect of the 10%mole addition is equivalent to 10%mole addition of carbon dioxide.

3. Asphaltene Precipitation Modeling using the Cubic Plus Association EoS

Equations of state are predictive because they include the effect of compressibility. Cubic equations of state are simple models that have been widely applied in industry. However, the major limitation of cubic EoS is that they cannot describe adequately the phase behavior of mixtures of molecules with large size disparities and they are unable to accurately calculate liquid densities. The reason for their poor prediction capability is because cubic EoS are typically fit to the critical point. It has been found that cubic EoS that are fit to liquid phase density have a better performance in reproducing phase behavior data. In the two models of CPA EoS evaluated in this study, the prediction of gas additions effects requires a new tuning process to develop a new set of parameters.

$$P=\frac{RT}{V_m-b}-\frac{\alpha(T)}{V_m(V_m+b)}-\frac{1}{2}\frac{RT}{V_m}\left(1+\rho\frac{\partial \ln g}{\partial \rho}\right)\sum_i x_i\sum_{A_i}(1-X_{A_i})$$

Repulsive Attractive Association

Cubic EOS
PR, SRK

SAFT (Statistical Association Fluid Theory).

Reservoir Research Institute (RERI) Cubic Plus Association Equation of State

Assumptions in the RERI model:

- AOP modeled as liquid-liquid or gas-liquid-liquid phase separation
- Live oil is characterized by considering pure and pseudo-hydrocarbon components (C1, C2, ….., C6, C7, …… Cn+)

- (Cn+): Hydrocarbon residue is divided in "heavy" component and asphaltene
- The "heavy" component stabilizes asphaltene in solution
- "Heavy" component contains heavy alkanes, heavy aromatics and all resins. It represents the secondary most polar component in petroleum. The polar-polar interactions of asphaltene constituents are suspected as the primary cause for the asphaltene aggregation.
- The polar-polar interactions between asphaltene molecules (self association) and between asphaltene and "heavy" molecules (cross-association) are described by the thermodynamic perturbation theory.
- Association parameters: $N_A = N_R = 4$; $k_{AA} = k_{AR} = 0.01$; $E_{AA}/k_B = 2000$ K

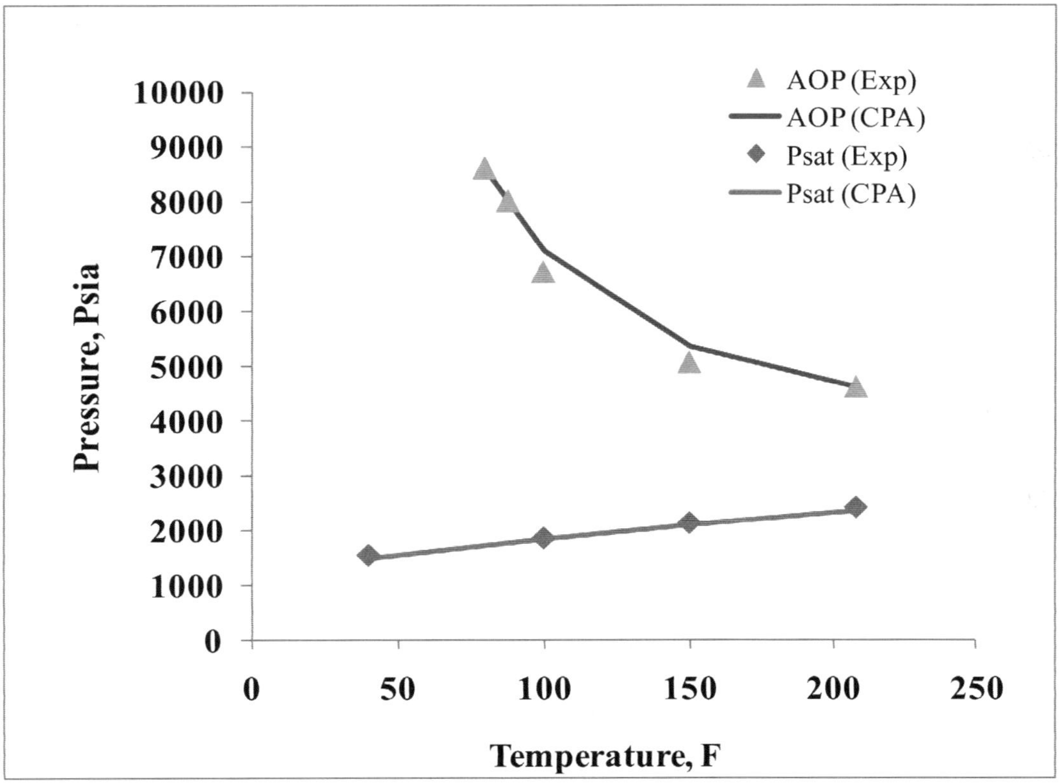

Figure 10. Asphaltene Phase Behavior prediction using CPA EoS from RERI.

Predictions using the RERI model require more than one asphaltene onset point for proper tuning of the asphaltene boundary.

DBR-Solids Cubic Plus Association (CPA) equation of state (EoS)

DBR-Solids is a state-of-the-art software application that models the precipitation of solids, specifically waxes and asphaltenes from petroleum. The asphaltene precipitation model is based in the Peng-Robinson equation of state plus and association term. The Model tuning function is capable of tuning to the onset point or tuning to the amount of asphaltene precipitated at a specified Temperature and Pressure.

For this study, the asphaltene onsets of precipitation measured at 208°F, 150°F and 100°F were used to tune the model. Figure 11 shows asphaltene phase envelopes predicted by tuning the model to one (purple), two (brown) and three (blue) experimental points.

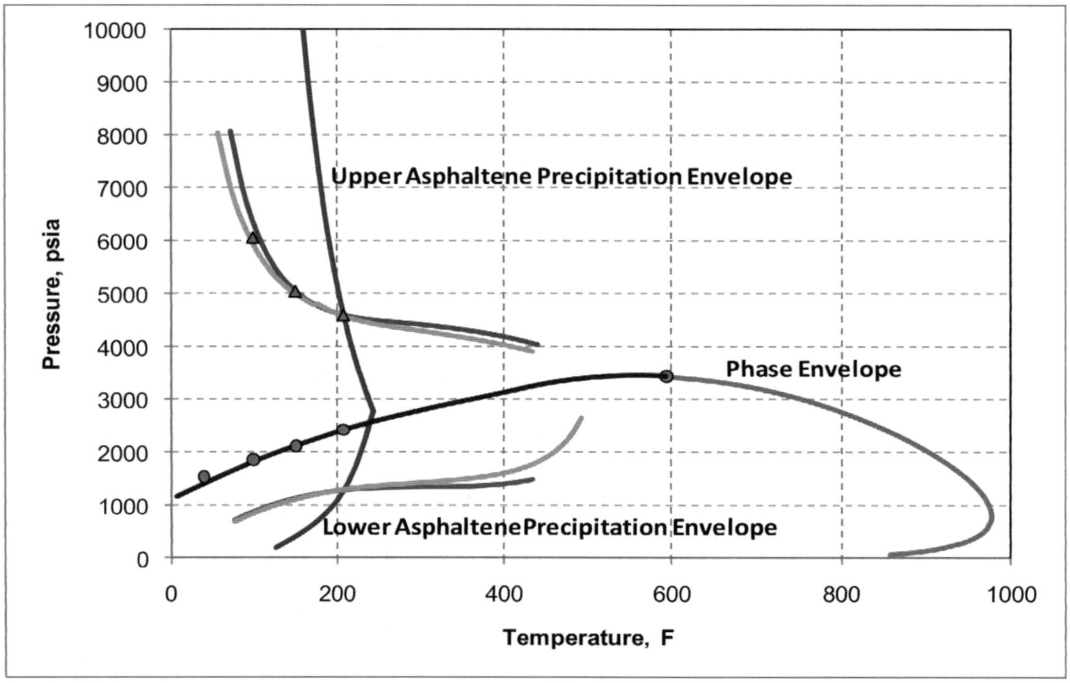

Figure 11. Asphaltene Phase Behavior prediction using CPA EoS DBR-Solids, Upper and Lower boundaries

Similar to RERI's model, predictions using the DBR-Solids model require to be tuned to more than one asphaltene onset point for proper prediction of the asphaltene boundary.

4. Asphaltene Precipitation Modeling using the PC-SAFT EoS

A modern equation of state is the Statistical Associating Fluid Theory (SAFT) family of models based on statistical mechanics. This equation of state can accurately model mixtures of molecules of different sizes. The detailed modeling principles and assumptions of this model can be found in references Gonzalez et al. [Gonzalez, 2005].

The PC-SAFT EoS compositional model as implemented in Multiflash^{TM} (Infochem) is used to predict asphaltene precipitation tendencies. An alternative version developed at Ecopetrol Colombia is available for the evaluation of the EoS at low temperatures.

The main assumption of the PC-SAFT model is that the asphaltene phase behavior can be explained based only on molecular size and van der Waals interactions. The associated interactions are already included in the formation of asphaltene pre-aggregates and no association parameters are required.

The equation of state parameters are calculated for each of the seven components and pseudo-components of the recombined sample using SARA, GOR, PVT data and compositional information. The asphaltene parameters are tuned to meet the onset of asphaltene precipitation measured by de-pressurization at reservoir temperature.

The PC-SAFT parameters are defined as:

- m, number of segments per molecule
- σ, diameter of each segment of a molecule in Å.
- ε/k, segment-segment interaction energy in Kelvin.

Fluid properties such us density of the live oil are calculated using PC-SAFT tuned model and compared to experimental data in Figure 12.

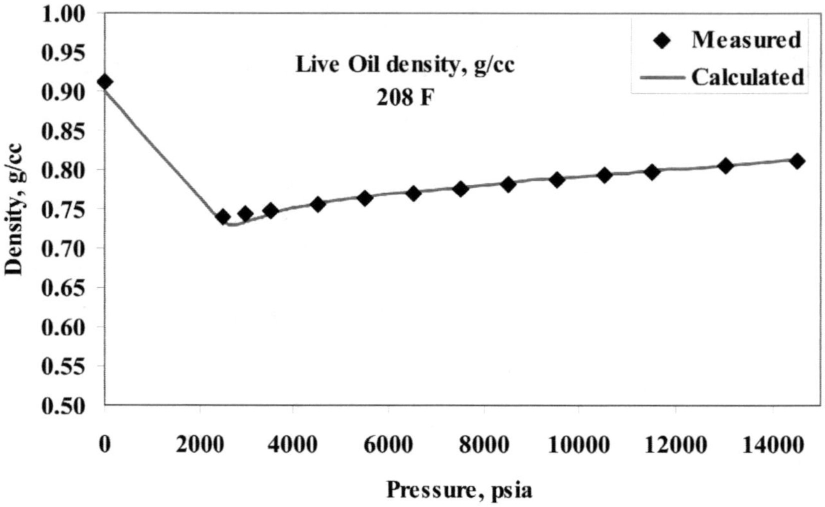

Figure 12. Recombine fluid density from constant composition expansion (CCE) test

The prediction of the Asphaltene Phase Envelope for the 4%Asph fluid is shown in Figure 13.

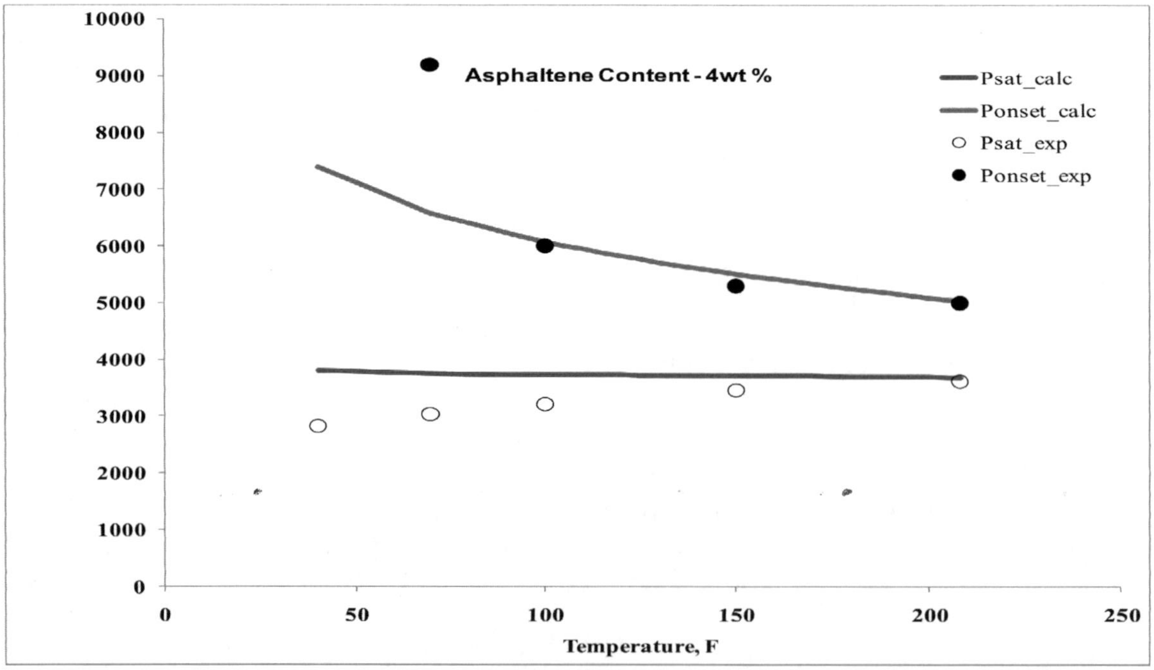

Figure 13. Prediction of the phase envelope for the 4%-asphaltene fluid at 208°F

5. Experimental Measurements of Gas Additions to GoM Reservoir Oil 'A'

The GoM fluid 'A' was under production with primary depletion for more than 20 years; the reservoir performance indicated minimal aquifer support and the reservoir pressure was expected to decline below the saturation pressure. An experimental study was carried out to evaluate the asphaltene phase envelope of fluid 'A' [Jamaluddin, 2001]. Summary of the reservoir 'A' fluid properties are provided in Tables 5 and 6.

Table 5. Reservoir fluid characterization: Reservoir Fluid 'A' [Jamaluddin et al., 2001].

Measured Depth	14,134 ft
Reservoir Temperature	296 ^{o}F
Reservoir Pressure	3,256 psia
Bubble Point Pressure at 255°F	3,045 psia
Gas-oil-ratio	900 scf/stb
Oil API Gravity	32°
Wax content	12% w/w
Cloud point	72°F
SARA Contents (ASTM D4124-97)	Wt%
Saturates	57.4
Aromatics	30.8
Resins	10.4
Asphaltenes (n-pentane insoluble)	1.4

The SARA Analysis for the liquid phase of Fluid 'A' indicates that the corresponding dead oil is Saturate in nature and that the asphaltene content is slightly above the limit suggested by de Boer to expect severe asphaltene problems (>1%) [deBoer, 1996]. Although the stock tank oil should looks relatively stable, the associated gas may cause asphaltene precipitation under certain conditions.

Table 6. Reservoir fluid 'A' composition

Components	Flashed liquid Mol %	Flashed Gas Mol %	Monophasic fluid Mol %
Nitrogen	0.00	0.77	0.49
Carbon Dioxide	0.00	17.67	11.37
Hydrogen Sulfide	0.00	5.00	3.22
Methane	0.00	42.49	27.36
Ethane	0.14	14.54	9.41
Propane	0.66	10.05	6.70
1-Butane	0.23	1.13	0.81
n-Butane	1.48	4.11	3.17
1-Pentane	1.17	1.26	1.22
n-Pentane	2.71	1.57	1.98
Pseudo C_6H_{14}	5.32	0.92	2.49
Pseudo C_7H_{16}	7.38	0.37	2.87
Pseudo C_8H_{18}	8.62	0.10	3.14
Pseudo C_9H_{20}	7.67	0.02	2.74
Pseudo $C_{10}H_{22}$	6.49	0.01	2.32
Pseudo $C_{11}H_{24}$	5.31	0.01	1.90
$C_{12}+$	58.82	0.01	18.82
Total	100.00	100.00	100.00
MW	229.21	31.69	102.04
Mole Ratio	0.3562	0.6438	
Molar Mass: $C_{12}+$	337.98	167.11	337.94
Density (g/cc): $C_{12}+$	0.906	-	0.906

The condition for asphaltene precipitation was measured using the Solid Detection System (SDS) experimental technique. Laboratory experiments using the transmittance of an optimized laser light in the near infrared (NIR) wavelength (~1600 nm) were reported by Jamaluddin et al. to define the pressure-temperature regions of asphaltene instability of the reservoir fluid. Depressurization tests were completed at 296, 260, 230, and 190°F.

6. Asphaltene Precipitation Modeling of Reservoir Oil 'A' using the PC-SAFT EoS

For prediction, the PC-SAFT parameters for each of the components (except for the asphaltenes) were determined from correlations with the average molecular weight. Asphaltene parameters were fitted to meet experimental asphaltene precipitation onset measurements. Table 7 summarizes composition, MW and the PC-SAFT EoS parameters.

Table 7. PC-SAFT characterization of the live sample at GOR = 900 scf/bbl

	X_{mole}	MW_n	m	σ (Å)	ε/k (K)
N_2	0.00495	28.01	1.2053	3.3130	90.96
CO_2	0.14583	44.01	2.0729	2.7852	169.21
Methane	0.27334	16.04	1.0000	3.7039	150.03
Light	0.21917	44.6	2.0546	3.6130	204.96
Saturates	0.23853	207.6	5.967	3.932	254.05
Aromatics + resins	0.11750	270.5	6.473	3.87	332.52
Asphaltenes	0.000676	1700	29.5	4.30	392.56

Figure 14 shows the agreement between experimental data and simulation results for fluid 'A'. The PC-SAFT equation of state accurately reproduces the bubble point pressures and the temperature dependence of the asphaltene precipitation pressure.

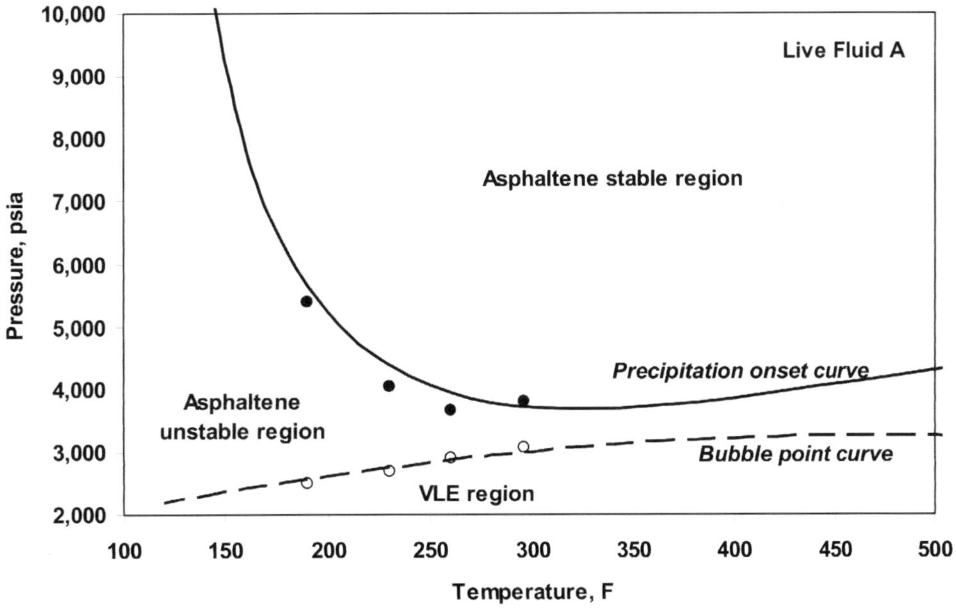

Figure 14. Onset of asphaltene precipitation and bubble points for reservoir fluid 'A'. Experimental data from Jamaluddin et al, 2001.

Figure 14 also shows that low temperatures within the well and flow lines increase the probability of asphaltene precipitation. This effect is even more important in deepwater production where seafloor temperatures range from 32-40°F (0 to 4°C). The SAFT EoS predicts the presence of upper and lower critical solution temperature (UCST and LCST) regions at low and high temperatures. In LCST region, an increase in temperature (at a fixed pressure) will result in asphaltene precipitation from the crude oil solution. The existence of the UCST and LCST regions can be explained in terms of contributions to Gibbs energy. At lower temperatures, entropy effects dominate and asphaltene solubility increases with increasing temperature. At the higher temperatures (i.e., as the critical temperature of the solvent is approached), enthalpy effects dominate due to increasing differences in the rate of free volume expansion of the components, which effects large changes in the cohesive energy density of the crude oil solvent. For reservoir fluid 'A', the onset pressure rapidly increases at temperatures below 220°F.

7. Prediction of gas addition effects on Reservoir Oil 'A' using the PC-SAFT EoS

To recreate the effect of is individual gas additions on asphaltene behavior, the PC-SAFT thermodynamic model was used to predict the precipitation boundaries when N_2, CO_2 and methane was added to reservoir fluid 'A'.

Simulation results showed that asphaltene precipitation onset pressure increases when 10 mol % of each of N2 and methane was added to the original fluid 'A' as shown in Figure 15 a) and b). A different behavior was observed upon the addition of 10 and 20 mol % CO_2. The boundary shows a crossover around 200°F; the addition of CO_2 increases asphaltene stability compared to the original oil. Above this temperature, the asphaltene solubility decreases (Figure 15 c)).

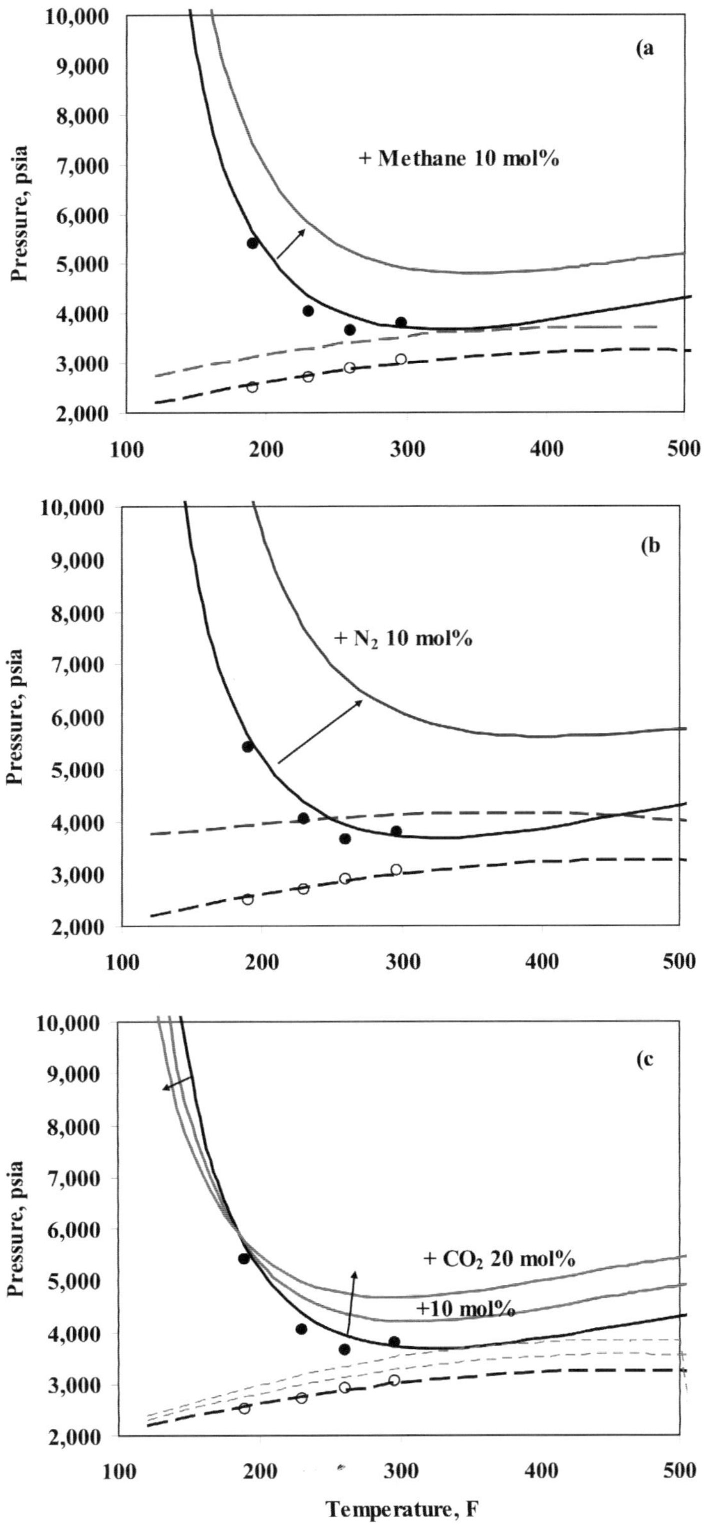

Figure 15. Fluid 'A' phase behavior after the addition of methane, N2 and CO2. Experimental data from Jamaluddin et al., 2001

The addition of increasing amounts of gas at different temperatures was also simulated as presented in Figure 16. Experimental measurements obtained during the addition of N_2 to Fluid A are included in the figure. Whereas the addition of methane and N_2 increases the asphaltene precipitation onset pressure of the mixture, the addition of CO_2 shows a different behavior at temperatures below the crossover point. Changes in the slope of the precipitation onset curves predicted by PC–SAFT at low temperature, confirms the special behavior of CO_2.

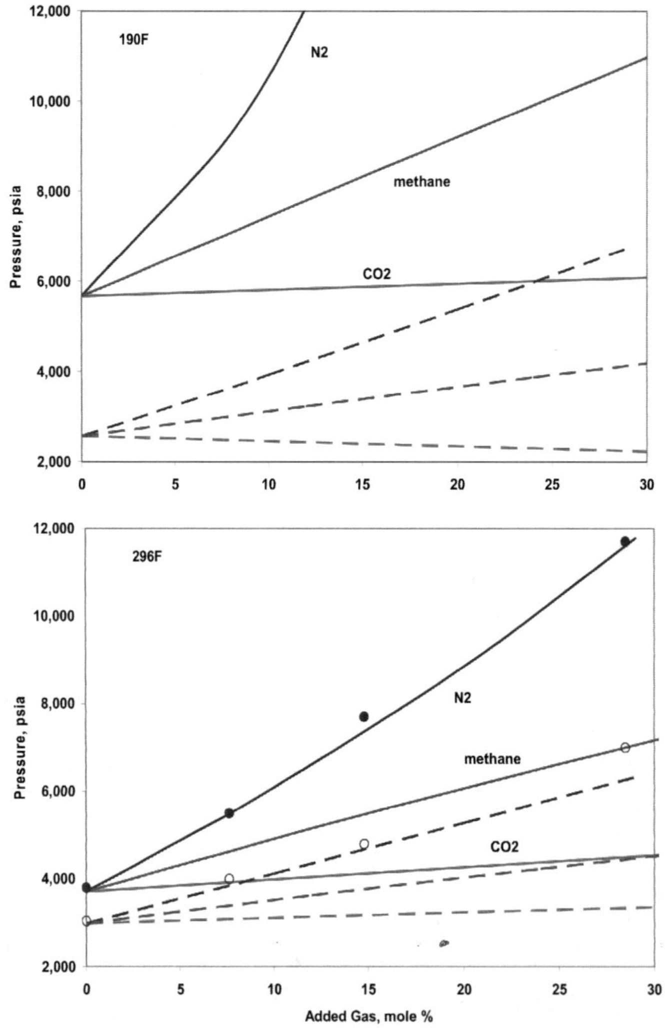

Figure 16. Addition of successive amounts of CO2, N2 and Methane to reservoir Fluid

8. Asphaltene precipitation due to light hydrocarbon

Light Hydrocarbon effect on a Simplify Live Oil

In the simplify system, methane, ethane or propane is added to 1g of n-C7 precipitated asphaltene dissolved in 100 ml of toluene. The Perturbed Chain version of the Statistical Associating Fluid Theory (PC-SAFT) EoS is used to predict bubble points and asphaltene precipitation onset for the mixture of asphaltene, toluene and light hydrocarbon gas. The PC-SAFT parameters for gases and toluene are taken from Gross and Sadowski [Gross, 2001] and the n-C7 monodisperse asphaltene parameters are those estimated by Ting for the simplify system (m = 80; σ = 4.05; ε/k = 350.8). The PC-SAFT calculates for this asphaltene a solubility parameter value of 21.71MPa$^{0.5}$ and a molar volume of 3,349 cm^3/mole. The simulated results are compared to methane-toluene-asphaltene experimental data previously evaluated by Ting et al. in 2003 (Figure 17).

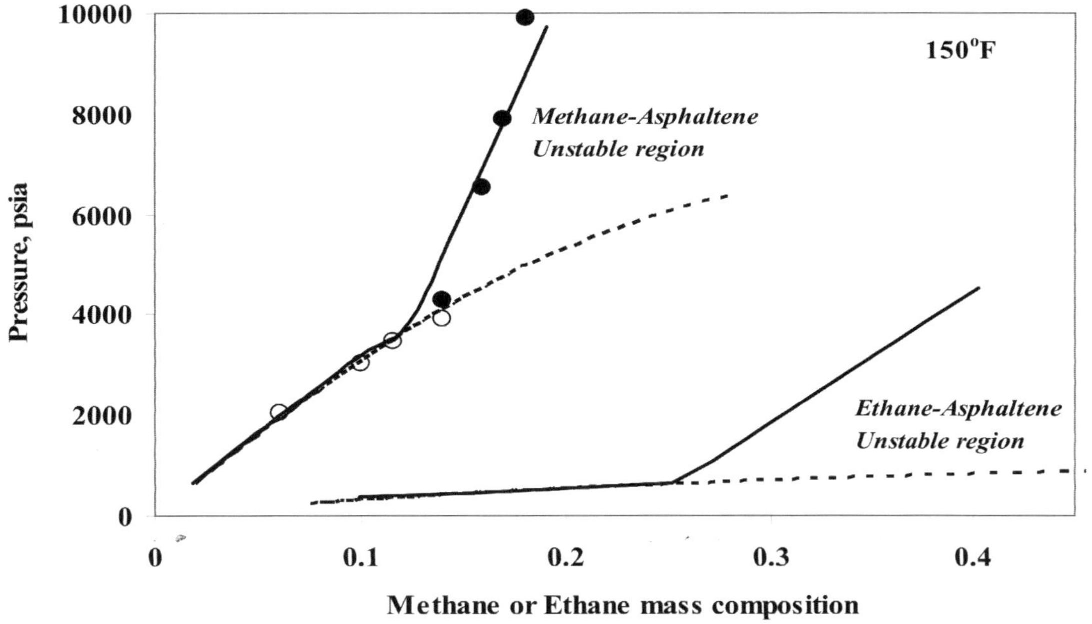

Figure 17. Methane and Ethane effect on Asphaltene Precipitation. Predicted onset and bubble point curves. Circles represent experimental values measured by Ting, 2003.

Figure 18 compares asphaltene instability and bubble point curves for solutions of asphaltene in toluene (1g of n-C7 precipitated asphaltene dissolved in 100 ml of toluene) titrated with methane, ethane and propane at different hydrocarbon concentrations.

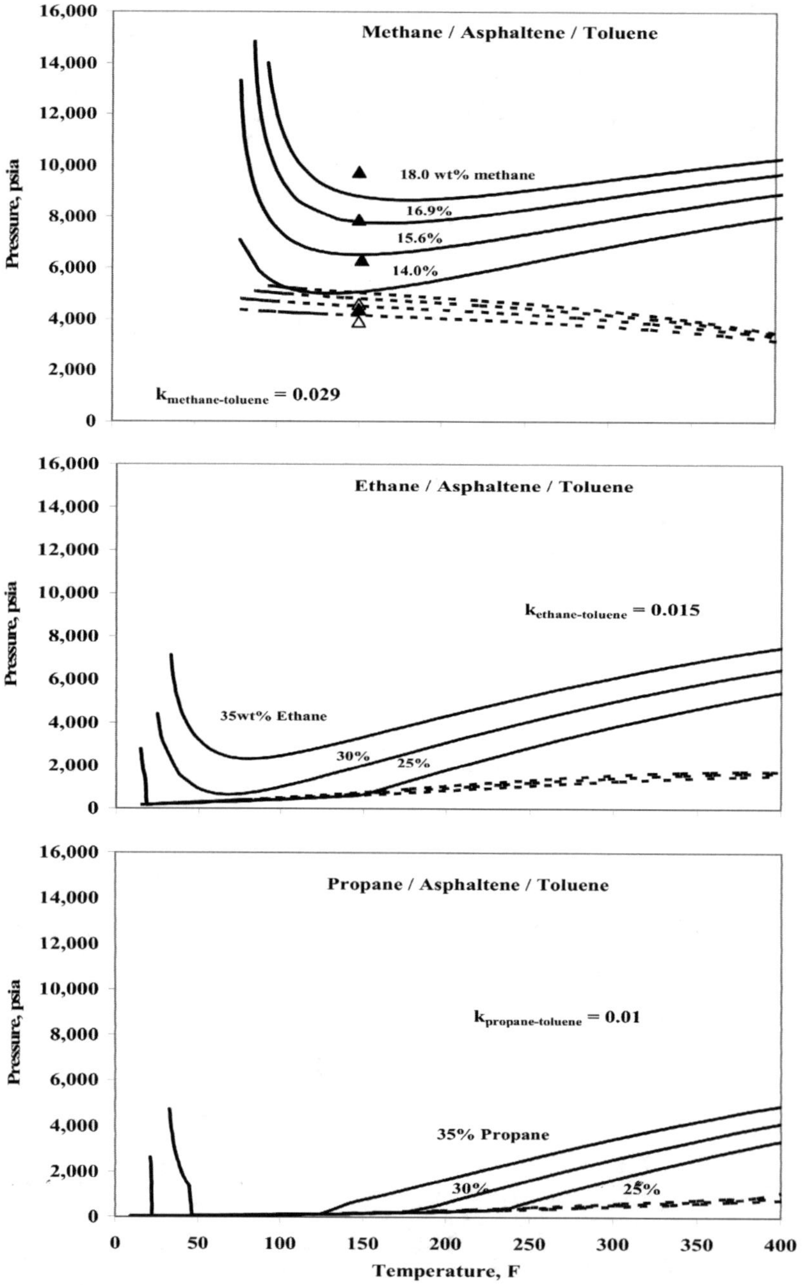

Figure 18. Methane, ethane and propane effect on asphaltene precipitation onset.

Light Hydrocarbon effect on a Reservoir Live Oil

Figure 19 compares the predictive asphaltene phase boundaries for a GoM reservoir fluid when C1, C2 or C3 is added. The asphaltene parameters are tuned to fit the onset with methane.

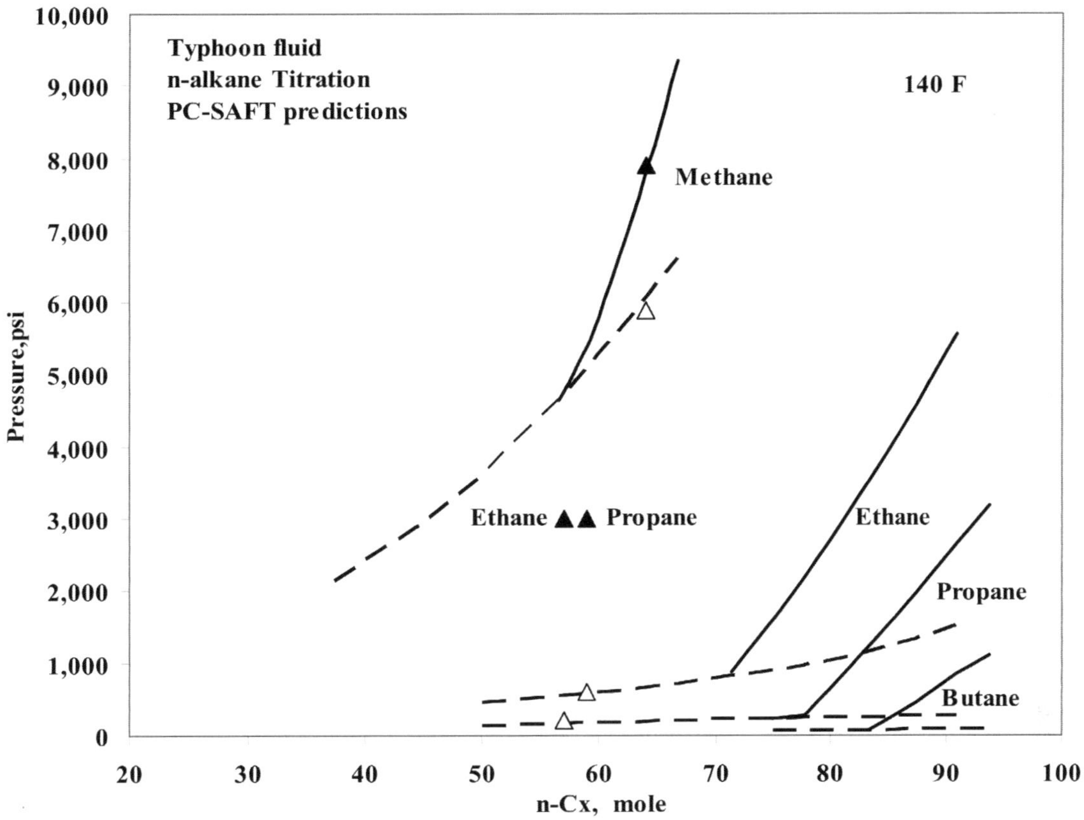

Figure 19. PC-SAFT predictions – C1, C2, C3 and C4 titration. Lines represent the PC-SAFT simulation of asphaltene precipitation onset (solid line) and bubble point (dashed line). Symbols represent onset (full) and bubble point (open) measurements.

In the prediction of the asphaltene onset, the results indicate that the ethane amount required to produce asphaltene precipitation is greater than methane concentration.

9. Asphaltene precipitation due nitrogen (N2) injection

N_2 is a low density gas that can enhance oil production processes such as miscible displacement, gravity drainage of a steeply dipping reservoir and gas cap displacement by substituting N_2 for natural gas in the pressure maintenance of a condensate field. N_2 is an inert, non-corrosive gas, which is optimal for equipment maintenance. However, from the chemistry point of view, N_2 exhibits a very low polarizability which makes it a strong precipitant agent for asphaltenes.

N_2 effect on a Simplify Oil System

Figure 20 shows the strong precipitant effect of nitrogen on the simplified system evaluated in the previous section. The predictions show that asphaltene instability occurs at lower concentrations for N_2 than for methane at the same pressure.

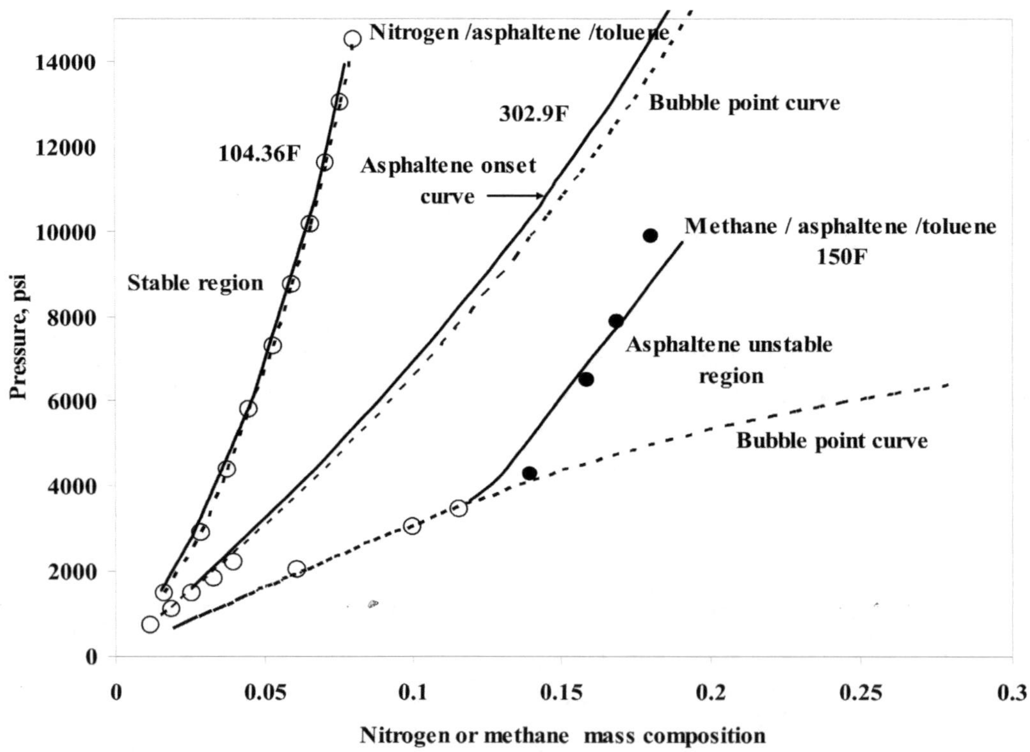

Figure 20. Nitrogen and methane effect on asphaltene precipitation onset

N2 effect on a Reservoir Live Oil

The effect of N2 injection on asphaltene precipitation onset has been simulated using reservoir live fluid 'A' from the GoM.

During isothermal depressurization of the reservoir oil 'A' with N2 injection at reservoir temperature of 296°F the saturation pressure and precipitation onset was determined for chosen injection gas concentrations of 5, 10, and 20 mole% (Figure 21).

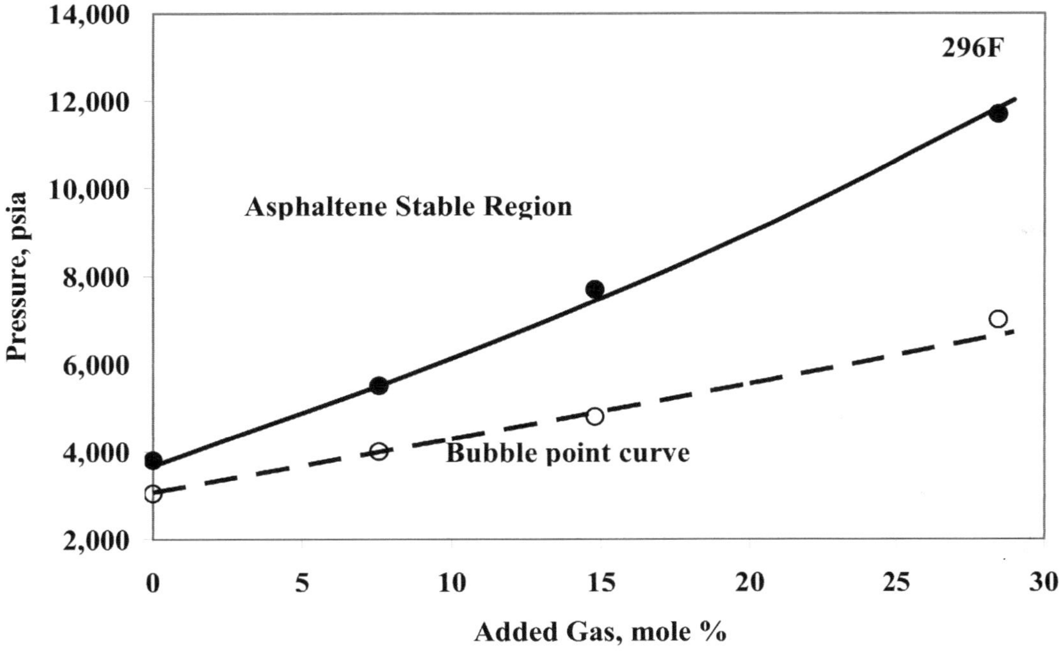

Figure 21. Nitrogen Injection effect on the asphaltene phase behavior. Symbols: experimental Pb and Ponset. Lines: PC-SAFT simulation.

The difference in pressures (Ponset - Pbbp) increases with the amount of N_2 added which is unfavorable for the asphaltene stability. Experimental and simulation results indicated that N_2 gas aggravated the asphaltene instability, and increased the bulk precipitation amount with increasing concentrations of N_2 in the reservoir fluid.

10. Asphaltene precipitation due CO2 injection

Carbon dioxide (CO_2) flooding has been used as a commercial process for enhanced oil recovery (EOR) [Yin et al. 2000]. In enhanced oil production, many light and medium reservoirs are subjected to miscible or near-miscible CO_2 flooding. In this process CO_2 acts as a solvent, lowering the viscosity of the oil and making it mobile, thus, significantly lowering the irreducible oil saturation. CO_2 dissolves in hydrocarbons to create a miscible zone, decreases surface tension of oil to increase mobility, "swells" residual oil so increasing oil recovery and enhances "gas lift" capability to increase well production rates. However, CO_2 flooding processes cause many changes in the flow and phase behavior of the reservoir fluids and can significantly favor the precipitation of asphaltenes [Kokal, 1995; Sarma, 2003]. Asphaltene precipitation has been cited as one of the most significant problems in the production of oil by CO_2 flooding from the Rangely field in Colorado [Wackowski et al, 1994] [Parra-Ramirez, 2002].

CO_2 effect on a Simplify Oil System

In the simplify live oil investigation, asphaltene precipitation onset, and bubble point pressures are predicted by PC-SAFT when CO_2 is added to 1 g of n-C7 precipitated asphaltene dissolved in 100mL of toluene. The n-C7 monodisperse asphaltene parameters are 80, 4.05A, and 350.8K respectively for m, σ and ε/k. The temperature-independent binary interaction parameters (k_{ij}) used in these calculations are presented in Figures 22 and 23. All binary interaction parameters between asphaltenes and other species were set to zero. The simulated results are compared to a methane-toluene-asphaltene system. CO_2 gas causes asphaltene instability at higher concentrations than methane.

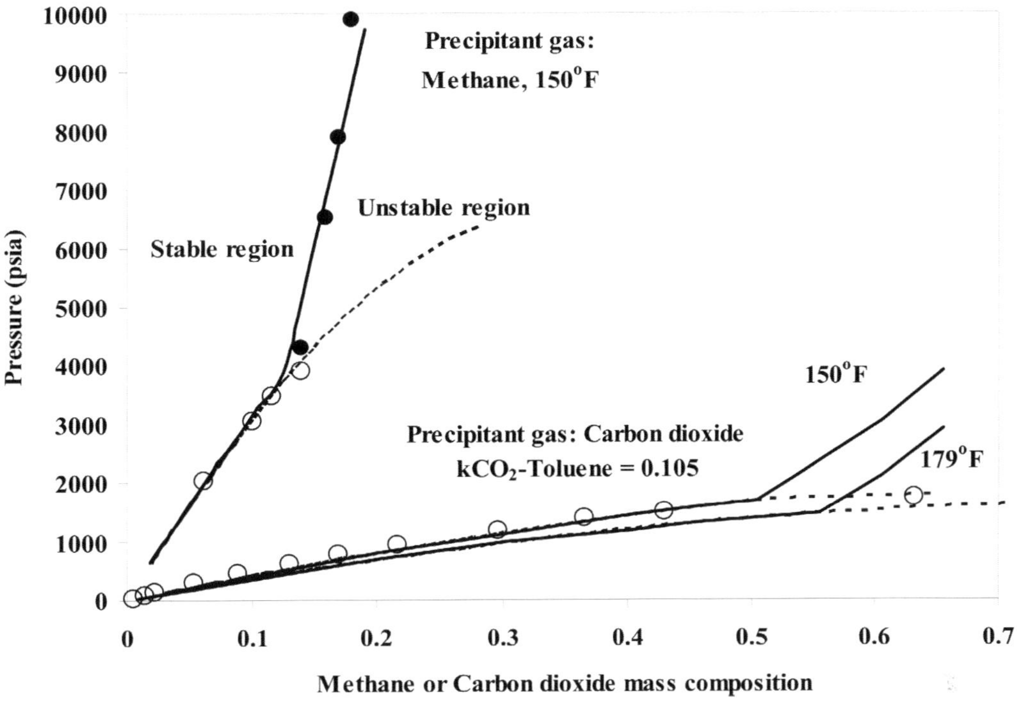

Figure 22. CO_2/toluene/asphaltene vs. methane/toluene/asphaltene. Open circles correspond to experimental bubble point data.

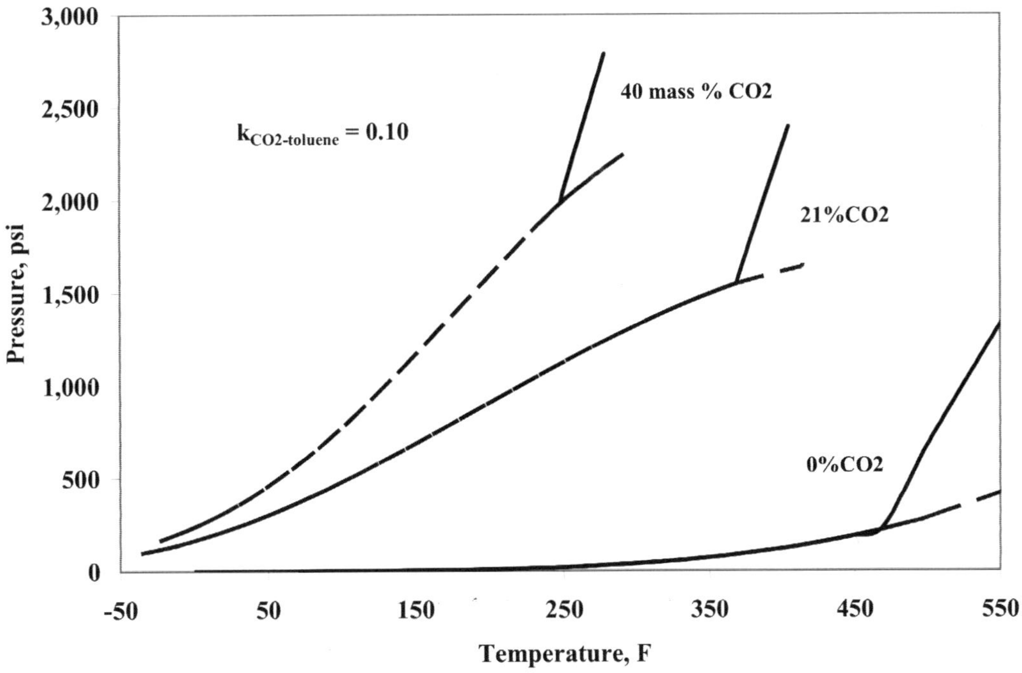

Figure 23. CO_2 - asphaltene phase behavior. System: CO_2/toluene/asphaltene

CO_2 effect on a Reservoir Live Oil

Several experimental measurements of the CO_2 effect on asphaltene precipitation onset have been reported in the last decades. In 1997, Srivastava et al. [1997] conducted static asphaltene precipitation tests for three samples of recombined crude oil (Weyburn reservoir fluid from Saskatchewan Canada) using a high pressure PVT cell at reservoir temperature (138 – 145 ºF). The reservoir fluid was reconstituted by recombining the separator oil and gas samples to a Gas-Oil-Ratio (GOR) of 107, 129 and 180 scf/bbl. The results from the mixtures indicate that the onset of asphaltene destabilization occurs at about 39-46 mol% CO_2 concentration for the three data sets.

In 2003, Takahashi, et al. evaluated asphaltene precipitation during CO_2 injection using a crude oil from a Middle East carbonate reservoir. The onset pressure for this oil was measured in a PVT cell using a light-scattering technique with near infrared by gradually decreasing the pressure of the mixture at each predetermined CO_2 concentration. Their results showed that the amount of precipitated asphaltene became noticeable when the CO_2 mol % exceeded more than 50 mol%.

In 2006, Verdier et al. studied the asphaltene phase behavior after injection of CO_2 into two dead oils (from South America and the Middle East) using a high pressure cell and a filtration technique. They found that, as expected, asphaltenes were more stable as pressure increased; however for CO_2 injection, asphaltenes were more stable as temperature decreased. The phase envelope of the Middle East crude was also investigated for a fixed CO_2 content (18 wt%) and the same conclusions were found. The higher the pressure, the more soluble are the asphaltenes and the higher the temperature, the lower is the solubility.

Conclusions

1. Results of experimental flow assurance assessments revealed that reservoir oils have high propensity for asphaltene precipitation due to addition of gas. The addition of nitrogen, carbon dioxide or methane significantly aggravates the asphaltene precipitation condition.
2. The comparison between the three gases showed that, when added in the same mole proportion, nitrogen was the stronger precipitant followed by methane.
3. N_2 produces a high asphaltene precipitation effect even at low concentrations.
4. The CPA models require to be tuned to more than one asphaltene onset point for proper prediction of the asphaltene boundary.
5. The prediction of gas addition effect on asphaltene phase boundary using CPA EoS requires a new tuning process and creates a new set of parameters for all components.
6. The prediction of gas addition effect on asphaltene phase boundary using the PC-SAFT EoS uses the same set of parameters for all components; only changes the amount of gas.
7. The SAFT EoS adequately predicts the onset of asphaltene precipitation when a light gas such as methane, ethane, CO_2, or N_2 is added to an oil mixture.
8. Simulation using the PC-SAFT EoS shows that the effect of CO_2 and ethane on asphaltene precipitation onset requires higher concentration of gas compared to the addition of methane.
9. The prediction for the model oil shows that a concentration of about 50 mass% of CO_2 is required to reach the asphaltene instability onset.
10. CO_2 addition to a crude oil can increase or decrease the stability of asphaltenes depending upon the characteristics of the fluid and temperature. The addition of CO_2 to a reservoir oil can cause asphaltene precipitation and deposition for temperatures above a certain

crossover point; however, for lower temperatures, the addition of CO_2 increases asphaltene solubility. This behavior is different to what we observe for methane and N_2.

11. The CO_2 behavior can be explained using the concept of the solubility parameter. Below the crossover point, the solubility parameter of CO_2 is greater than the solubility parameter of the oil. Therefore, the addition of CO_2 under those conditions increases the solubility of asphaltene. Above the crossover temperature, the addition of CO_2 reduces the solubility parameter of the mixture.
12. PC-SAFT predicts the presence of the upper and lower critical solution temperature (UCST and LCST) regions in the asphaltene phase envelope.

Acknowledgement

The authors are thankful to Professor Walter Chapman at Rice University for helpful discussions. The authors thank the technical committee of "The 1st International Conference on Upstream Engineering and Flow Assurance" for inviting the presentation of this work.

Acronyms

AOP: Asphaltene onset pressure

CPA: Cubic Plus Association

DBR: Schlumberger DB Robinson company

EoS: Equation of state

GoM: Gulf of Mexico

GOR: Gas-to-oil ratio

HPM: High Pressure Microscope

LCST : Lower critical solution temperature

NIR: Near Infrared

PSA: Particle Size Analysis

PC-SAFT: Perturbed chain form of the statistical associating fluid theory

PVT: Pressure-Volume-Temperature

RERI: Reservoir Research Institute

SAFT: Statistical associating fluid theory

SDS: Solid Detection System

STO: Stock tank oil

SARA: Saturates, aromatics, resins and asphaltene

UCST : Upper critical solution temperature

WAT: Wax Appearance Temperature

Nomenclature

δ : Solubility parameter

k_{ij}: Binary Interaction Parameter

T : Temperature

RI : Refractive index

R : Universal gas constant

References

1. De Boer, R.B.; Leerlooyer, K.; Elgner, M.R.P. and Bergen, A.R.D. *Screening of Crude Oils for Asphaltene Precipitation*. SPE Production and Facilities, 24987 (1995).
2. Gonzalez, D.L., Ting, D., Hirasaki, G.J., Chapman, W.G., *Prediction of Asphaltene Instability under Gas Injection with the PC-SAFT Equation of State*. Energy & Fuels, 2005. **19**: p. 1230-1234.
3. Gonzalez, D.L., Vargas, F.M., Hirasaki, G.J., Chapman, W.G., *Modeling Study of CO2-Induced Asphaltene Precipitation*. Energy & Fuels, 2008. **22**(2): p. 757–762.
4. Jamaluddin, AKM.; Joshi, N., Iwere, F. and Gurpinar, O. *An Investigation of Asphaltene Instability under Nitrogen Injection*. SPE 74393, 2001.
5. Kokal, S. L.; Sayegh, S. G. *Asphaltenes: The Cholesterol of Petroleum*. Presented at the Middle East Oil Show, Bahrain, SPE 29787, 1995.
6. Lim, F., Munoz, E., Browning, B., Joshi, N., Jackson, C., Smuk, S., *Design and Initial Results of EOR and Flow Assurance Laboratory Fluid Testing for K2 Field Development in the Deepwater Gulf of Mexico*. OTC 19624, 2008.
7. Parra-Ramirez Martha J. *Precipitation and Characterization of Asphaltenes*. PhD. The University of Utha, 2002.
8. Sarma, H. K. Can we ignore asphaltene in a Gas Injection Project for Light-Oils? SPE 84877, 2003.
9. Srivastava, R. S. and Huang, S. S. *Asphaltene Deposition during CO2 flooding: a Laboratory Assessment*. SPE 37468, 1997.
10. Takahashi, S; Hayashi, Y; Yazawa, N. and Sarma H. *Characteristics and Impact of Asphaltene Precipitation during CO2 Sandstone and Carbonate Cores: An Investigative Analysis through Laboratory Tests and Compositional Simulation*. SPE 84895, 2003.
11. Ting, D., Hirasaki, G.J., Chapman, W.G., *Modeling of asphaltene phase behavior with the SAFT equation of state*. Petroleum Science and Technology 2003. **21**(3-4): p. 647–661.

12. Verdier, S., H. Carrier, S.I. Andersen and J-L Daridon, *Study of Pressure and Temperature Effects on Asphaltene Stability in Presence of CO2*. Energy & Fuels, 2006, 20, 1584-1590.

13. Wackowski, R.K. and L. Masoner. *Rangely weber sand unit CO2 project update: operation history*. SPE 27755, pp. 203-216, 1994.

14. Wang, J.X., Buckley, J.S., *A Two-Component Solubility Model of the Onset of Asphaltene Flocculation in Crude Oils*. Energy & Fuels, 2001. **2**(5): p. 1004-1012.

15. Yin, Y. R.; Yen, A. T. *Asphaltene Deposition and Chemical Control in CO2 Floods*. Presented at the 2000 SPE/DOE Improved Oil. Recovery Symposium, Tulsa, OK, SPE 59293, 2000.

A Generalized Partial Molar Volume Algorithm Provides Fast Estimates of CO_2 Storage Capacity in Depleted Oil and Gas Reservoirs

Ernesto Valbuena and Maria Barrufet
Petroleum Engineering Department, Texas A&M University

Abstract
Understanding fluid behavior is critical for the success of any gas injection project into a reservoir whether this is to recover additional oil or to sequester and store greenhouse gases including CO_2. Fluid phase behavior of mixtures resulting from the injected streams and reservoir fluids determine the design of production and injection schemes and facilities.

This manuscript presents an analytical method to estimate the ultimate CO_2 storage capacity in depleted oil and gas reservoirs by implementing a volume-constrained thermodynamic equation of state (EOS) and using average reservoir pressure and fluid compositions. This method can handle all impurities contained in the injection stream by defining and applying a generalized partial molar volume calculation.

The developed algorithm provides fast and thermodynamically consistent estimates of storage capacity and enables the selection of candidate storage reservoirs, schedule injection strategies, and design of surface facilities including compressors and tubulars.

Results from this analytical method are in excellent agreement with those from a commercial reservoir simulator. A total of 24 numerical runs were conducted to evaluate scenarios with large pressure and compositional gradients while injecting. The reservoir used was heterogeneous, had a five-spot injection pattern and local grid refinement in the neighborhood of wells. CO_2 storage capacity was predicted with an average difference of 1.26 wt% between analytical and numerical methods; the average oil, gas, and water saturations at the end of injection were also matched within 2.35% difference. Additionally, the analytical algorithm performed several orders of magnitude faster than numerical simulation, with an average of 5 seconds per run.

Introduction
Geological storage in aquifers and depleted oil and gas reservoirs is an attractive option to mitigate carbon dioxide (CO_2) emissions, a greenhouse gas (GHG) that has been continuously increasing in the past 3 decades. More than 70% of GHG emissions are composed of CO_2, which reached 30 Gt in 2009.[1]

Carbon capture and sequestration (CCS) programs consists of various stages, first the CO_2-based stream is subjected to an impurities removal procedure to improve the storage efficiency. Next, the concentrated CO_2-stream is compressed and transported to a storage location[2-4], and finally this stream is injected into a geological formation, usually a saline aquifer or a depleted oil and gas reservoir. Particularly, interest now exists in using depleted reservoirs because extensive knowledge of static and dynamic properties of the reservoir (acquired during the developing phase) is available to optimize the efficiency of the project and increase the final storage capacity and profits. In addition, oil and gas depleted reservoirs have a higher storage density in comparison with aquifers[5].

According to EIA[1], coal-fired electrical power plants produced nearly 45% of CO_2 worldwide emissions in 2009. A large plant of this type, generating 500 MW, emits approximately 2.9 Mt-CO_2 per year. Given that coal is mainly used in power plants to generate electricity, these locations are large sources of CO_2 and become the most important target for CCS processes.

The concentration of CO_2 in a typical coal-fired power plant flue gas is low, nearly 14% (dry volume)[2]. The rest of the flue gas is normally composed by nitrogen (~80%), oxygen, carbon monoxide, and other components in smaller proportions. These impurities should be removed to increase the CO_2 concentration in the gas stream, hence maximize the storage capacity in the target formation. However, this is an expensive process and the cost for CO_2 capture and compression, depending on the level of separation, can range from USD 12 to USD 60 per ton of CO_2.[3]

Numerical simulation is a common practice to forecast CO_2 storage processes. However, numerical simulation is time consuming, expensive, and requires detailed input data to develop representative models. This investigation proposes a fast and accurate analytical method to estimate the ultimate CO_2 storage capacity (UCSC) and fluid saturations in depleted oil and gas reservoirs.

Common Assumptions in Designing a CCS Process

CO_2 injected in a formation will partially mix with oil and gas, a smaller fraction will dissolve in the water (brine), and the rest will remain in a dense CO_2 supercritical phase; in addition, a small fraction may react in mineralization forming carbonates. Mineralization rates in a reservoir volume base are extremely slow and highly uncertain, only after a few hundred years of storage mineralization and dissolution are noticeable[6], therefore these components are neglected in this study. Mineralization may occur during injection in the nearwellbore region reducing injectivity[7]; fluid compatibility laboratory studies can be performed to assess this potential. The solubility of CO_2 in brine is also negligible when compared to the amount that can be dissolved in the hydrocarbons and stored in a supercritical phase for CCS processes in depleted oil and gas reservoirs. Valbuena[8] validated that for a wide range of pressures, temperatures, and salinities, less than 1% of injected CO_2 dissolves in brine. Hence, the analytical method developed in this study neglects the solubility of CO_2 and brine, although it could easily be incorporated into the analysis.

The UCSC is independent from the number of wells, spacing, or well stimulation (e.g. hydraulic fracturing, acidizing, etc.) techniques[9]. Furthermore, Barrufet et al.[5] demonstrated that injection rates have no effect on the final storage capacity. Injectivity, injection rates, and number of wells will only determine the duration of the project; thus, the UCSC in depleted oil and gas reservoirs is controlled by the difference between initial and final average pressures, temperature, and fluid composition and characterization, of both reservoir hydrocarbons and injected stream.

CO_2 capture and compression costs range from USD 12 to USD 60, indicating that impurities removal from the CO_2-based stream is the most expensive process in CCS projects[3]. Often, certain level of impurities is allowed in the injection stream due to increasingly high costs for separation. Nogueira and Mamora[10] experimentally demonstrated that storage capacity and displacement efficiency are reduced in the presence of impurities in the CO_2-based stream; nonetheless, the storage reduction is not just a linear function of CO_2 concentration in the injection stream, though as found by Barrufet et al.[5]

This suggests that an economic analysis between separation costs to a certain level of CO_2 concentration and storage efficiency is a key factor for the success of CCS projects. Therefore, the partial molar volume (PMV) concept in this study has been generalized, to allow the change of the amount of more than one component in the mixing process between hydrocarbons in place and injected impure CO_2-based stream.

Aquifers are common candidates for CCS projects due to their large volume for storage. Conversely, storage efficiency (defined as the ratio between the maximum storage volume and the actual injected volume)[11] in aquifers is low, below 5%, exhibiting low storage density given their initial high pressure[7,12]. Although the storage efficiency definition is ambiguous because pressure, temperature, and CO_2 pressure-volume-temperature (PVT) behavior is not evidently stated in the ratio, it is clear that storage in aquifers may be inefficient. CCS processes in depleted oil and gas reservoirs provide large pressure differential between depletion and target pressures leading to high storage densities. Moreover, often enough detailed reservoir data are available, wells are already drilled, and if enhanced oil recovery (EOR) has been implemented most of the surface facilities will already be in place.

CCS Process in Depleted Oil and Gas Reservoirs

A brief description of the operations for this type of reservoirs will help to understand the process and the data required to forecast injectivity and storage of CO_2 (or any other injected fluid) in a particular formation using reservoir simulation and modeling tools.

During the production life of the reservoir, seismic surveys, core and fluid samples, formation evaluation logs, well tests, and measurement of produced fluids allow reservoir characterization, a field development plan, and the design of facilities. The reservoir is produced until economical limit (low pressure or rates), and the wells are abandoned to conclude operations. CCS projects are developed in this stage.

Estimations and measurements of original fluids in place (oil, gas, and water), reservoir size, temperature, porosity, formation and water compressibilities, average initial and depletion reservoir pressures, and fluid laboratory studies are performed during production phase to characterize the reservoir. Fluid analyzes are used to determine average fluid composition at depletion.

The analytical method presented in this study estimates the UCSC in depleted oil and gas reservoirs using average pressures and fluid compositions and reservoir pore volume (PV). Good estimates of PV can be obtained during production and injection using material balance methods, production data, and average reservoir pressure[13]. This method is independent of relative permeability functions which could only impact the injection times.

Figure 1 shows the trend of selected reservoir properties while the reservoir is produced and later when CO_2 is injected with no production taking place. As fluids are withdrawn during production, average reservoir pressure and mass or moles of hydrocarbon fluid in place decrease. The PV shrinks with decreasing pressure due to rock expansion, and if no water is produced the water saturation increases with decreasing pressure as water expands. PV and water saturation changes depend on rock and water compressibilities, available from the reservoir characterization phase. When water is produced at any stage, material balance and compressibility concepts are used to estimate water saturation.

Once depletion pressure is reached, production ceases and injection begins. The injected fluid causes the average reservoir pressure to increase if no fluids are produced; hence the quantity on fluid in place (hydrocarbon and injected CO_2-stream) increases. PV and water saturation display the opposite behavior observed in the production phase. Target pressure is usually set below initial reservoir pressure to ensure hydraulic integrity and avoid rock fracturing.

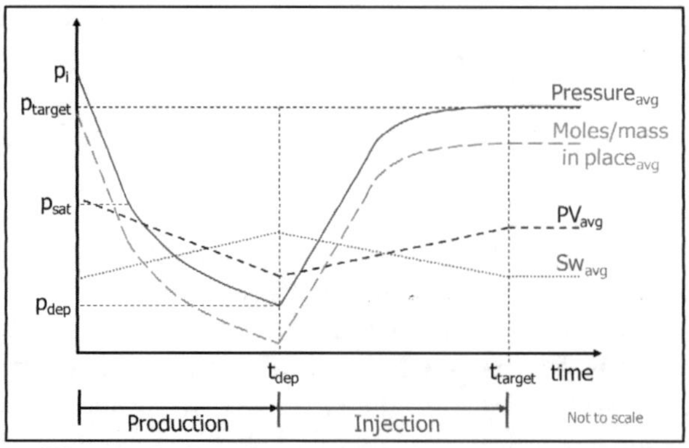

Figure 1 Selected properties of depleted oil and gas reservoirs subjected to CCS projects. Average pressure, moles or mass in place, pore volume, and water saturation interdependency.

Water is usually considered inert, thus there are no physical interactions with the hydrocarbons or the injected fluid. Consequently, the hydrocarbon pore volume (HCPV), i.e. the space occupied by hydrocarbons and injected fluid, can be calculated using compressibility principles[13], independent from fluid composition. The relationship between pressure and moles/mass in place is performed through thermodynamic vapor-liquid equilibrium (VLE) calculations using an EOS.

Analytical Approach to Determine the UCSC

Intuitively, one could approach the CO_2 storage problem neglecting mixing between hydrocarbons in place and injected fluid. This piston-like approach involves compression of the hydrocarbon fluid to the target final pressure and subsequent filling of the remaining volume with injected gas to target pressure, with no mixing between the two fluids. Using this approach may lead to errors higher than one order of magnitude in the UCSC estimations depending on reservoir PV, type of fluids, pressure, and, temperature.[8]

The proposed analytical method in this study allows the estimation of UCSC and average fluid saturations in depleted oil and gas reservoirs used for CO_2 storage purposes considering thermodynamic VLE between hydrocarbons in place and injected CO_2-stream.

The focus of the analytical approach is the volume-pressure-moles relationship from depletion to target states. The injection process begins at depletion pressure p_{dep}, where n_{dep} moles of hydrocarbon fill the volume $HCPV_{dep}$. Subsequently, injection is initiated and pressure increases as the quantity of injected fluid (moles) increases; thus, the $HCPV$ increases due to pressure increase and water volume shrinkage. **Figure 2a** shows that this process is inconveniently complex to solve since all three variables (pressure, volume, and number of moles) are implicitly related. To simplify this analysis the concept of State Functions can be applied.

The State Function principle proposes that fluid properties at the end of the process are independent from the path followed to reach the final state characterized by the final pressure, volume, composition, and, temperature. Hence, the analysis of reservoir fluid volume change (due to pressure and mixing with injected fluids) can be divided in two stages, as shown on **Figure 2b**.

The first stage starts at depletion and ends at a fictitious intermediate state by increasing the pressure to p_{target} while holding constant the initial number of moles (n_{dep}). The pressure increase causes the hydrocarbon fluid to compress to a fictitious volume Vn_{dep}, estimated from EOS calculations. The second stage starts at the fictitious state (Vn_{dep}, n_{dep}, p_{target}) and ends at ($Vn_{dep+\Delta n}$, $n_{dep+\Delta n}$, p_{target}). This path is performed at constant pressure, with Δn moles of fluid added such that the hydrocarbon and injected gas mixture volume, $Vn_{dep+\Delta n}$, equals $HCPV_{target}$ independently determined using the compressibility of rock and water, and final water saturation.

The second stage involves the analysis of reservoir fluid volume change with the injected CO_2-based stream at constant pressure. Solution of Eq. (1) provides the amount of stored CO_2-based fluid Δn (the unknown).

$$V_{n_{dep}+\Delta n} = V_{n_{dep}} + \int_{n_{dep}}^{n_{dep}+\Delta n} \bar{V}_x \, dn_x \quad \text{(1)}$$

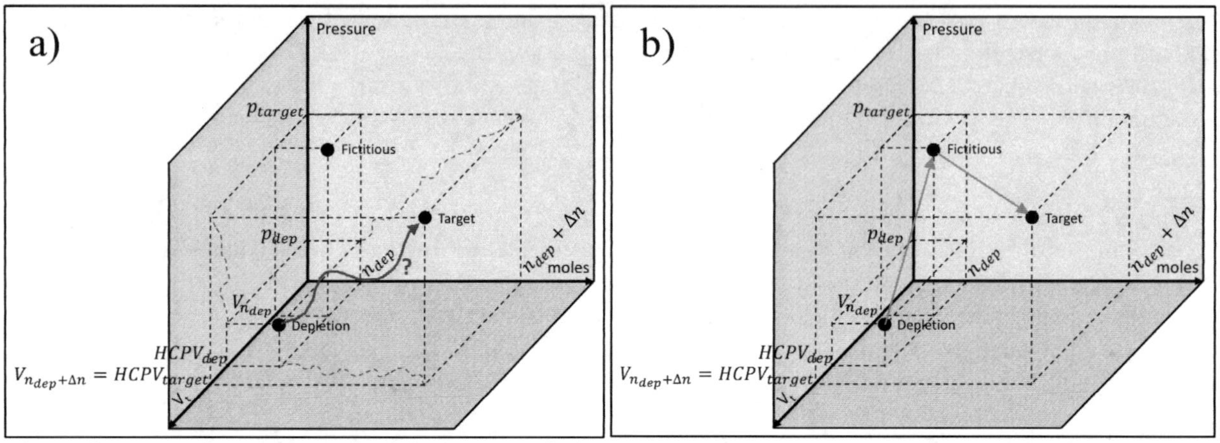

Figure 2 Thermodynamic relationships between fluid volume, pressure, and moles during injection. a) Direct path from depletion to target state is complex to solve with all variables changing at the same time. b) Two-stage approach using State Function concept simplifies analysis.

Appendix A provides the derivation of Eq. 1 using fundamental thermodynamic relations and the Peng-Robinson (PR) EOS[14]. The argument of the integral in Eq. 1 is the generalized PMV, defined as the change of fluid volume (hydrocarbon and injected fluid mixture) with respect to the change of one or more component in the mixture (i.e. injected CO_2-based stream). Conventional PMV represents the change of volume of a fluid mixture with respect to the change of only one of its components[15]; given the multi-compositional nature of the CO_2-based injection stream, a generalization of the PMV was developed for this study. The expression for $\bar{V}_x$ is presented in Appendix B.

Finally, average oil, gas, and water saturations at depletion and target states are estimated using compressibility relations and VLE calculations with the corresponding fluid compositions, pressure, and temperature.

This analytical method was implemented in an algorithm using Visual Basic for Applications (VBA). The required input data are: average initial and depletion reservoir pressures, temperature, initial PV and water saturation, average fluid composition at depletion (end of production), produced water volume, and rock and water compressibility. Controls for the process are target pressure, composition of injected fluid, and selected EOS, e.g. PR[14]. The results obtained include: total reservoir moles and molar volume, average fluid (oil, gas, and water) saturations, and pore volumes at depletion and target states, amount of CO_2 (or CO_2-mixture) stored, and final reservoir fluid composition.

Validation

Results from the analytical method were validated with reservoir numerical simulation. A total of 24 reservoir models were created to represent a wide diversity of scenarios varying reservoir pressures, temperature, initial water saturation, cases with and without water production, presence of single- and two-phase fluids at depletion and target states, and reservoir and injected fluid compositions. **Table 1** presents the cases developed for comparison and validation.

Reservoir static properties are representative of the Cupiagua field in Colombia. The selected section of the stratigraphic column is formed by layered sandstone units, which display sheet-like packages of shoreface bodies[16]. The areal extension and thickness of the reservoir model are 1,740 acres (about twice the area of Central Park, NY) and 270 ft respectively. Rock and water compressibilities are 4E-6 and 3E-6 psi^{-1} respectively. Porosity ranges from 1.54% to 18.22%, with an average value of 13%, and 4.2% standard deviation. Horizontal permeability in the reservoir ranges from 0.21 to 85.4 mD. The Dykstra-

Parsons coefficient for this model is 0.79, indicating high heterogeneity. The anisotropy is given by a vertical to horizontal permeability ratio of 0.1.

Sensitivities to reservoir temperatures were performed between 100 and 285 °F. Initial pressures were calculated for each case to maintain thermodynamic equilibrium, ranging between 5,600 and 9,000 psia. Depletion and target pressures were designed to evaluate cases with small and large pressure differential, ranging from 500 to 6,000 psia between depletion and target states, and the presence of single- and two-phase fluids. Three compositional fluid descriptions were also used to validate the analytical method for different fluid types: black oil (B-O), volatile oil (V-O), and gas condensate (G-C). Fluid models were constructed by calibration of the PR-EOS with experimental PVT data. Initial reservoir water saturation sensitivity was performed by altering the water oil contact depth, generating cases with and without water production. Six injection gas compositions were used with mixtures of CO_2, nitrogen (N_2), and carbon monoxide (CO); CO_2 molar composition ranged between 100% and 60%, N_2 from 0% to 40%, and CO from 0 to 10%.

Table 1—SENSITIVITY CASES. 24 SCENARIOS COVER DIFFERENT RESERVOIR AND INJECTED FLUIDS, TEMPERATURES, SINGLE- AND TWO-PHASE STATES, AND WATER PRODUCTION.

Sensitivity Cases						
Case	Res. Fluid	Inj. Fluid Comp. (CO_2, N_2, CO)	T, °F	Depletion State*	Target State*	Water Prod.
1	V-O	(1, 0, 0)	100	S	S	No
2	V-O	(0.9, 0.1, 0)	100	T	S	No
3	V-O	(0.8, 0.2, 0)	150	T	S	No
4	V-O	(0.6, 0.4, 0)	150	S	S	No
5	V-O	(0.8, 0.15, 0.05)	200	T	S	Yes
6	V-O	(0.75, 0.15, 0.1)	200	T	T	No
7	V-O	(1, 0, 0)	250	S	S	No
8	V-O	(0.9, 0.1, 0)	250	T	S	Yes
9	G-C	(0.8, 0.2, 0)	270	T	T	No
10	G-C	(0.6, 0.4, 0)	270	S	S	No
11	G-C	(0.8, 0.15, 0.05)	270	T	S	Yes
12	G-C	(0.75, 0.15, 0.1)	285	T	T	No
13	G-C	(1, 0, 0)	285	S	S	No
14	G-C	(0.9, 0.1, 0)	285	T	S	Yes
15	G-C	(0.8, 0.2, 0)	285	T	T	No
16	G-C	(0.6, 0.4, 0)	285	S	S	Yes
17	B-O	(1, 0, 0)	250	S	S	No
18	B-O	(0.9, 0.1, 0)	200	T	S	Yes
19	B-O	(0.8, 0.2, 0)	200	T	T	Yes
20	B-O	(0.6, 0.4, 0)	250	S	S	No
21	B-O	(0.8, 0.15, 0.05)	250	T	S	No
22	B-O	(0.75, 0.15, 0.1)	250	T	T	Yes
23	B-O	(1, 0, 0)	285	S	S	No
24	B-O	(0.9, 0.1, 0)	285	T	S	Yes

* (S) Single-phase hydrocarbon fluid. (T) Two-phase hydrocarbon fluid.

The behavior of the generalized PMV $\bar{V}_x$ was also validated through comparison with a commercial reservoir fluid analysis software package. To evaluate $\bar{V}_x$ from commercial software, the numeric first derivative of the total volume with respect to added moles was calculated in a spreadsheet.

Discussion of Results

Due to the diversity and length of these 24 cases, a complete analysis and validation of the analytical method will be explained only for Case 3 from Table 1. Overall analysis of results from the 24 cases is summarized later.

Case 3 is represented by a volatile oil reservoir at a high initial pressure (above saturation point) and 150 °F temperature. The computed phase envelope and properties are shown in **Figure 3**. Initial reservoir pressure was 8,847 psia. The reservoir was produced until an average pressure of 4,004 psia (below the bubble point of 5,700 psia). Next, injection was performed to repressurize the reservoir to 8,790 psia average pressure. Injected fluid composition was 80% CO_2 and 20% N_2. No water was produced during depletion stage.

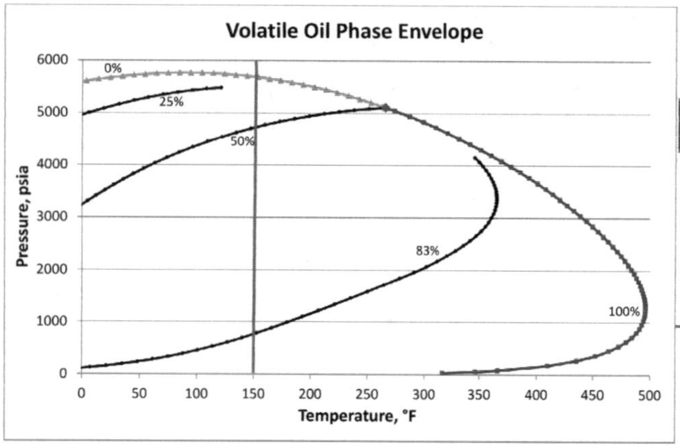

Component or Pseudo-comp.	M_w, lb/lb-mol	T_c, °R	P_c, psia	ω
CO_2	44.010	548.46	1071.33	0.28686
N_2	28.013	227.16	492.31	0.04000
C_1	16.093	342.60	667.05	0.01672
C_2- C_3	34.940	590.00	676.20	0.14952
C_4- C_6	63.213	777.78	520.02	0.26301
C_7- C_{10}	105.660	1027.22	423.52	0.40695
C_{11+}	254.193	1367.80	231.55	0.86200

Figure 3 Volatile oil phase envelope and fluid properties used in Case 3.

Cumulative oil production during depletion was 31 MMSTB, representing 16.4% recovery factor. The reservoir hydrocarbon fluid fell in two-phase during depletion; liquid and gas phases have different mobilities, thus fluid composition was altered from its initial condition. Subsequently, the CO_2-based stream was injected to repressurize the reservoir to 8,790 psia average pressure, a 4,786 pressure differential. Final amount of injected fluid from numerical simulation was 633.56 MMlb-mol, corresponding to 506.85 MMlb-mol of CO_2 and 126.71 MMlb-mol of N_2. **Table 2** displays fluid composition and moles at initial, depletion, and final states. Notice that only the number of moles of CO_2 and N_2 change from depletion to final state, injected fluid contains only these two components.

Table 2—CASE 3 FLUID COMPOSITION AND MOLES AT INITIAL, DEPLETION, AND FINAL STATES FROM NUMERICAL SIMULATION. FLUID IS IN TWO-PHASE AT DEPLETION AND SINGLE-PHASE AT FINAL STATE.

Initial Fluid Composition			Depletion Fluid Composition		Final Fluid Composition		
Comp. or Pseudo-comp.	Z_i, %	n_{init}, MMlb-mol	Z_i, %	n_{dep}, MMlb-mol	Z_i, %	n_{final}, MMlb-mol	Δn, MMlb-mol
CO_2	4.596	91.28	4.582	71.83	26.287	578.68	506.85
N_2	0.000	0.00	0.000	0.00	5.756	126.71	126.71
C_1	61.719	1,225.73	60.764	952.67	43.276	952.67	0.00
C_2- C_3	14.418	286.33	14.515	227.58	10.338	227.58	0.00
C_4- C_6	4.815	95.63	4.919	77.12	3.503	77.12	0.00
C_7- C_{10}	7.507	149.09	7.840	122.92	5.584	122.92	0.00
C_{11+}	6.944	137.90	7.381	115.72	5.257	115.72	0.00

Figure 4 illustrates the CO_2 distribution (molar composition) at the end of the injection process from numerical simulation. The CO_2 plume is mainly located near-wellbore and reaches some extent into the reservoir. The CO_2 distribution is not homogeneous due to the heterogeneity of the reservoir, gravity effects, and fluid characteristics (e.g. diffusion coefficients). The objective of this evaluation is to ensure that the zero-dimensional analytical model accurately predicts average composition and saturations with respect to numerical simulation, even for a case where large compositional gradients are observed.

Since the analytical model is zero-dimensional, that is average pressures are used for computations, an important case to analyze is when noticeable pressure gradients exist in the reservoir. Case 3 exhibits a maximum pressure gradient of 400 psi at the end of production and 130 psi at the end of injection, mainly observed nearwellbore, see **Figure 5**. Pressure gradients depend on gravity, phase distribution, and well production/injection controls.

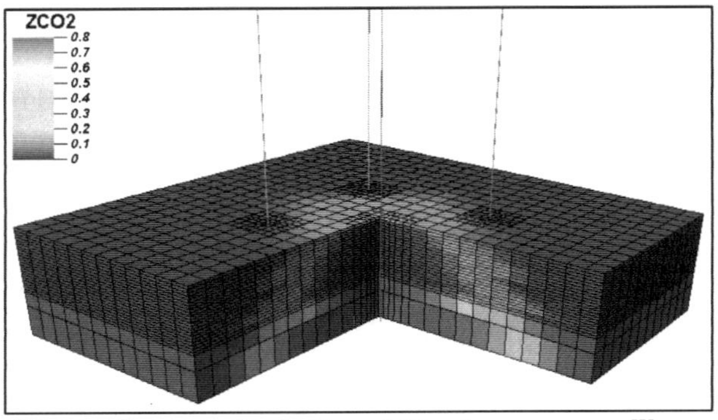

Figure 4 CO_2 molar composition at the end of injection. CO_2 plume nearwellbore, vertical and lateral growth depends on gravity, permeability heterogeneity, and fluid characteristics.

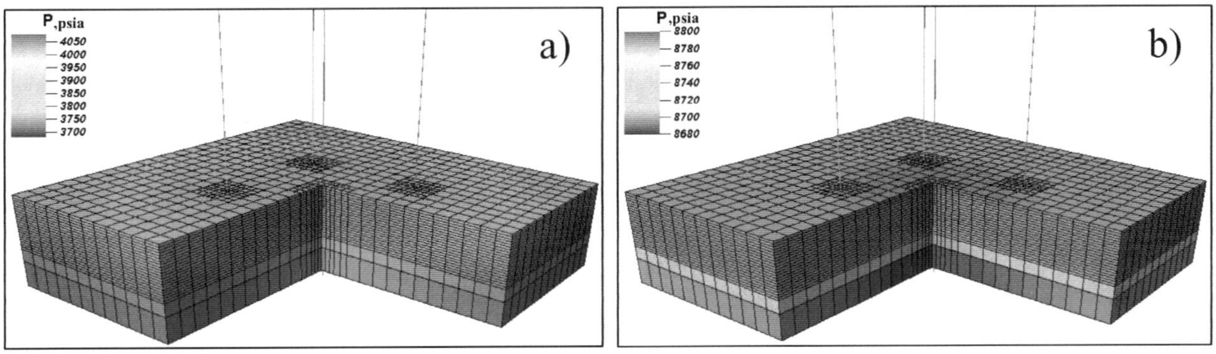

Figure 5 Pressure gradients in Case 3. a) Depletion stage presents a 400 psi maximum pressure gradient. b) Final state (end of injection) presents a 130 psi maximum pressure gradient.

Likewise, the analytical method forecasts average fluid saturations, which in real scenarios are heterogeneously distributed. **Figure 6** shows the fluid distribution at the end of production and during the injection process; as expected, distribution of saturations is heterogeneous as multiple fluid phases are present in the model.

Results from the numerical model were used as base of comparison to validate analytical method results. The analytical model predicted storage of 495.52 MMlb-mol of CO_2 and 123.88 MMlb-mol of N_2 for Case 3 model. This is equivalent to 9.88 Mt of CO_2, given the injection fluid composition. This amount corresponds to nearly 9 years of CO_2 emissions from a mid-size 200 MW coal-fired power plant.

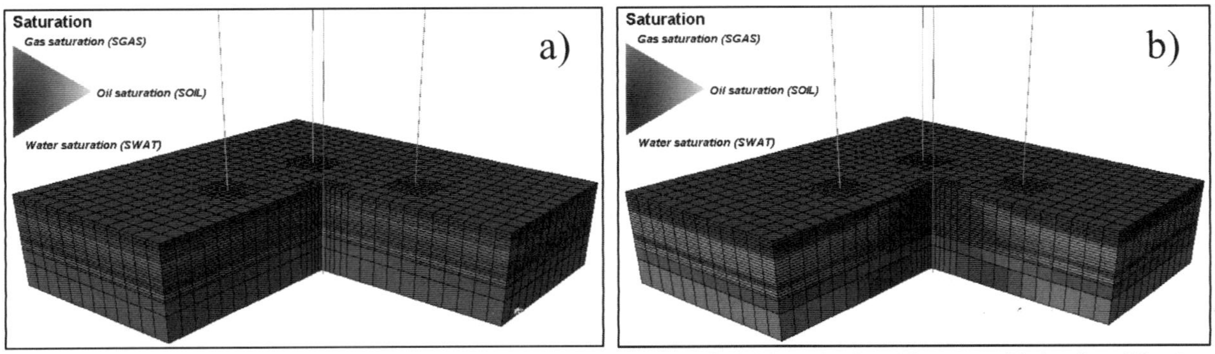

Figure 6 Oil, gas, and water saturations at the end of production (a) and after 10 years of injection (b). a) Hydrocarbon fluid falls in two-phase across the reservoir. Oil and gas distribution is mainly driven by permeability differences in reservoir layers and gravity. b) CO_2 plume in nearwellbore, heterogeneous distribution. Fluid-phase at the end of the process (not shown in this figure) is 100% liquid.

Table 3 presents the comparison of results between the analytical model and numerical simulation. The difference between analytical and numerical simulation results in UCSC is 2.23%. Average oil, gas, and water saturations from the analytical model match with those observed in numerical simulation. In addition, the analytical model performs 233 times faster than numerical simulation.

Table 3—CASE 3-COMPARISON OF RESULTS FROM NUMERICAL SIMULATION AND ANALYTICAL MODEL. ANALYTICAL MODEL RESULTS MATCH THOSE FROM NUMERICAL SIMULATION AND PERFORMS NEARLY 233 TIMES FASTER.

Case 3—Comparison of Results			
Property	Numerical Simulation	Analytical Model	Difference, %
Depletion PV, $ft^3 x10^9$	3.04	3.04	0.00
Final PV, $ft^3 x10^9$	3.10	3.10	0.00
n_{dep}, lb-mol x 10^9	1.5678	1.5719	0.26
v_{dep}, ft^3/lb-mol	1.5460	1.5419	0.26
$S_{o,dep}$, %	49.42	50.74	1.32
$S_{g,dep}$, %	30.22	28.90	1.32
$S_{w,dep}$, %	20.36	20.36	0.00
Δn, lb-mol x 10^9	0.6336	0.6194	2.23
v_{target}, ft^3/lb-mol	1.1318	1.1370	0.46
$S_{o,final}$, %	80.31	80.31	0.00
$S_{g,final}$, %	0.0	0.0	0.00
$S_{w,final}$, %	19.69	19.69	0.00
CPU time, sec	932	4	233x

A supplementary analysis of the behavior of the generalized PMV ($\bar{V}_x$) concept allowed to validate the fundamentally derived expression in Appendix B. **Figure 7** shows the comparison of the PMV calculation between the analytical model and the calculations performed with the output of the PVT commercial package. Results show that $\bar{V}_x$, argument of the integral in Eq. 1, is not constant and may be non-monotonic. Data points calculated from the PVT commercial package present scattering due to limited significant digits in the molar volume output. The complex behavior of $\bar{V}_x$ depends on reservoir fluid compositions, components characteristics, pressure, temperature, and injected fluid composition, and it is unique for each scenario. Integration of $\bar{V}_x$ with respect to Δn provides the solution for Eq. 1.

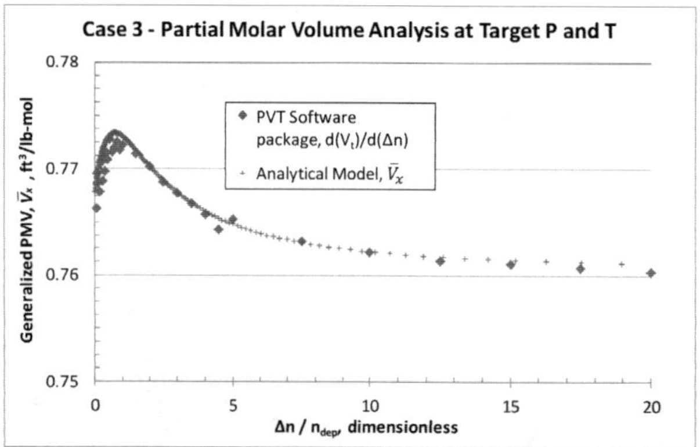

Figure 7 Partial molar volume comparison from analytical model and calculations from commercial PVT software package. Results present excellent agreement. Limited significant digits in commercial software lead to scattered data.

Results from the analytical model matched those from reservoir simulation in the 24 comparison cases with an average of 1.26 wt% as indicated from the near unity slope in **Figure 8a**. Maximum difference in the UCSC between methods was 3.60 wt%. Average oil, gas, and water saturations were also matched with a maximum difference of 2.35% at target final state. In addition, the analytical model performed several orders of magnitude faster than numerical simulation in all cases, with an average CPU time of 5 seconds versus 1600 seconds from numerical simulation as illustrated in **Figure 8b**. CPU time in the numerical method is highly dependent on reservoir size/number of cell in the model, while the analytical method CPU time is independent from those parameters.

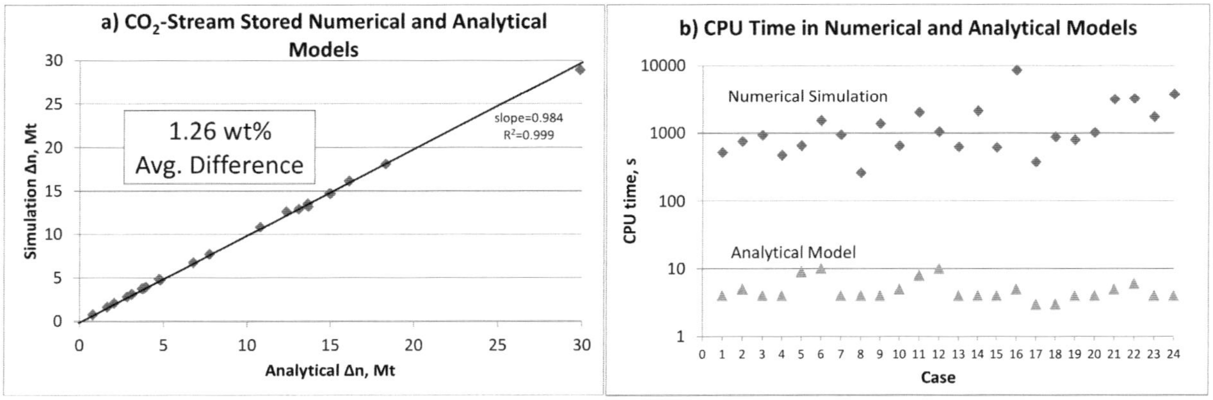

Figure 8 Performance of proposed analytical method against compositional numerical simulation. a) Analytical model results match reservoir simulation within 1.26 wt% average difference. b) Analytical model performs orders of magnitude faster than numerical simulation in all 24 cases.

Cases with larger pressure differential between depletion and target state achieved UCSCs from 12 and 30 Mt-CO_2 (Figure 8a) for this relatively small reservoir (around twice the areal extension of Central Park, NY), storing between 10 and 25 years of CO_2 emissions from a 200 MW coal-fired power plant.

Conclusions

The proposed analytical method allows fast and accurate estimations of ultimate CO_2 storage capacity in depleted oil and gas reservoirs. The average difference between analytical and numerical simulation methods was 1.26 wt%, with a maximum of 3.60 wt% in all studied cases.

Average oil, gas, and water saturations at depletion and target states were accurately estimated. Maximum observed difference between numerical and analytical results was 2.35% in saturations at final state.

Implementation of the analytical method in an algorithm performed with an average CPU time of 5 seconds versus 1600 seconds from numerical simulation.

Generalized PMV allowed to complete the calculations required in cases with impure CO_2-based injection streams. The PMV is not constant during injection and it may be non-monotonic. Complex behavior pf PMV is unique for each case, depending on reservoir and injection fluid compositions, pressure, and temperature.

Nomenclature

CCS = carbon capture and sequestration
CO_2 = carbon dioxide
EOS = equation of state
PMV = partial molar volume
PR = Peng-Robinson
PVT = pressure-volume-temperature
UCSC = ultimate CO_2 storage capacity
VLE = vapor-liquid equilibrium
B-O = black oil

$G\text{-}C$ = gas condensate
$V\text{-}O$ = volatile oil
$HCPV$ = hydrocarbon pore volume from rock-water compressibility, ft^3 or bbl
$(a\alpha)_m$ = parameter of EOS for mixtures (quadratic mixing rule)
b_m = parameter of EOS for mixtures (linear mixing rule)
f_i = mole fraction of component "i" in CO_2-based injection stream
n_{dep} = number of moles at depletion, lb-mol
Δn = cumulative injected moles, lb-mol
p = pressure, psia
PV = pore volume, ft^3 or bbl
R = gas constant, 10.7316 ft^3-psi-°R^{-1}/lb-mol
S_g = gas saturation
S_o = oil saturation
S_w = water saturation
T = temperature, °F ([°F] = 32 + [°C] x 9/5)
υ = molar volume, ft^3/lb-mol
$\bar{V}_x$ = generalized partial molar volume of pseudo component x; where x could be a mixture with constant composition, e.g. CO_2-based injection stream (see Appendix B), ft^3/lb-mol
$V_{n_{dep}}$ = fictitious volume occupied by n_{dep} moles of fluid at target p and T, ft^3
$V_{n_{dep}+\Delta n}$ = final volume of fluid at p and T from EOS, ft^3

Conversions of Units

acres = area (1 acre = 43,560 ft^2 = 4,046.86 m^2)
ft = feet, length (1 ft = 0.3048 m)
°F = degrees Fahrenheit, temperature (°F = °C × 9/5 + 32 = K × 9/5 – 459.67)
°R = degrees Rankine, temperature (°R = °F + 459.67 = K × 9/5)
lb = pound, mass (1 lb = 0.453592 kg)
MW = Megawatt, energy rate (1 MW = 10^6 W = 10^6 J/s = 10^6 kg-m^2/s^3)
mD = milliDarcy, permeability (1000 mD = 1 D = 9.869233×10^{-13} m^2)
MMSTB = million stock tank barrels, volume (1 MMSTB = 10^6 STB = 158,987.29 m^3)
MMlb-mol = million pound mole, quantity of substance (1 MMlb-mol = 10^6 lb-mol)
psi = pound per square inch, pressure (1 psi = 6894.7573 Pa)
t = tonne, mass (1 t = 10^3 kg)
USD = United States dollar, currency
Gt = Gigatonne, mass (1 Gt = 10^9 ton = 10^{12} kg)
Mt = Megatonne, mass (1 Mt = 10^6 ton = 10^9 kg)
wt% = weight percent, percentage

References

1. EIA US. Total Carbon Dioxide Emissions from the Consumption of Energy. 2011; http://tonto.eia.doe.gov/cfapps/ipdbproject/IEDIndex3.cfm?tid=90&pid=44&aid=8. Accessed January 18, 2011.
2. Metz B, Davidson O, Coninck Hd, Loos M, Meyer L. *IPCC Special Report on Carbon Dioxide Capture and Storage* 2005.
3. Dooley J, Dahowski R, Davidson C, et al. *Carbon Dioxide Capture and Geologic Storage. A Core Element of a Global Energy Technology Strategy To Address Climate Change.* Columbus, Ohio, United States 2006.

4. EPA US. Geologic Sequestration. 2011; http://www.epa.gov/climatechange/emissions/co2_geosequest.html. Accessed March 14, 2011.

5. Barrufet MA, Bacquet A, Falcone G. Analysis of the Storage Capacity for CO_2 Sequestration of a Depleted Gas Condensate Reservoir and a Saline Aquifer. *Journal of Canadian Petroleum Technology.* 2010;49(8).

6. Sifuentes WF, Giddins MA, Blunt MJ. Modeling CO_2 Storage in Aquifers : Assessing the key Contributors to Uncertainty. *Offshore Europe*. Aberdeen, UK: SPE; 2009.

7. Gallo YL, Couillens P, Manai T. CO_2 Sequestration in Depleted Oil or Gas Reservoirs. *SPE International Conference on Health, Safety and Environment in Oil and Gas Exploration and Production*. Kuala Lumpur, Malaysia: SPE; 2002.

8. Valbuena E. *Analytical Estimation of CO_2 Storage Capacity in Depleted Oil and Gas Reservoirs Based on Thermodynamic State Functions*. College Station, TX: Petroleum Engineering, Texas A&M University; 2011.

9. van der Meer LGH, Egberts PJP. A General Method for Subsurface CO_2 Storage Capacity Calculations. *Offshore Technology Conference*. Houston, Texas, U.S.A., 5–8 May 2008.: SPE; 2008.

10. Nogueira M, Mamora DD. Effect of Flue Gas Impurities on the Process of Injection and Storage of CO_2 in Depleted Gas Reservoirs. *SPE/EPA/DOE Exploration and Production Environmental Conference*: SPE; 2005.

11. van der Meer LGH. The CO_2 Storage Efficiency of Aquifers. *Energy Conversion and Management.* 1995;36(6-9):513-518.

12. Okwen RT, Stewart MT, Cunningham JA. Analytical Solution for Estimating Storage Efficiency of Geologic Sequestration of CO_2. *International Journal Of Greenhouse Gas Control.* 2010;4(1):102-107.

13. Dake LP. *Fundamentals of Reservoir Engineering*. First Edition ed. Amsterdam, The Netherlands: Elsevier; 1978.

14. Peng D-Y, Robinson DB. A New Two-Constant Equation of State. *Industrial & Engineering Chemistry Fundamentals.* 1976;15(1):59-64.

15. Michelsen M, Mollerup J. *Thermodynamic Models: Fundamentals & Computational Aspects*. First Edition ed. Denmark: Tie-Line Publications; 2004.

16. Ramon JC, Fajardo A. Sedimentology, Sequence Stratigraphy, and Reservoir Architecture of the Eocene Mirador Formation, Cupiagua Field, Llanos Foothills, Colombia. *AAPG Memoir.* 2006 2006(88/SEPM Special Publication):p. 433-469.

Appendix A

This appendix presents the derivation of Eq. 1 from fundamental thermodynamic principles, which allows the solution for the proposed analytical method. The derivations presented following are intended to perform the analysis of the second path in Figure 2b, from fictitious to target states.

The total volume V occupied by the fluid mixture at target pressure, at the end of the injection, is a function of pressure, temperature, and the total number of moles of fluid; that is:

$$V = f(p, T, n) \quad \text{(A.1)}$$

Total number of moles of fluid is given by the summation of the moles of hydrocarbons in place before the start of the injection (at depletion stage) and the injected fluid.

$$n = n_{dep} + \Delta n \quad \text{(A.2)}$$

The change of the total volume, as the injection fluid mixes with the hydrocarbon evaluated at final target pressure and reservoir temperature, can be expressed as:

$$dV = \left(\frac{\partial V}{\partial p}\right)_{T,n} dp + \left(\frac{\partial V}{\partial T}\right)_{p,n} dT + \sum_{n_x} \left(\frac{\partial V}{\partial n_x}\right)_{p,T,n_j} dn_x \quad \text{(A.3)}$$

Notice that reservoir temperature is considered constant, which allows to drop the second term on the right-hand side. Additionally, taking advantage of the State Function, the path between depletion and target states was divided in two steps. In the first step, the pressure is increased at constant number of moles, which only causes fluid compression. The second step is developed holding constant target pressure (allowing to drop the first term) and increasing the number of moles. For the second step, integrating Eq. A.3 with respect to the number of moles injected at target pressure gives:

$$\int_{n_{dep}}^{n_{dep}+\Delta n} dV = \int_{n_{dep}}^{n_{dep}+\Delta n} \left(\frac{\partial V}{\partial n_x}\right)_{p,T,n_j} dn_x \quad \text{(A.4)}$$

Where, n_{dep} is the number of moles of the reservoir fluid before injection and Δn is the cumulative number of injected moles at the end of the process. The solution of the left-hand side integral of Eq. A.4, and the application of $\bar{V}_x$, the definition for the generalized partial molar volume (PMV) developed in Appendix B, provides the final form:

$$V_{n_{dep}+\Delta n} = V_{n_{dep}} + \int_{n_{dep}}^{n_{dep}+\Delta n} \bar{V}_x \, dn_x \quad \text{(A.5)}$$

Appendix B

This appendix presents the generalized partial molar volume (PMV) equation derived in this investigation. Since the CO_2-based injection stream is multicompositional, this study requires a generalization of the PMV concept considering more than one component changing in the mixture.

Using the PR-EOS[14] as example, one can derive pressure with respect to the change of moles and with respect to volume. These derivations are somewhat tedious, however their manipulation yields the generalized PMV expression for $\bar{V}_x$:[8]

$$\bar{V}_x = \left(\frac{\partial V}{\partial n_x}\right)_{p,T,n_j} = -\frac{\left(\frac{\partial p}{\partial n_x}\right)_{V,T,n_j}}{\left(\frac{\partial p}{\partial V}\right)_{T,n}} = -\frac{R_1 + A_1}{D_1} \quad \text{(B.1)}$$

where R_1 and A_1 are the repulsion and attraction terms derivatives with respect to the change in moles of injected mixture n_x (CO_2-based stream with impurities), and D_1 is the partial derivative of p with respect to V, expressed as follows:

$$R_1 = \frac{RT}{v - b_m} + \frac{RT \sum_{i=1}^{N_c} f_i b_i}{(v - b_m)^2} \quad \text{(B.2)}$$

$$A_1 = -\frac{2\Psi}{(D_{v})} + \frac{2(a\alpha)_m v \sum_{i=1}^{N_c} f_i b_i}{(D_v)^2} - \frac{2(a\alpha)_m b \sum_{i=1}^{N_c} f_i b_i}{(D_v)^2} \quad \text{(B.3)}$$

$$D_1 = -\frac{RT}{(v - b_m)^2} + \frac{2(a\alpha)_m (v + b_m)}{(D_v)^2} \quad \text{(B.4)}$$

with:

$$D_v = v^2 + 2b_m v - {b_m}^2 \quad \text{(B.5)}$$

and:

$$\Psi = \sum_{i=1}^{N_c} \sum_{j \neq i=1}^{N_c} [f_i x_j (a\alpha_{ij})] + \sum_{i=1}^{N_c} [f_i x_i a_i \alpha_i] \quad \text{(B.6)}$$

The variables in Eqs. B.2 to B.5 come from PR-EOS[14]. v is the molar volume of the fluid, R is the gas constant, parameters $a\alpha$ and b, evaluated with Eqs. B.7 to B.10, depend on the critical properties of the fluid (subscript "m" represents the fluid mixture, "i" represents the i-component, and "j" the j-component), f_i is the fraction of component "i" in the injection stream, and x is the liquid fraction if the mixture is in liquid phase and can be substituted if the mixture is in single-vapor phase or even two-phase.

$$(a\alpha)_m = \sum_{i=1}^{N_c} \sum_{j=1}^{N_c} x_i x_j \sqrt{a_i \alpha_i a_j \alpha_j} (1 - k_{ij}) \quad \text{(B.7)}$$

$$b_m = \sum_{i=1}^{N_c} x_i b_i \quad \text{(B.8)}$$

with:

$$\alpha = \left[1 + m\left(1 - \sqrt{T_r}\right)\right]^2 = \left[1 + m\left(1 - \sqrt{\frac{T}{T_c}}\right)\right]^2 \quad \text{(B.9)}$$

and:

$$\begin{aligned} m &= 0.37464 + 1.54226\,\omega - 0.2699\,\omega^2 \quad \text{for} \quad \omega < 0.49 \\ m &= 0.379642 + 1.48503\,\omega - 0.164423\,\omega^2 + 0.01667\,\omega^3 \quad \text{for} \quad \omega > 0.49 \end{aligned} \quad \text{(B.10)}$$

Where k_{ij} represent the binary interaction coefficients between components "i" and "j", T_r and T_c are reduced and critical temperatures respectively, and ω is the Pitzer acentric factor.

Integrated Equation of State Modelling for Flow Assurance

Nuno Pedrosa, Richard Szczepanski and Xiaohong Zhang
Infochem, Unit 4 The Flag Store, 23 Queen Elizabeth Street, London
SE1 2LP, UK

Introduction

Flow assurance involves ensuring the transportation of hydrocarbon fluids from the reservoir to the processing facilities. It encompasses multiphase flow, thermal and hydraulic analysis, production chemistry, corrosion and operability. Most subsea developments will include production from multiple wells and maybe several field regions via a common tie-back to a production platform; in these circumstances multiphase flow and the risk attendant on operating at low temperatures inevitable give rise to flow assurance challenges.

An understanding of phase behavior and physical properties is fundamental to achieving successful flow assurance strategies. It underpins all other modeling and analysis. It is necessary to handle multiple fluid phases, typically gas, oil and aqueous phases, plus many possible solid phases that can hinder production. The solids of principal interest are hydrates, waxes, asphaltenes and inorganic scales, although other solid phases such as diamondoids and naphthenates may be equally problematic in some developments.

In this work we describe an integrated approach to modeling phase behavior and thermodynamic properties for flow assurance applications. We focus on a standard engineering cubic equation of state and show how it can be extended in a systematic way to handle the complete range of fluid phases encountered in flow assurance studies. We demonstrate a number of applications ranging from relatively simple oil and gas systems, through situations including hydrates, wax and asphaltenes. The integrated approach allows us to handle situations where multiple solid phases may exist and to examine the interactions between the phases.

There exists an important distinction between solid phase *precipitation* (or formation) and *deposition* of the solid in a flowline or in the reservoir. Modeling deposition involves a close interaction between the thermodynamic modeling and the thermal/hydraulic modeling. We describe some of the important issues raised when incorporating our thermodynamic wax model in a wax deposition simulator.

Thermodynamic modeling strategy

An enormous number of thermodynamic models for oil and gas applications has been developed and published over the past 50 years or so. An up-to-date survey is provided by Kontogeorgis and Folas [1]. However, few models have had a lasting impact that is comparable to that of cubic eos.

Cubic eos were established as practical engineering models by the work of Soave [2] and Peng and Robinson [3]. They have proved to be very successful tools for the oil industry and derivatives of these eos still form the basis of almost all modeling work. The recent development of the SAFT family of eos [1] has made progress in some specialized areas but for most applications cubic eos remain dominant because of their accuracy, versatility and computational speed.

In their original for these cubic eos suffer from several well-known limitations:

- the prediction of liquid phase density is usually poor;
- generalized correlations for the eos parameters do not represent well the vapor pressure of polar or associating compounds
- Phase behavior of mixtures including polar and associating compounds is not well-predicted.
- The eos are not applicable to electrolytes.

It is important to tackle each of these shortcomings in order to provide a reliable basis for flow assurance calculations.

Our thermodynamic modeling strategy is based on the Soave modification of the Redlich-Kwong (RKS) eos. The Peng-Robinson eos could equally well have been used but our choice is influenced by the evidence that the RKS eos gives better predictions of the fugacity of light gases such as methane.

The RKS eos

For the RKS eos the residual Helmholtz energy is given by the equation

$$A^r = -NRT\ln\left(1-\frac{b}{v}\right) - N\frac{a}{b}\ln\left(1+\frac{b}{v}\right) \tag{1}$$

The parameters a and b depend on the composition of the mixture and the properties of the pure components in the mixture:

$$\begin{aligned} a &= \sum\sum n_i n_j \sqrt{a_i a_j}(1-k_{ij}) \\ a_i &= a_{ci}(1+\kappa_i t_i)^2 \\ \kappa_i &= c_0 + c_1\omega_i + c_2\omega_i^2 \\ t_i &= 1-\sqrt{T_{ri}} \\ T_{ri} &= T/T_{ci} \\ a_{ci} &= \Omega_a R^2 T_{ci}^2 / p_{ci} \end{aligned} \qquad \begin{aligned} b &= \sum n_i b_i \\ b_i &= \Omega_b R T_{ci} / p_{ci} \end{aligned} \tag{2}$$

Where n_i is the number of moles of component i, T_{ci} is the critical temperature, p_{ci} is the critical pressure, R is the gas constant and k_{ij} is the binary interaction parameter for components i and j. Ω_a and Ω_b are constants that are fixed by the criticality conditions. The coefficients c_0, c_1 and c_2 were fixed by Soave [2] based on vapor pressure data and the equation for a_i can be regarded as a generalized vapor pressure correlation.

The generalized vapor pressure correlation is reasonable for $T_r > 0.7$ but fails badly at lower temperatures, especially for water. The errors in liquid densities are unacceptably large in all cases. Note that a 2-parameter cubic eos cannot reproduce the critical temperature, pressure and volume simultaneously so there will always be a large density deviation at the critical point.

Improved vapor pressure correlation

We replace the generalized correlation for the a parameter with an extended Mathias-Copeman [5] equation

$$a_i = a_{ci}(1+\kappa_{i1}t_i + \kappa_{i2}t_i^2 + \kappa_{i3}t_i^3 + \kappa_{i4}t_i^4 + \kappa_{i5}t_i^5)^2 \tag{3}$$

For each component the coefficients κ_{ij} are fitted to vapor pressure data from the triple point to the critical point. This ensures that an accurate representation of the vapor pressure is obtained for irrespective of the nature of the component.

Peneloux volume correction

The well-known volume translation method proposed by Peneloux [6] is used to improve the liquid volume predictions:

$$V = V^{RKS} - \sum n_i c_i \tag{4}$$

where the, possibly, temperature-dependent volume shift parameters c_i are fitted to the experimental saturated liquid density at one or more temperatures. The volume translation has no effect on the vapor pressure because the fugacity of each phase is changed by the same amount.

We denote the eos with these simple modifications as RKSA (RKS Advanced).

The vapor pressure of all substances is better reproduced. The deviations in liquid density are within 5% except in the critical region where they are still very large for the reason mentioned above.

Mixtures

The classic van der Waals 1-fluid mixing rule for the a parameter is both simple and effective for hydrocarbon mixtures. The binary interaction parameter k_{ij} is generally adjusted to improve the prediction of binary VLE data and makes the equation predictive for multicomponent mixtures. Densities in the single-phase supercritical region are also reasonably well predicted.

Petroleum fluid characterization

A procedure for characterizing petroleum fluids is an essential ingredient of flow assurance modeling By this we mean a procedure that allows us to use a PVT lab report to construct a compositional model of a fluid. Our recommended procedure is based on using the most reliable measured data reported by the lab. It is summarized briefly below:

- The primary information is taken from the GC analysis of the fluid reported in weight % and additional directly measured data, such as the molecular weight and density of the stock tank oil , is also taken into account.
- The information is used to construct an extended compositional distribution that accounts for the material in the heaviest fraction.
- The distribution is used to assign properties to single carbon number fractions extending to the highest significant fraction. The Raizi and Al-Sahhaf correlations[7][8] are applied for the molecular weight and specific gravity,followed by the Twu correlations for the critical properties[9].Lumped pseudocomponents are then obtained by grouping the single carbon number fractions to produce the desired number of pseudocomponents.
- Eos properties such as critical temperature, critical pressure etc. are obtained using the well-established Lee-Kesler correlations [10].

Tuning the model

The objective of characterizing a fluid is to obtain the best possible compositional model based on the analytic and other data as described above. In practice it is usually necessary to tune the model by adjusting the initially-assigned petroleum fraction properties. Although the initial model is close to the reported bubble points it is possible to improve the match by adjusting the volatility of the petroleum fractions.

For condensates it is desirable to reproduce the liquid dropout measurements below the retrograde dew point. Figure 1 shows an example of a liquid dropout curve with a tuned eos model. The data are taken from Pedersen [11].

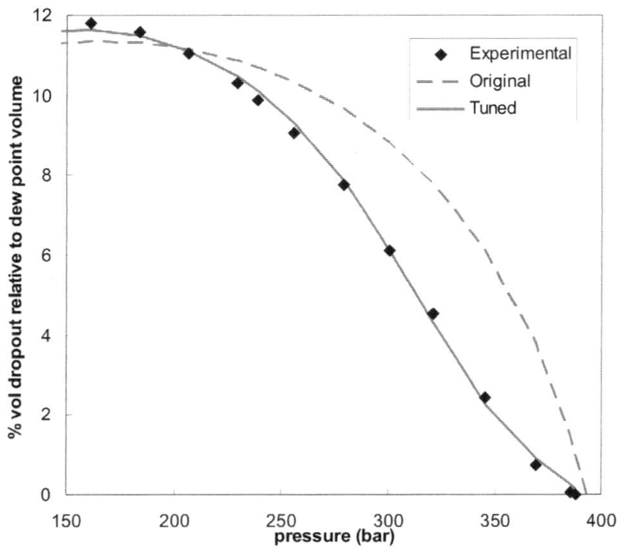

Figure 1 Liquid dropout curve for a gas condensate. Data from Pedersen and Christensen [11]

More complex systems

As already stated, flow assurance applications demand modeling of not only gas and oil phase but must also encompass an aqueous phase that will usually contain electrolytes and several possible solid phases. Our modeling approach is to enhance the basic eos model by adding appropriate terms to the Helmholtz energy. Hence our general fluid phase model has the following structure

$$A^r = A^r_{SRK} + A_{assoc} + A_{electrolyte} \tag{5}$$

where A^r is the total fluid phase residual Helmholtz energy, A_{assoc} is the contribution from chemical association and $A_{electrolyte}$ is the term that accounts for contributions in electrolyte solutions. The association term makes it possible to model highly non-ideal aqueous phases containing hydrate inhibitors such as alcohols and glycols. It is also the basis of our approach to modeling asphaltenes [12]. The electrolyte term allows us to handle mixed-solvent electrolyte solutions. More details of the extended model are included in subsequent sections of this paper.

Aqueous systems and hydrates

The suppression of gas hydrates in oil and gas production is of major importance for cost and safety reasons, especially with the continuing trend towards deep-water developments. The conventional engineering approach is to add inhibitors such as methanol or glycol to alter the thermodynamic equilibrium so that hydrates cannot form under production conditions. Although there has been much work recently on low-dosage kinetic inhibitors, there are many situations where it remains essential to use thermodynamic inhibitors [13].

The CPA (Cubic Plus Association) model is based on the RKS equation combined with a simplified version of the association theory developed by Wertheim [14, 15] that explicitly represents the effect of hydrogen bonding in fluid phases. The model has be]en described in previous works [16-18] and will not be presented here. The CPA method is good in predicting the mutual solubility of hydrocarbons in water. The consequence of this is that the partition of hydrate inhibitor for example methanol is better described when using the CPA model.

The modeling considerations up to now concerned the fluid phases: gas, liquid and aqueous. When predicting hydrate behavior the following solid phases have to be taken into account as well:

- Hydrates;
- Ice;
- Inorganic salt scales.

The most successful model used in industry to model hydrates is the Van der Waals – Platteeuw [19] model with Kihara spherical core potential. It was later one simplified by Parrish and Prausnitz [20].

This model can predict accurately the hydrate dissociation temperatures for fluids with a large number of components but there are still unsolved issues that researchers are tackling. One of them is multiple occupancy of the sites. The hydrate cavities can hold several small guest molecules such as hydrogen, helium and to a less extent nitrogen and argon. In a recent work by Martin [21], a simplified version of the Van der Waals – Platteeuw method has been used to describe this behavior for hydrogen and helium successfully.

Other limitations / assumptions of the Van der Waals – Platteeuw model are:

- The presence of the guest molecule does not distort the cavity
- The interactions between encaged molecules can be ignored
- Spherical symmetry is assumed

Despite these assumptions, the model has proven to be quite accurate giving predictions errors of less than 1 K for the hydrate dissociation pressure.

An example of the capabilities of the CPA model with the Van der Waals – Platteeuw hydrate model is illustrated in Figure 2. The experimental data is from a synthetic natural gas prepared by Ng and Robinson [22].

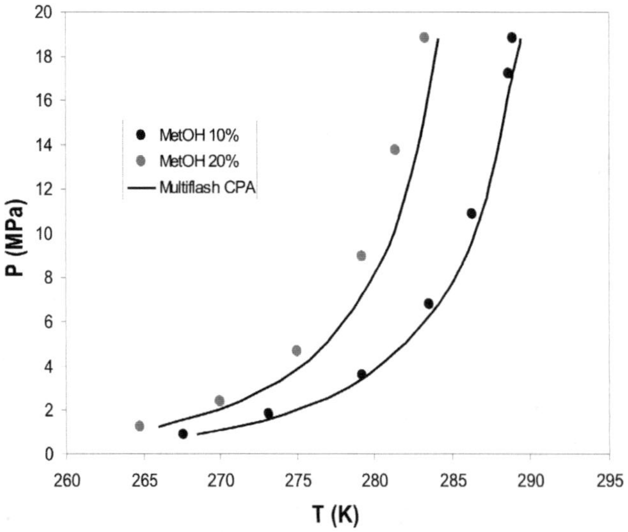

Figure 2 Prediction of methanol hydrate inhibition with CPA model. Experimental data from [22]

At very low temperatures, water can form ice. The ice phase can be also treated thermodynamically and included in the list of possible phases that the system may present.

The ice phase is modeled using the freeze-out model where the fugacity of the solid phase is calculated using properties from the solid transition such enthalpy of fusion and change of volume and heat capacity at the fusion point.

The last set of components, which can also form a solid phase, are the salts that may be present in the water. The presence of these components may occur due to several reasons:

- already present in the formation water;
- added with the mud used to stabilize the well;
- when sea water is used in the process.

Another consequence of having salts in solution is that the water activity is reduced, hence, the salts have a hydrate inhibition inhibition effect. The model used to account for the presence of salts is based on the Debye-Hückel theory [23] which determines the effect of the electrolytes on the water activity. With this method it is possible to accurately predict the solubility of gases in salted water.

An example of the inhibition effect is illustrated in Figure 3 a synthetic gas mixture. The experimental data and gas composition can be found in the work by Tohidi at al. [24].

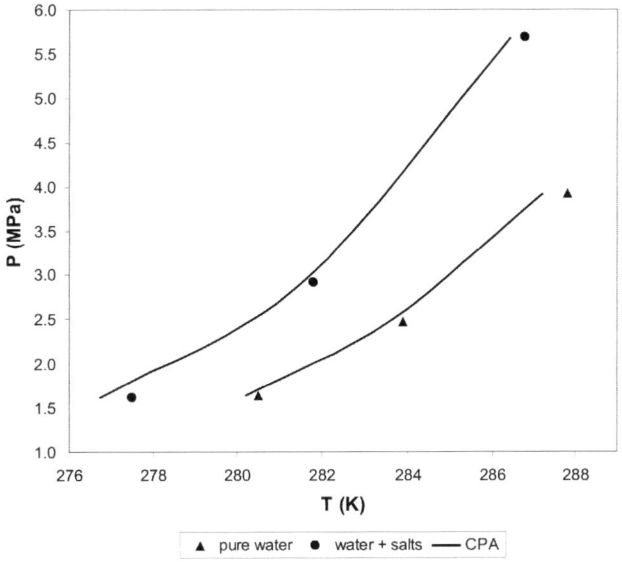

Figure 3 Experimental data from Tohidi, et al.[24].

The set of CPA, Van der Waals – Platteeuw and solid freeze-out models make a extensive framework where the vapour-liquid equilibria can be modeled simultaneously with hydrates, ice, salts and solid salts using the same equation of state.

Using the same gas from Tohidi at al. [24] a complete phase diagram can be created that shows phase loci for hydrate II, gas, liquid hydrocarbon, water and ice.

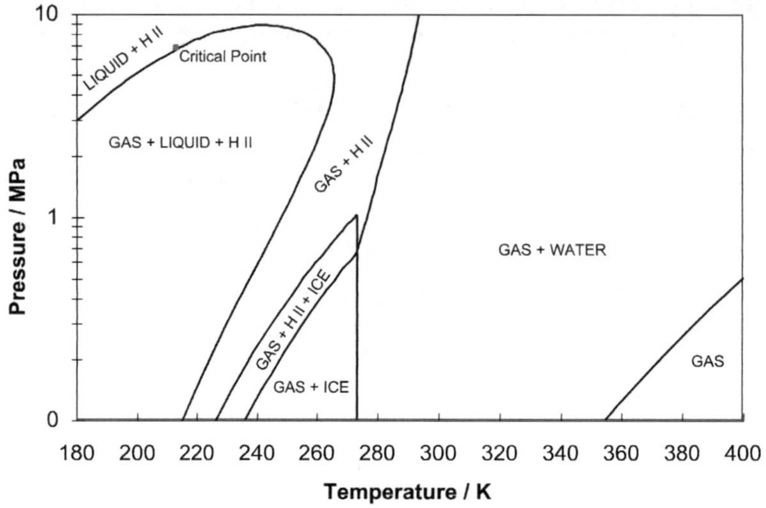

Figure 4 Phase envelope diagram for gas with composition from [24].

Waxes

Transportation of waxy crude in cold environment such as subsea pipelines presents another flow assurance and operability challenges. When the waxy crude temperature drops below the Wax Appearance Temperature (WAT), the waxes will start to crystallize and deposit on the pipe walls if a radial heat flux to the surrounding exists. This phenomenon, known as wax deposition, can result in significant operational and remedial costs, reduced or delayed production. For the oil industries, it is very crucial to identify the thermodynamic conditions of the wax precipitation and predict the wax deposition rates for the design and optimization of pipeline operation systems, as well as to implement appropriate strategies for prevention and remediation.

Waxes are solid crystals consisting predominately of n-paraffins that precipitate out of oil solution when temperature falls. This type of waxes has straight-chain with varying chain length. For given any carbon number, the linear straight-chain n-paraffins have the highest melting points, are most likely to precipitate first and therefore is most interest in operational situations. For temperatures of operational interest, i.e., above 0 °C, waxes consist predominantly of the C20+ n-paraffins. The solid wax phase is actually solid solutions of n-paraffins; a mixture of n-paraffins of different lengths, not a single n-paraffin.

A number of engineering models have been proposed for calculating oil-wax equilibria, for example, the works of Won[25] , Hansen et al[26], Erickson et al.[27], Pedersen[28], Rønningsen et al.[29], Lira-Galeana et al.[30], Pan et al.[31], and Coutinho et al [32-38]. Among these models, the most predicative and accurate model is the one developed by Coutinho et al and it is the one used in this work.

The key factors of this wax model are the solution behavior of the n-paraffin in the oil phase, the thermodynamical properties of the fusion of the n-paraffins and the thermodynamics of solid solutions of n-paraffins, the wax phase. By breaking the problem down, it is possible to set up a model in which the accuracy of the individual assumptions and associated parameters can be reliably assessed by comparison with high quality experimental data and the problem of internal cancellation of errors can be minimized. The model is a predictive local composition model, can accurately predict the waxing behavior of diesel and jet fuels [32, 33] and crude oils [34, 35]. The model exists in two variants, the Wilson [36] and Uniquac [37] wax models. The Wilson model is simpler to apply, as it treats the wax phase as a single solid solution of n-paraffins. The Uniquac model is more realistic in that it predicts that the wax phase splits into a number of coexisting solid solution phases;

experimental evidence confirms this to be the case [39]. However, for the engineering applications, the Wilson model has the advantage of being easier to use. With a correction factor, the Wilson model is capable of giving the results as good as those from Uniquac model.

Fluid characterization and results

In order to apply the wax model for calculating the phase equilibrium between the wax and fluid phases, a procedure for characterizing the crude oils is required together with the n-paraffin analysis. To characterize the oil, we split the oil into two distributions, one is for the n-paraffins and one for all other components referred as non n-paraffins. The n-paraffins are represented by a component from the non n-paraffin distribution with the same molar density as that of the n-paraffin in question. The other components are characterized with the method described previously for petroleum fluids using the Raizi and Al-Sahhaf correlations [7, 8] for molecular weight and specific gravity, and the Twu correlations for the critical properties[9].

For the characterization and wax calculations, the input data requirement includes:

- A compositional GC analysis of the fluid or distillation data , eg. TBP distillation.
- One of the following n-paraffin analysis:
    - Measured n-paraffin distribution to as high carbon number as possible.
    - The UOP-46 wax content that is interpreted as giving the total C20+ n-paraffin content in the stabilized oil.
    - Estimate the wax content but this will give the least reliable predictions and, in any case, should not be used for condensates.
- The wax model can be tuned to match additional wax precipitation data. The best measurement is the amount of wax precipitated as a function of temperature. Typically only the wax appearance temperature (WAT) is measured but this is not a good choice for tuning the model because any measurement requires a finite amount of wax to be precipitated before it can be detected.
- If the measured bubble points are available, the data can be used to "tune" the petroleum fraction properties. This is optional but will improve predictions of the saturation lines for live oils as described in section "Tuning the model".

After characterizing the fluid, the wax phase boundary and the amounts of waxes precipitated can be calculated. Here presented is the oil 3 from the paper of Rønningsen et al [40, 41] which not only has the WATs measured but also the amounts of wax precipitated as a function of temperature. As shown in Figure 5, the predicted amount of wax precipitated are in good agreement with the experimental data, whereas the predicted WAT is higher than the measured WATs by both DSC(Differential Scanning calorimetry) and CPM(Cross polar microscopy). This is quite reasonable as the predicted WAT is the point where the first wax crystal appears, whereas the measured WAT is the temperature at which the technique is able to detect the presence of a finite amount of wax crystals.

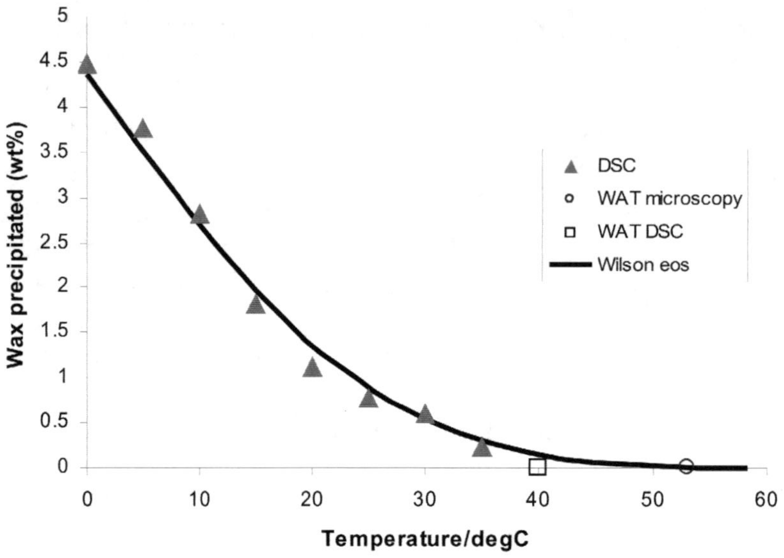

Figure 5: The amount of waxes precipitated as a function of temperature at 1 atm, Data from Rønningsen et al[41] .

Wax deposition model

The avoidance or remediation of wax deposition is one key aspect of flow assurance. Which depends on the reliable prediction on the tendency and the likelihood of the wax deposition. The ability to accurately predict the rate of the wax deposition depends on a number of factors among which the most important ones are the driving force calculated by the thermodynamic equilibrium between the wax and fluid phases, and the method for evaluating the rate of wax deposition.

The Infochem wax deposition model [42] is based on flow-loop experiments [43]. Our investigations suggest that using a compositional treatment in modeling the wax deposition is essential to give physically realistic prediction of the wax deposition profiles. The model takes account of the removal of wax by shearing using an expression that has been devised to be compatible with observations from flow-loop studies. The model is capable of giving correct trend of the flow-loop experimental measurements and realistic predictions of the wax deposition profiles with the compositional treatment for the n-paraffins.

For modeling wax deposition, lumping the n-paraffin into a few pseud-components has drastically effect on the wax deposition profiles. The example presented in Figure 6 is the simulation of the wax thickness profiles for a single-phase unburied pipeline reported by Lindeloff and Krejbjerg [44]. Three different characterizations for the n-paraffins are considered in the simulations to demonstrate the lumping effect, they are the continuous n-paraffin distribution with n-paraffins up to N76+, 7 n-paraffin pseudo-components and 3 n-paraffin pseudo-component. In Figure 6, the effect of using seven and three pseudocomponents is clearly shown with spikes occurring at the points along the pipeline where the temperature falls below the value needed to cause a particular pseudocomponent to precipitate.

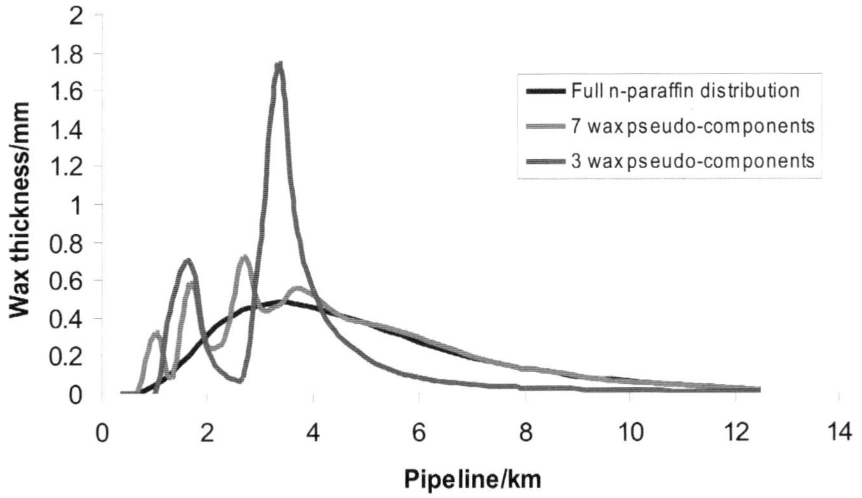

Figure 6: Comparison of wax deposition profiles between the cases with full n-paraffin distribution and a few pseud-components, case A in Lindeloff et al. [44]

However applying the continuous n-paraffins in the simulations can easily end up with a number of components approaching or over 100. As the phase equilibria of both the flowing fluids and the wax deposited must be repeatedly calculated along with the physical properties of the phases present, the simulations are computationally intensive taking several hours for a typical real pipeline. To make the model into a practical engineering tool, the calculations have to be performed much more quickly. Our method to achieve this is to convert the phase equilibrium and physical property data into empirical expressions that are fitted to the rigorous model. Figure 7 gives a comparison between the rigorous and approximate forms of the model. The visual impression is that there is little appreciable difference in the results. The differences in the predictions are normally within about 1%, but the impact on run time is enormous. Taking this example which uses 74 components, the rigorous version took several hours to run, whereas, using the same computer, the approximate version took about 20 s for the physical property correlations to be regressed, followed by the actual simulation of the pipeline which took about 1 min.

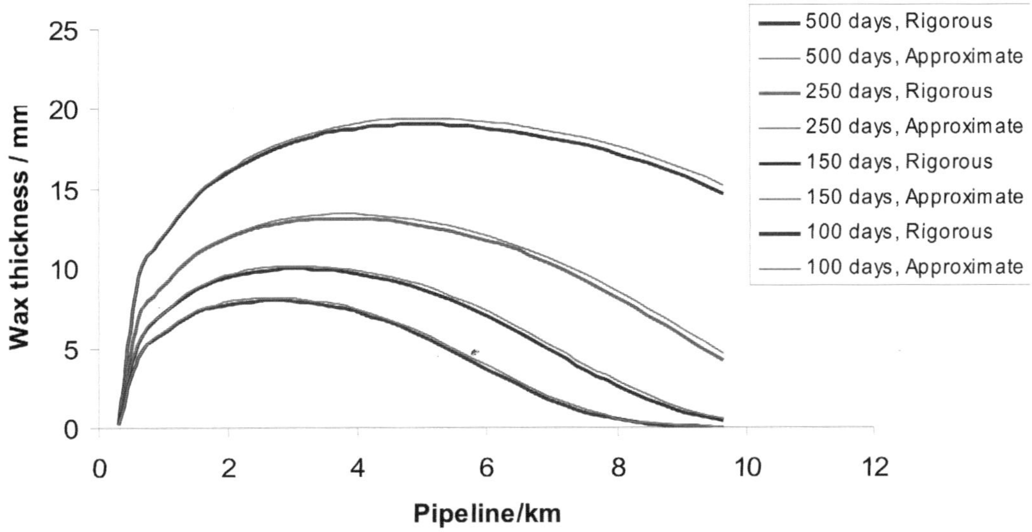

Figure 7: Comparison of the rigorous and approximate phase equilibrium methods

Asphaltenes

Asphaltenes, as the heaviest, most highly polarizable and polydisperse petroleum fractions in crude oils, are the another major flow assurance challenge for the oil industries. The asphaltene precipitation and deposition can occur at various stages from the production, transportation to refining. Unlike the waxes, their actual chemical structures are very difficult to define due to the lack of volatility, the complex and continuous variation in their chemical structures and self-aggregation. The solubility of the asphaltenes in crudes is usually affected by reduction of pressure, temperature and/or oil composition change as a result of commingling with other crudes or gas injection. This may lead to asphaltene precipitation and deposition, decline in permeability, blockage of well and surface facilities, finally production decrease or termination which has substantial economical impact. The ability to understand and predict asphaltene phase behavior is therefore essential for both upstream and downstream processing so that appropriate strategies can be implemented for prevention and remediation.

The asphaltene phase behavior of petroleum mixtures is quite complex and depending on the hypothesis off the mechanism of asphaltene precipitation and stabilization, the models are mainly classified into two categories; the colloidal approach and the true solution approach. The colloidal approach is mainly represented by the solid colloidal model by Leontaritis and Mansoori [45], reversible micellization model by Victorov and Firoozabadi [46, 47] and McMillan-Mayer-SAFT model by Wu and Prausnitz[48]. On the other hand, the true solution approach includes the solubility models [49, 50] and equations of state such as Cubic-Plus-Association (CPA) [51-54] and perturbed-chain statistical associating fluid theory (PC-SAFT) [55-57].

This section is to demonstrate the potential of the true solution approach, the capability and extensions of the SRK equations of state plus association (CPA) to model asphaltene stability in live oils and compare the predictions with the various types of measured asphaltene precipitation data over wide ranges of temperature and pressure. In general the model can give excellent results for a wide range of different types of precipitation envelopes [51, 54] and is able to describe the gas, oil, aqueous and asphaltene phases simultaneously, making it possible to calculate the phase stability and phase equilibrium in a general and unified way.

Fluid characterization and model parameters

The experimental data required by the CPA asphaltene model to predict the asphaltene phase boundary are: gas chromatography (GC) compositional analysis for live or stabilized oil, the amount of asphaltenes and resins in stock tank oil (SARA).

Before modeling the asphaltene phase, the crude oils are characterized using the standard Lee-Kesler and Riazi correlations as described in section “Petroleum fluid characterization”. In the characterization, the asphaltenes are defined as the heaviest fractions followed by the resins, the second heaviest ones. The measured bubblepoints of live oils are optional. But if bubble point data are available the physical properties of pseudo-components such as acentric factors or the Mathias Copeman parameters are tuned to reproduce the experimental bubblepoints.

For the association model parameters accounting for the cross-association between asphaltenes and resins, three types of data can be applied:

- Measured asphaltene precipitation upper onset points, ideally at two different temperatures.
- The titration onset data of n-heptane (amount of n-heptane for initializing the onset of asphaltene precipitation).
- The reservoir conditions (only valid for live oils). Using reservoir condition; based on

the de Boer[58] screening technique, it is assumed that the reservoir fluid is nearly saturated with asphaltenes at reservoir condition.

After the parametrization of the association model parameters, the same model parameters are used to predict the entire asphaltene phase boundary of the original reservoir, the amounts of asphaltene in solution as a function of pressure at other temperatures and the effect of gas injection on the asphaltene phase instability.

The example used to demonstrate the asphaltene model is a live oil from the Magwa area of the Greater Burgan field in south Kuwait[59]. This is a fairly paraffinic crude with little asphaltenes, 0.5wt% in stock tank oil but the field suffered severe asphaltene problems ranging from plugging of tubulars and flowlines to separators.

The predicted asphaltene phase envelopes is summarized in Figure 8 together with the measured asphaltene onset data, the predicted and measured bubblepoints. The asphaltene upper and lower onset data in the figure were established from the measured asphaltene content as a function of pressure at various temperatures. As shown in Figure 8, the predicted asphaltene phase boundary is in very good agreement with the asphaltene upper and lower onset data over the entire temperature range applied in the measurements.

Knowing the asphaltene precipitation boundary is critical for implementing appropriate strategies for preventing the asphaltene precipitation and deposition along the production line. As shown in Figure 8 , the initial reservoir condition is just above the asphaltene upper boundary but when production occurs from the reservoir state to surface condition, asphaltene precipitation and deposition becomes more probable as the pressure drops.

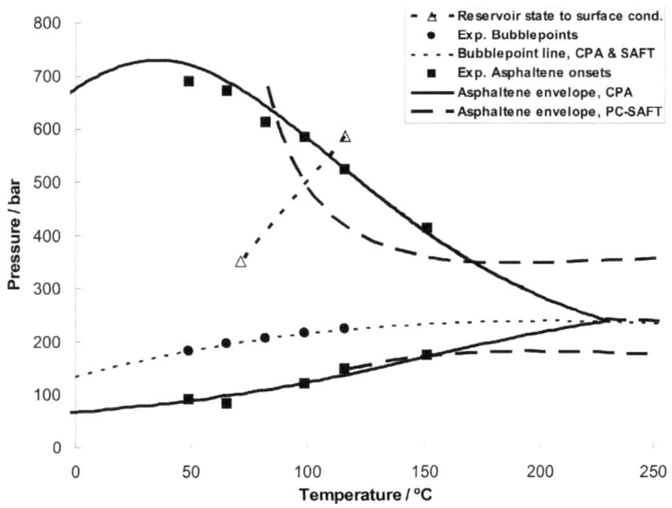

Figure 8: Asphaltene phase envelope along the saturation line together with experimental data [59].

The asphaltene model can also model very accurately the effect of gas injection in the asphaltene stability. The next example is a reservoir fluid from an offshore carbonate oil field in the Arabian Gulf, recently published by Yonebayashi et al [60]. The field exhibits asphaltene precipitation and deposition problem inside tubing of some production wells.

The detailed gas titration experiments (GTE) with a gas mixture containing 40mol% of methane at four different temperatures were carried out. Based on the asphaltene onset pressures as a function of the molar amounts of injected gas at a given temperature by GTE, it is possible to estimate the asphaltene onsets for the original reservoir fluid by extrapolating the onset pressure to the zero amount of gas at the given temperature. The estimated possible asphaltene onsets data for the Well 3 are presented in Figure 9 together with four of the several measured asphaltene onsets for the cases with gas injection.

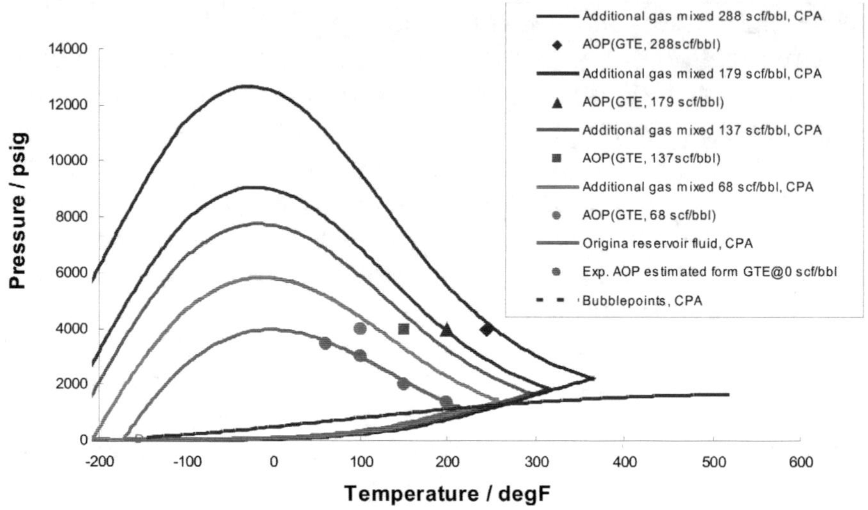

Figure 9: Asphaltene phase envelopes of the original fluid and the cases with gas injection.

Multiple solids

The example presented here that makes use of the whole framework is a light oil with a molecular weight of 87 g/mol. From the the SARA analysis the amount of asphaltenes is 0.7 % and the amount of resins is 12.4 % by mass.

The amount of wax present was estimated from the single carbon distribution and consequently the n-paraffin distribution was estimated. Figure 10 shows how an equation of state coupled with a few extra terms can fully model a fluid and all the solids that may precipitate from it.

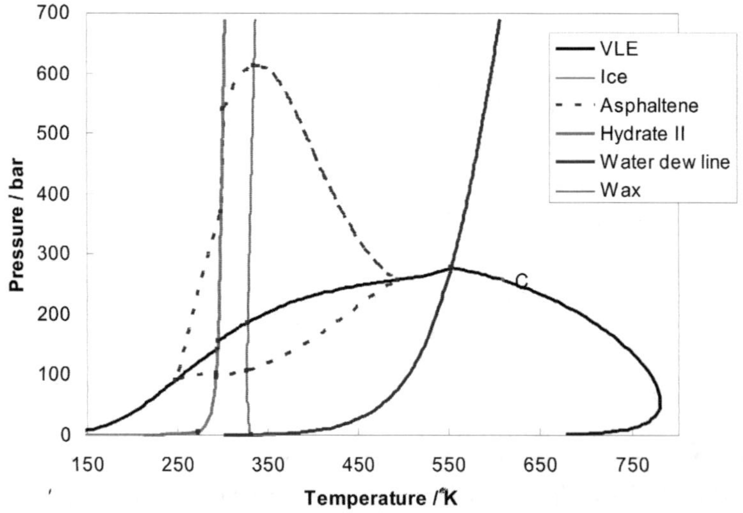

Figure 10 Phase behaviour for an oil showing all the possible solid and liquid phase lines

The plot was made using the RKSA model, described previously, as base with the following extra terms:

- Association to improve water phase behavior description;
- Freeze-out model to model ice formation;

- Van der Waals and Plateew term to account for hydrate formation
- simplified association model to describe asphaltene precipitation behavior;
- Wilson solid solution model to describe wax appearance in the oil.

The results present in Figure 10 also show how the various phases can affect the behavior of other phases.

- When the hydrate phase boundary line crosses the VLE line, it shows a small depression due to the depletion of gas from the fluid by the hydrate solid phase.
- When the water dew point point line crosses the VLE line it also shows that the solubility of the gas components is slightly altered in that point.
- The most remarkable effect happens when the hydrate line crosses the asphaltene line: as the hydrate phase removes the light components from the fluid, it stabilizes the asphaltene in solution making the the asphaltene onset pressure much lower.

Conclusions

In this work an integrated equation framework has been presented. It allows modeling of full range of flow assurance applications just by using a common equation of state. To model the individual solids separate terms are included to account for the particular phase behavior of each solid.

With all the separate terms included, the multiphase equilibrium algorithm supports all flash calculations for any combination of phases: gas, hydrocarbon liquid, aqueous, hydrates, wax, asphaltenes and solid scale phases.

It is also possible to infer about any interactions the various possible phases may have and this will have an impact on the calculated overall behavior.

References

[1] Kontogeorgis, G.M. And Folas, G.K., Thermodynamic models for industrial applications, Wiley (2010).

[2] Soave, G., Equilibrium constants from a modified Redlich-Kwong equation of state, Chem. Eng. Sci., **27**, 1197-1203 (1972)

[3] D.Y. Peng, D.B. Robinson, A new two-constant equation of state,Ind. Eng. Chem. Fundam. 15 59-64, (1976)

[4] Lungaard, L and Mollerup, J., The influence of gas phase fugacity and solubility on correlation of gas-hydrate formation pressure, Fluid Phase Equilibria, **70**, 199-213 (1991).

[5] Mathias P.M., Copeman T.W., Fluid Phase Equilib., 13, 91-108, (1983).

[6] Peneloux, A., Rauzy, E. and Freze, R., Fluid Phase Equilib., **8**, 7-23 (1982).

[7] Riazi, M. R., and Daubert. T. E., Hydrocarbon Processing, 59, 115 (1980).

[8] Riazi, M.R. and Al-Sahhaf, T.A., Fluid Phase Equilibria, 117, 217-224 (1996).

[9] Twu, C.H., Fluid Phase Equilibria, 16 137-150 (1984).

[10] M. G. Kesler and B. I. Lee, Hydrocarbon Processing, 153-158, (1976).

[11] K. S. Pedersen and P. L. Christensen, Phase Behavior of Petroleum Reservoir Fluids, Taylor and Francis, (2007).

[12] Zhang, X.; Pedrosa, N. Moorwood, T.; Energy & Fuels, 2012, Article ASAP (DOI: 10.1021/ef201383r)

[13] Sloan, E. D.; Koh, C. A. 'Clathrate Hydrates of Natural Gases', 3rd. Ed., CRC Press, New York (2008).

[14] Wertheim, M. S. J. Stat. Phys. 35, 19 (1984).

[15] Wertheim, M. S. J. Stat. Phys. 42, 459 (1986)

[16] Kontogeorgis, G. M.; Voutsas, E. ;Yakoumis, I.; Tassios, D. P. Ind. Eng. Chem. Res.

35 4310-4318 (1996).
[17] Yakoumis, I.; Kontogeorgis, G. M.; Voutsas, E.; Tassios, D. P. Fluid Phase Equilibria 130 31-47 (1997).
[18] Voutsas, E.; Kontogeorgis, G. M., Yakoumis, I.; Tassios, D. P. Fluid Phase Equilibrium 132 61-75 (1997).
[19] van der Waals, J. H; Platteeuw, J. C. Clathrate Solutions, Adv. Chem. Phys. 2, 21 , 1-55 (1959).
[20] Parrish, W. R.; Prausnitz, J. M. Ind. Eng. Chem. Proc. Des. Dev., 11, 26-35, (1972).
[21] Martin, A. J. Phys. Chem B, 114, 9602-9607 (2010).
[22] Ng, H.-J.; Robinson, D. B. Gas Processors Association Research Report 66, (1983)
[23] P. Debye and E. Hückel .Physikalische Zeitschrift 24: 185-206 (1923).
[24] Tohidi, B., Danesh, A., Burgass, R.W., and Todd, A.C "Measurement and Prediction of the Amount of Gas Hydrates", BHR Group Conference – Multiphase 95, Cannes, France, 7-9 June (1995).
[25] Won, K. W. Fluid Phase Equilib., 53, 377 (1989).
[26] Hansen, J. H.; Fredenslund, Aa.; Pedersen, K. S.; Rønningsen, H. P., AIChE J. 34, 1937 (1988)
[27] Erickson, D. D.; Niesen, V. G.; Brown, T. S. SPE 26604, Society of Petroleum Engineers' Annual Technical Conference and Exhibition, Houston, Oct 3-6, (1993)
[28] Pedersen, K. S. SPE Prod. Facil., 2, 46 (1995)
[29] Rønningsen, H. P.; Sømme, B. F.; Pedersen, K. S. In Proceedings of Multiphase '97, BHR Group, (1997)
[30] Lira-Galeana, C.; Firoozabadi, A.; Prausnitz, J. M. AIChE J., 42, 239 (1996)
[31] Pan, H.; Firoozabadi, A.; Fotland, P. SPE Prod. Facil., 11, 250 (1997)
[32] Coutinho, J.A.P., Energy Fuesl, 14, 625 (2000).
[33] Coutinho, J.A.P.; Dauphin, C.; Daridon, J.-L., Fuels, 79, 607 (2000).
[34] Daridon, J.-L.; Pauly, J.; Coutinho, J.A.P.; Montel, F, Energy Fuels, 15, 730 (2001).
[35] Coutinho, J.A.P.; Daridon, J.-L., Energy Fuels, 15, 1454 (2001).
[36] Coutinho, J. A. P.; Stenby, E. H. Ind. Eng. Chem. Res., 35, 918 (1996)
[37] Coutinho, J. A. P. Ind. Eng. Chem. Res., 37, 4875 (1998).
[38] Coutinho, J.A.P.; Edmonds, B; Moorwood, T; Szczepanski, R and Zhang, X, Energy & Fuels, 20, 1081-1088 (2006).
[39] Broadhurst, M.G., J. Res. Nat. Bur. Stand., 66A, 241 (1962).
[40] Rønningsen, H.P.; Bjørndal, B; Hansen, A.B.; Pedersen, A.B., Energy Fuels, 5, 895 (1991).
[41] Pedersen, K.S.; Skovborg, P.; Rønningsen, H.P., Energy Fules, 5, 924 (1991).
[42] Edmonds, B.; Moorwood, T., Szczepanski, R. and Zhang, X., Energy & Fuels, 22, 729-741 (2008).
[43] Hernandez Perez, O. C. Investigation of Single-Phase Paraffin Deposition Characteristics. M.S. thesis, University of Tulsa: Tulsa, OK, (2002).
[44] Lindeloff, N.; Krejbjerg, K. Energy Fuels, 16, 887 (2002).
[45] Leontaritis, K. J. and Mansoori, G. A., in Soc. Pet. Eng. Intl. Symp. On Oilfield Chem. San Antonio, TX, (1987).
[46] Victorov, A. I. and Firoozabadi, A., Amer. Inst. Chem. Eng. J. 42, 1753-1764 (1996).
[47] Pan, H. and Firoozabadi, A, Thermodynamic Micellization Model for Asphaltene Aggregation and Precipitation in Petroleum Fluids, SPE Production & Facilities, May (1998).
[48] Wu, J; Prausnitz, J.M.; Firoozabadi, A., AIChE J., 44, 1188-1199 (1998).
[49] Hirschberg, A., de Jong, L.N., Schipper, B.A., and Meijer, J.G., SPE11202 (1984).
[50] Wang, J.X.; Buckley, J.S., Energy Fuels, 15, 1004 (2001).

[51] Edmonds, B.;Moorwood, R. A. S.; Szczepanski, R.; Zhang, X.; Heyward, M.; Hurle, R.; "Measurement and prediction of asphaltene precipitation from live oils", Third international symposium on colloid chemistry in oil production, asphaltenes and waxes deposition (ISCOP'99), Mexico (1999).
[52] Zhidong Li and Abbas Firoozabadi, Energy Fuels, 24, 1106–1113 (2010).
[53] Zhidong Li and Abbas Firoozabadi, Energy Fuels, 24, 2956–2963 (2010).
[54] Zhang, X.; Pedrosa, N. and Moorwood, T: Energy and Fuels, DOI: 10.1021/ef201383r. ASAP (2012).
[55] Ting, D.; Hirasaki, G. J.; Chapman, W. G.; Petroluem Science & Technology, 21, 647-661 (2003).
[56] Gonzalez, D. L.; Vargas, F. M.; Hirasaki, G. J.; Chapman, W. G. Energy and Fuels, 22, 757–762 (2008).
[57] Vargas, F. M.; Gonzalez, D. L.; Hirasaki, G. J.; Chapman, W. G. Energy & Fuels, 23, 1140 (2009).
[58] de Boer, R.B., Leerlooyer, K.,Eigner, M.R.P., van Bergen, A.R.D., SPE24987 (1995).
[59] G.S.Kabir, G.S.,SPE, KOC/Chevron; Jamaluddin, A.K.M., SPE53155 (1999).
[60] Hideharu Yonebayashi, INPEX; Tsutomu Masuzawa and Chawki Dabbouk, ZADCO; and Daisuke Urasaki INPEX/JODCO, SPE125643 (2009).

(92a) Unconventional Natural Gas How Much At What Cost?

Stephen A. Holditch, Petroleum Engineering, Texas A& M University, College Station, TX

Unconventional gas reservoirs (UGRs) are low quality reservoirs that have to be stimulated to produce gas at commercial flow rates. Well stimulation can occur by drilling horizontal wells and then hydraulically fracture treating the horizontal well. In this presentation, I will discuss the prospects of finding unconventional gas reservoirs in virtually every basin in the world and how the industry can use existing technology to begin to unlock those resources. In fact, most of the reservoirs undergoing development in 2011 in the United States have been thought of as source rocks for the past century. The source rocks are now the most important reservoirs, thanks to technology.

The presentation will first cover how natural resources are distributed in nature and how gas production from fields in North America increases every time there is an increase in gas price or there is a technology breakthrough. We will discuss the possibilities of producing gas from UGRs in virtually every hydrocarbon basin in the world during the 21stCentury. Finally, I will discuss how the economic recovery of gas from UGRs is directly affected by new developments in technology, gas prices and finding and development costs.

(92b) Natural Gas Hydrates: An Emerging Massive Energy Source

Michael J. Economides, Chemical and Biomolecular Engineering, University of Houston, Hoston, TX, Xiuli Wang, XGas, Houston, TX and Harold Brannon, Baker Hughes, Tomball, TX

The quest for energy resources is an ongoing process and while over the last two decades there has been a significant move towards alternatives to the traditional fossil fuels, natural gas is poised to become the premier fuel of the world economy both for its obvious environmental benefits but also because of its abundance. Shale gas, mostly the result of the deployment of technology to an energy source that was known but considered until recently inaccessible, has provided a large volume of new gas and, especially, almost unilaterally has doubled the world expected ultimate recovery. The success of shale gas has triggered by analogy a renewed interest in natural gas hydrates whose volume dwarfs any other energy source on earth.

If techniques for their commercial recovery materialize, natural gas hydrates deposits have the potential to provide a huge amount of natural gas for human use. Hydrate accumulations have been found worldwide on the slopes of continental shelves and arctic permafrost where there is natural gas and water plus the required temperature and pressure. Considerable research has already been carried out for the characterization, composition and location of known deposits of natural gas hydrates.

Notions of potential exploitation have included the sublimation of natural gas from solid hydrate primarily either with depressurization to below the hydrate formation pressure for a given temperature or by heating the deposit to a temperature above the hydrate formation temperature via a thermal stimulation process. Another technique has been the potential injection of a chemical compound which would depress the hydrate formation pressure. Such compounds might include alcohols or glycols.

We present here an exploitation scheme for natural gas hydrates that involves hydraulically fractured horizontal wells with multiple treatments, using a suite of wells that would provide mechanisms of injection and/or production. Pressure reduction or temperature increase or, even the injection of chemical compounds is possible, depending on the deposit to be produced.

(92c) Production Strategy for Marine Hydrates Reservoirs

Jyoti Phirani, UT austin, Austin, TX and **Kishore K. Mohanty**, Petroleum & Geosystems Engineering, The University of Texas at Austin, Austin, TX

Natural gas hydrate reservoirs have been found in large quantities in marine and arctic sediments. This research focuses on finding the optimum production strategy of natural gas production from marine reservoirs using horizontal wells. A 3-dimensional, compositional, thermal simulator has been developed in the past, which can simulate hydrate formation and dissociation in porous medium. Three components (hydrate, methane and water) and four phases (hydrate, gas, aqueous-phase and ice) are considered in the simulator. A combination of depressurization and warm water flooding of these reservoirs is studied to find the production strategy. The production strategy is dependent on the injection water temperature, injection water pressure, production well pressure, reservoir confinement, initial hydrate saturation, size of aquifer attached to hydrate reservoirs, dip of the reservoir, orientation and positions of the wells.

(92g) Massively Parallel Simulation of Field-Scale Oceanic Gas Hydrate Deposits

Matthew T Reagan, George J. Moridis, Jeffrey N. Johnson, Lehua Pan, Katie L. Boyle and Craig M. Freeman, Lawrence Berkeley National Laboratory, Berkeley, CA

The quantity of hydrocarbon gases trapped in natural hydrate accumulations is enormous, leading to significant interest in the evaluation of their potential as an energy source. It has been shown that large volumes of gas can be readily produced at high rates for long times from some types of methane hydrate accumulations by means of depressurization-induced dissociation, and using conventional technologies with horizontal or vertical well configurations. However, these systems are currently assessed using simplified or reduced-scale 3D or even 2D production simulations. In this study, we use the massively parallel TOUGH+HYDRATE code (pT+H) to assess the production potential of a large, deep-ocean hydrate reservoir and develop strategies for effective production. The simulations model a full 3D system of over 24 km^2extent, examining the productivity of vertical and horizontal wells, single or multiple wells, and variations in reservoir properties. Systems of up to 2.7M gridblocks, running on hundreds of supercomputing nodes, are required to simulate such large systems at the highest level of detail. The simulations reveal the challenges inherent in producing from deep, relatively cold systems with extensive water-bearing channels and connectivity to large aquifers. Also highlighted are new frontiers in large-scale reservoir simulation of coupled flow, transport, thermodynamics, and phase behavior, including mesh construction, parallel numerical solvers, and large-scale visualization.

(92d) Hydraulic Fracture Height Growth - Real

Kevin Fisherand Norman Warpinski, Pinnacle Technologies/Halliburton, Houston, TX

By some reports, hydraulic fracturing accounts for up to 30% of North America's oil and 90% of its natural gas reserves. However, until recently, real data on hydraulic fracture growth mechanisms has been limited.

The first tool used to try to understand hydraulic fracture growth, and still the industry's favorite predictive tool, was linear elastic fracture modeling. Additional technologies, including temperature logs, radioactive and chemical tracers, along with mapping technologies such as tiltmeters and microseismic monitoring, have been increasingly utilized in recent years to verify and calibrate frac models by measuring where and how fractures grow. These technologies, along with mineback studies, core-throughs, laboratory studies, and modeling, are helping to provide a better understanding of fracture growth mechanisms and conditions under which vertical fracture propagation is helped or hindered.

Recently, much public discourse has taken place about fracture growth in unconventional reservoirs and whether fractures could potentially grow all the way back to the surface, creating communication pathways for frac fluids or produced hydrocarbons to pollute groundwater supplies. This paper will present real fracture growth data mapped during thousands of fracturing treatments in unconventional reservoirs along with aquifer depths in the vicinity of the fractured wells. Additionally, the authors will present an in-depth discussion of fracture growth limiting mechanisms augmented by the mineback tests and other studies which were performed to visually examine hydraulic fractures. These height growth limiting mechanisms, which are supported by the mapping data, provide insight into why hydraulic fractures can be longer but more vertically constrained than originally modeled. This information can be used to improve models, optimize fracturing, and provide definitive data for regulators and interest groups.

(92f) Optimization of Marine CNG Transport for Stranded Gas

M. Nikolaou, Chemical and Biomolecular Engineering, University of Houston, Houston, TX

While liquefied natural gas (LNG) is the dominant technology for marine transport of natural gas, compressed natural gas (CNG) is an economically attractive alternative for transportation of relatively small amounts of natural gas over short distances (up to 6 Bcm/year transported over 2000 km, respectively). Because the main capital expenditure in a CNG project is on marine transport vessels, accounting for more than 3/4 of potential CNG projects, careful design of CNG transport fleets and compatible distribution schedules is important. Design through numerical solution of an all-encompassing optimization problem would be feasible but complicated and intuitive unappealing, in that it would provide little insight into the guiding principles of CNG transport system design. In this work, a structured optimization framework is applied to potential marine CNG transport. A case study for transport from a source in the Trinidad/Venezuela area to island countries in the Caribbean is used to illustrate related concepts. Two general patterns emerge in the results: A "hub-and-spoke" pattern for servicing of the largest four consumption markets (Puerto Rico, Dominican Republic, Jamaica, and The Bahamas) and a cyclical "milk-run" pattern for the remaining (smaller) markets. Furthermore, a general guiding principle emerges that favors medium rather than large-size CNG transportation vessels, such that a closer approximation of the ideal (but infeasible) optimum of a two-way CNG pipeline can be achieved.

Adsorption of Ethoxylated Alkyl Phenols from High Salinity Brines

David A. Goodwin Jr

Department of Chemical, Biological and Materials Engineering, University of Oklahoma, Norman OK
dagoodwin@ou.edu

***Keywords**: Nonionic Surfactants, Adsorption Isotherm, High Salinity.*

Abstract

Anionic surfactants are generally used in oil field applications to lower the interfacial tension between crude oil and brine because they are less expensive and have lower adsorption on reservoir minerals. One of the main challenges in applying nonionic surfactant in is a significant loss of surfactant by adsorption onto rock or clay surfaces. At high salinities, however, ionic surfactants may precipitate, and the electrostatic repulsion that reduces their adsorption on rock surfaces may become insignificant. Because nonionic surfactants possess low critical micelle concentration and high brine tolerance, they are of interest in this research. Nonionic surfactants have the potential to decrease surfactant adsorption and reduce the potential for phase separation in high salinity brines. In this study, the adsorption behavior of polyethoxylated octylphenol and polyethoxylated nonylphenol on silica in various aqueous solutions was investigated. Surfactant adsorption isotherms were used to study the effects of increasing salinity concentrations on the mechanism of adsorption and the overall adsorption density. The results show that increasing salinity causes increased adsorption; however, the similarity between adsorption isotherms and the single curve obtained after normalization of the isotherms at various salinities suggests a consistent mechanism of adsorption that is independent of salt concentration. These similarities also suggests that there may be an upper limit to nonionic surfactant adsorption so that after some maximum electrolyte concentration little additional adsorption occurs.

Introduction

Until recently oil production has been declining in the continental United States; however, a large amount of crude oil remains within the mature reservoirs because conventional methods of oil production are unable to extract it (1). Developments in enhanced oil recovery coupled with high oil prices and an ever increasing energy demand make applying EOR techniques to recover the trapped crude oil in mature reservoirs economically feasible (2).

Enhanced oil recovery by surfactant flooding is a method of improving oil recovery by reducing the surface tension forces that bind the oil to the rock. However, adsorption-desorption processes cause a reduction in bulk solution concentration and a chromatographic separation of surfactant and solution as it moves through the reservoir. Loss of surfactant due to adsorption on the surface of reservoir rock is also an important consideration when determining the economic feasibility of implementing an EOR system (3).

Minimizing the surfactant loss to adsorption is economically desirable. Anionic surfactants are often used in surfactant flood based EOR; however, the brine environment of many reservoirs is not amenable to these surfactants (4). Nonionic surfactants are an option for surfactant flooding based EOR in reservoirs with high TDS content because they are less sensitive to electrolyte concentration than ionic surfactants (5).

Several studies (6-9) have determined adsorption isotherms for polyethoxylated nonionic surfactants on hydrophilic silica surfaces; however, the majority of the surfactants chosen for study contained small to medium sized oxyethylenic groups, an EO number less than 20, with the exception of Partyka et al.(7) and Somasundaran et al.(6) who each developed an adsorption isotherm for an ethoxylated phenol with an EO number of 40. Furthermore, none of the studies have thoroughly investigated the effect salinity has on the overall adsorption and shape of the adsorption isotherm. A few generalizations about the adsorption of ethoxylated alkylphenols can be drawn from the previous studies. Most notable is that the adsorption density is directly proportional to the EO number of the surfactant. Thus the longer the oxyethylenic chain the lower the adsorption density. Additionally, the area per ethylene oxide unit remains constant regardless of the size of the surfactant's associated hydrophilic and hydrophobic portion, and the area per molecule is proportional to the square root of number of EO groups (6,9). Most authors report the adsorption isotherm of surfactants with EO chains follows the traditional s-shaped Langmuir isotherm. However, it is important to note that first phase representing monomeric adsorption is extremely abbreviated and most adsorption occurs near the critical micelle concentration (6-9).

The shape of the adsorption isotherm suggests a mechanism of adsorption that is highly dependent on the aggregation of surfactant molecules at the solid-liquid interface. Several studies have been conducted to determine the exact mechanism of adsorption (8-11) A four part mechanism for adsorption of polyethoxylated alcohols with small EO numbers has been proposed in which at low surface coverage, less than 0.1 monolayers, monomeric adsorption occurs. These adsorbed monomers act as nucleation sites for the adsorption of micelle-like aggregates. As the CMC is approached the aggregates fuse, and beyond the CMC a classic bilayer is formed. Other studies (11-12) demonstrated that adosprtion isotherms for polyethoxylated alkylphenols, when normalized to the CMC and maximum adsorption, collapse to a single curve. Taken together these results suggest a consistent mechanism of adsorption, and a relationship between the adsorption at solid-liquid interface and bulk aggregation.

However, Partyka et al.(7) and Somasundaran et al. (6) show that the adsorption isotherm for polyethoxylated phenols with an EO number of 40 is fundamentally different, suggesting a different mechanism of adsorption than described previously. These adsorption isotherms do not show the distinct break characteristic of molecular aggregation on the solid-liquid interface, and a slope less than one leading upto the plateau, which occurs near the CMC. Similar behavior has been observed for the adsorption of polyethoxylated alcohols on a hydrophobic silica surface and was shown to be described well by the Frumkin adsorption isotherm (12).

In this paper, the adsorption behavior of two polyethoxylated nonionic surfactants on hydrophilic silica was investigated to determine if the presence of inorganic electrolytes affects the mechanism of adsorption of polyethoxylated phenols with large oxyethylenic groups, as suggested in the results of Partyka et al. (7).

Materials

Nonionic Surfactants

The two polyethoxylated phenol nonionic surfactants used in this study were supplied by Huntsman Chemical. Table 1 provides a description of the surfactants structure.

Table 1

Surfactant	Alkyl Chain Length	EO Number
H-C_8EO_{40}	8	40
H-C_9EO_{55}	9	55

Adsorbent Materials

The silica used in this study was provided by Evonik. Aerosil 300 is a nano sized fumed silica with a BET measured specific surface area of 300 m^2/g and Aeroperl is a micron sized, porous fumed silica with a BET measured specific surface area of 300 m^2/g. Aerosil 300 does not have any significant internal surface area. Aeroperl surface area is predominantly internal.

Methods

The following reviews the methods that will be used to complete the experiments required to investigate the effect a high salinity environment has on the adsorption of nonionic surfactants.

Adsorption Isotherms

Distinct adsorption isotherms are developed for each nonionic surfactant at various concentrations of sodium chloride. All other variables are held constant, such that any differences between the two adsorption isotherms will be directly related to the difference in salinity levels.

Surfactant solutions are made at a variety of concentrations, ranging from 10 to 100,000 micromoles per liter. Twenty milliliters of each surfactant solution is pipetted into a 40 mL vial containing 0.3 grams of Aerosil or Aeroperl. Samples are then equilibrated at 30 °C in a water bath for two days. After the samples are equilibrated, the concentration was determined using HPLC and UV spectroscopy.

Results and Discussion

Adsorption isotherms are developed for H- C_9EO_{55} and H- C_8EO_{40} in solutions of distilled water, 0.15 M NaCl, and 1.5 M NaCl in order to determine the effect salt has on the adsorption of polyethoxylated phenol nonionic surfactants.

In each case, the adsorption isotherms show the same shape as those determined in previous studies. The adsorption density increases linearly with a slope less than one without the break which is characteristic of aggregate formation at the water-silica interface. Figures 2 -4 show the adsorption isotherms of H-C_8EO_{40}., and Figures 5-7 show the adsorption isotherms of H-C_9EO_{55}. Two adsorption isotherms are plotted on Figures 3, 4, 6, and 7. One isotherm corresponds to surfactant adsorption on Aerosil and the other to adsorption on Aeroperl. The figures show that adsorption is higher on the Aerosil silica than it is on the Aeroperl silica, which is most likely the result of smaller pores in the Aeroperl limiting the adsorption of the large surfactant molecules onto the internal surface area of the Aeroperl.

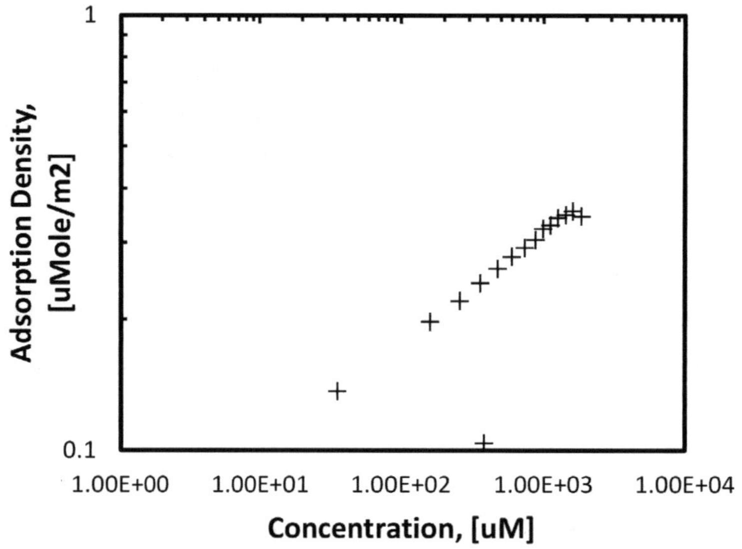

Figure 1: H – C_8EO_{40} Adsorption Isotherm in Di Water with Aerosil 300

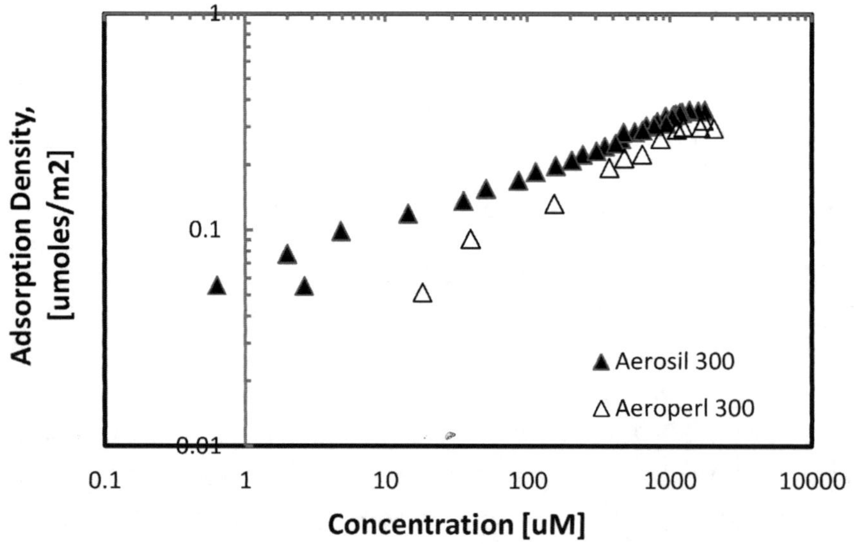

Figure 2: H – C_8EO_{40} Adsorption Isotherm in 0.15 M NaCl Solution

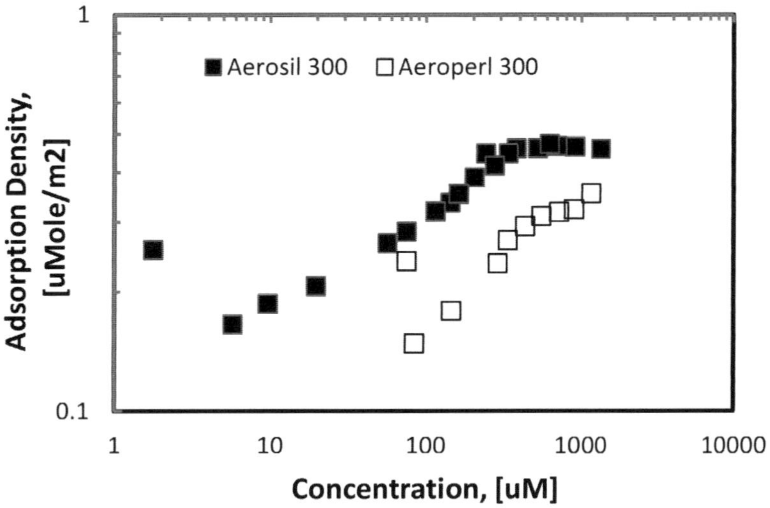

Figure 3: H – C_8EO_{40} Adsorption Isotherm in 1.5 M NaCl Solution

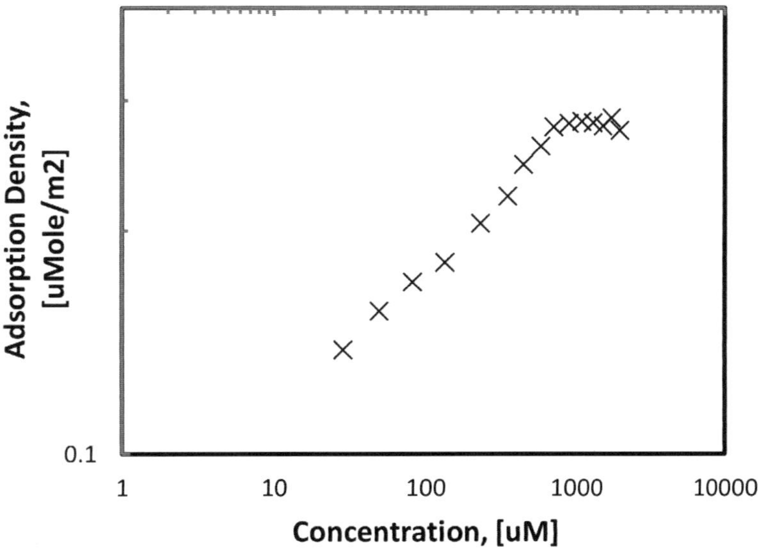

Figure 5: H – C_9EO_{55} Adsorption Isotherm in Di Water with Aerosil 300

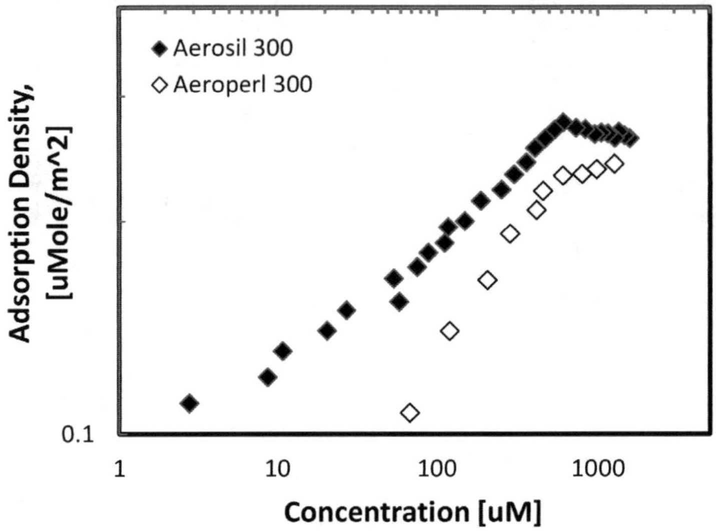

Figure 6: H – C_9EO_{55} Adsorption Isotherm in 0.15 M NaCl Solution

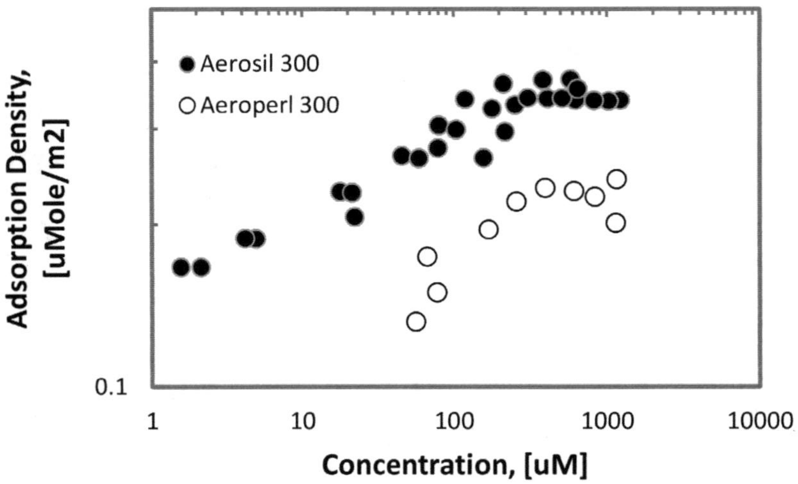

Figure 7: H – C_9EO_{55} Adsorption Isotherm in 1.5 M NaCl Solution

When Figures 8 and 9 are plotted together, , it is apparent that the increased salinity causes the total adsorption to increase, presumably due to the same electrolyte-surfactant interactions that cause the CMC and cloud point to decrease. The total adsorption densities and area per molecule at the silica-water interface are summarized in Table 2.

Table 2: Maximum Adsorption Density and Minimum Area per Molecule for H-C_8EO_{40} and H-C_9EO_{55}

Solution	C_8EO_{40}	C_9EO_{55}
	Maximum Adsorption Density, [μmoles/m^2]	Maximum Adsorption Density, [μmoles/m^2]
Di water	0.35	0.28
0.15 M NaCl	0.35	0.28
1.5 M NaCl	0.46	0.34
	Minimum Area per Molecule, [$Å^2$/molecule]	Minimum Area per Molecule, [$Å^2$/molecule]
Di water	460	600
0.15 M NaCl	460	600
1.5 M NaCl	350	450

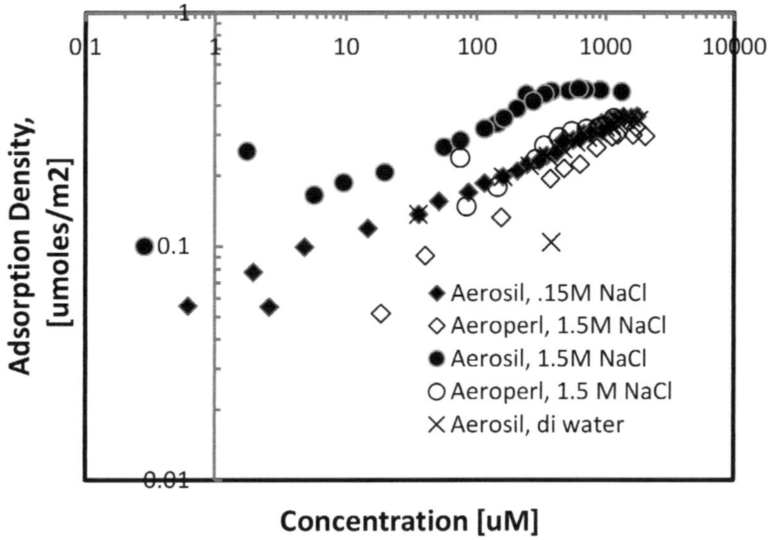

Figure 8: H – C_8EO_{40} Combined Adsorption Isotherms

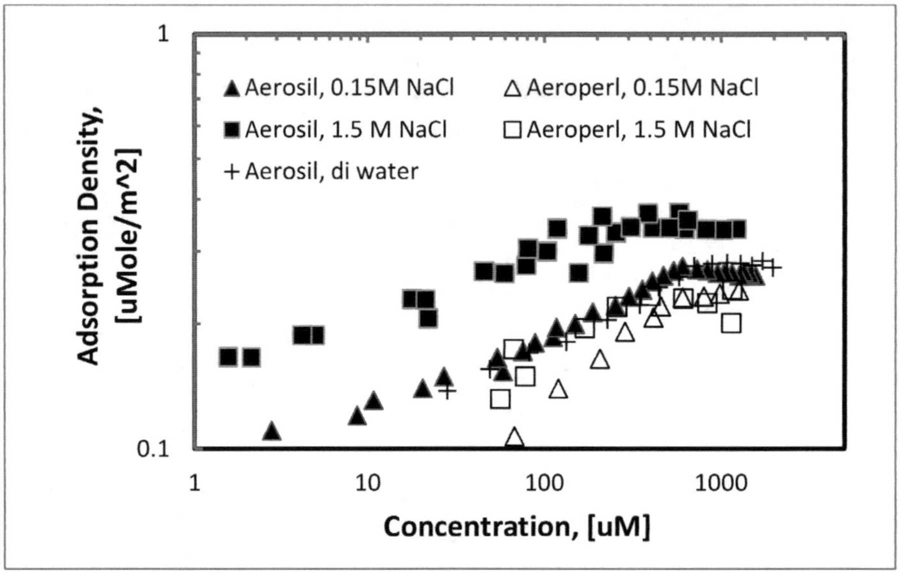

Figure 4: H – C_9EO_{55} Combined Adsorption Isotherms

Though the total adsorption density changes with changing salt concentration, the general shape of the adsorption isotherms does not change with salt concentration. To compare the curve shapes, adsorption isotherms can be normalized by dividing the concentrations by the critical micelle concentration and dividing the adsorption density by the adsorption density at the critical micelle concentration. This transformation results in single curve isotherms, for a given surfactant. This is shown in Figure 5, and Figure 6. This result implies that the salinity does not affect the mechanism of adsorption.

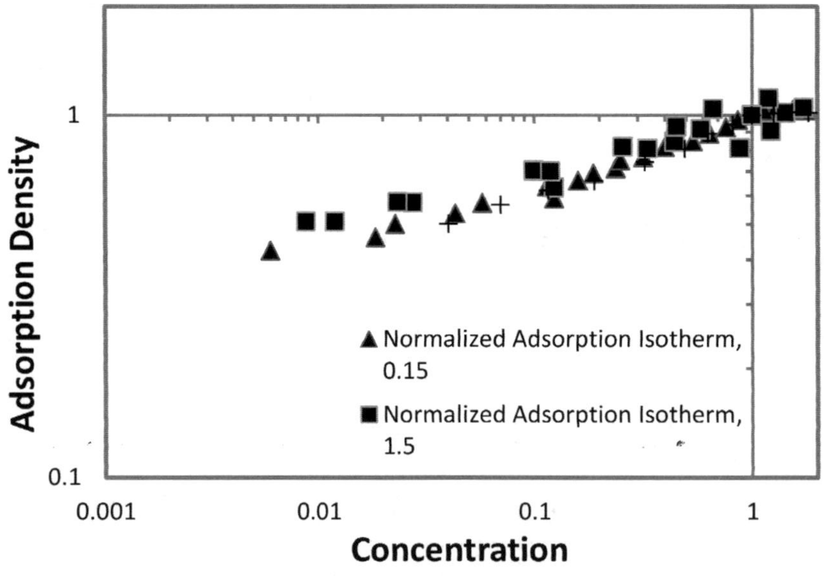

Figure 5: Normalized Adsorption Isotherms for H – C_9EO_{55}

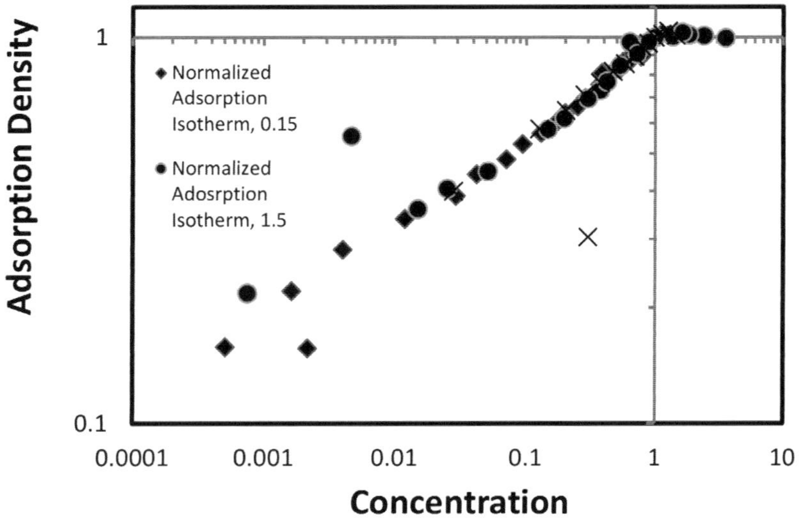

Figure 6: Normalized Adsorption Isotherms for H – C_8EO_{55}

Conclusion

The results clearly indicate that increasing salt concentration increases the adsorption of polyethoxylated phenols on silica. This result is most likely due to the interactions of the sodium and chloride ions with the water molecules surrounding the surfactant. These same interactions are responsible for the lowering of the critical micelle concentration and cloud point of the surfactant. This suggests the adsorption of surfactant and the aggregation of surfactant in bulk solution is related. Furthermore, the adsorption isotherms increase linearly with concentration until the critical micelle concentration is reached, and there is no break in the curve indicating a critical admicelle concentration, and suggesting a mechanism of competitive adsorption. This mechanism is most likely due to the large size of the polyethylene oxide chains, which prevents admicelle formation. Therefore the adsorption density does not depend on the surfactant – surfactant interaction, but only on the size of the polyethoxylate chain and its interaction with the silica surface. It is likely that the area per molecule at the silica-water interface is directly proportional to the square root of the number of oxyethylenic groups, like at the liquid-air interface.[13] The similarity between the shape of the adsorption isotherms for the nonylphenol and octylphenol and the collapse of the normalized isotherms to a single curve suggests a consistent mechanism of adsorption that is not influenced by salinity.

Current work is being conducted on how the cmc and cloud point change with salt concentration and whether there is a correlation with the adsorption. Adsorption isotherms will also be developed for several more surfactants to assess how the mechanism of adsorption changes with surfactant structure, including isotherms for surfactants with lower EO numbers.

References

1. Mohan, H.; Biglarbigi, K.; Carolus, M.; Van Wagener, D. (2011), *"Assessing the EOR Potential of the United States".* SPE Enhanced Oil Recovery Conference, Kuala Lumpur, Malaysia, July 19-21.
2. Manrique, E.; Thomas, C.; Ravikiran, R.; Izadi, M.; Lantz, M.; Romero, J. (2010), *"EOR: Current Status and Opportunities".* SPE Improved Oil Recovery Symposium, Tulsa, Oklahoma, USA, April 24-28.
3. Lui, Q.; Dong, M.; Zhou, W.; Ayub, M.; Zhang, Y.P.; Huang, S.(2003), *"Adsorption-Desorption-Related Interfacial Tension Behaviour in Chemical Flooding and Its Effects on Oil Recovey".* Canadian International Petroleum Conference, Calgary, Alberta, Canada, June 10-12.
4. Lewis, S.J.; Verkruyse, S.A.; Salter, S.J. (1986), *"Selection of Nonionic Surfactants for Minimized Adsorption and Maximized Solubilization".* SPE Enhanced Oil Recovery Symposium, Tulsa, Oklahoma, USA, April 20-23.
5. Verkuyse, L.A.; Salter, S.A. (1985), *"Potential Use of Nonionic Surfactants in Micellar Flooding".* SPE Oilfield and Geothermal Chemistry Symposium, Phoenix, Arizona, USA, March 9-11.
6. Somasundaran, P.; Snell, E.D.; Xu, Q.(1991), *"Adsorption Behavior of Alkylarylethoxylated Alcohols on Silica"*. Journal of Colloid and Interface Science 144, 165-173.
7. Partyka, S.; Zaini, S.; Lindheimer, M.; Brun, B. (1984), *"The Adsorption of Non-Ionic Surfactants on a Silica Gel"*. Colloids and Surfaces 12, 255-269.
8. Lindheimer, M.; Keh, E.; Zaini, S.; Partyka, S. (1989), *"Interfacial Aggregation of Nonionic Surfactants onto Silica Gel: Calorimetric Evidence"*.Journal of Colloid and Interface Science 138, 83-91.
9. Levitz, P.; Van Damme, H.; Keravis, D. (1984), *"Fluorescence Decay Study of the Adsorption of Nonionic Surfactants at the Solid-Liquid Interface. 1. Structure of the Adsorption Layer on a Hydrophilic Solid"* .Journal of Physical Chemistry 88, 2228-2235.
10. Rutland, M.W.; Senden, T.J. (1993), *"Adsorption of the Poly(oxyethylene) Nonionic Surfactant $C_{12}E_5$ to Silica: A Study Using Atomic Force Microscopy".* Langmuir 9, 412-418.
11. Levitz, P.(1991), *"Aggregative Adsorption of Nonionic Surfactants onto Hydrophilic Solid/Water Interface. Relation with Bulk Micellization".* Langmuir 7, 1595-1608.

12. Kumar, N.; Garoff, S.; Tilton, R.D. (2004), *"Experimental Observations on the Scaling of Adsorption Isotherms for Nonionic Surfactants at a Hydrophobic Solid-Water Interface".* Langmuir 20, 4446-4451.
13. Sedev, R. (2001), *"Limiting Area per Molecule of Nonionic Surfactants at the Water/Air Interface"*. Langmuir 17, 562-564.

(107b) CFD Modeling of Drop Coalescence - Improving Separator Efficiency

Santosh Appathurai, Chemical Engineering, Purdue University, West Lafayette, IN

Drop coalescence is commonly observed in numerous industrial processes of note, as well as in natural phenomena. The industrial processes typically tend to involve emulsions that either need to be stabilized or separated, depending on the specific application or end use. Of particular interest to the petroleum industry are multiphase systems involving dispersions of one fluid in another immiscible fluid medium. In several aspects of upstream engineering and flow assurance, one encounters such multiphase flows, and the need to remove the dispersed water from oil (or oil from water, during late life production). Some of these include, flow conditioning, water management, and coalescence-induced separation units. Drop coalescence is studied at a basic "doublet-system" level – where the underlying physics of two liquid drops coalescing in both a dynamically passive, as well as active external medium, are uncovered at very small (nanometer) length scales, using CFD. These computational results were also validated with controlled experiments. The physical insight obtained from such a grassroots study is then extended to improve the existing population balance models on the coalescence efficiency in separators. Additionally, a novel way of extending droplet shapes encountered during coalescence to multiphase flow measurement is also presented.

(107c) CFD Modeling of Electrically-Enhanced Oil-Gas Separators

Krishnaraj Sambath, Chemical Engineering, Purdue University, West Lafayette, IN

Oil and gas production is often accompanied by water and the separation of these phases becomes a necessity. Chemical de-emulsification and heat treatment have been widely applied methods to address this separation issue and these processes are time consuming, require injection of chemicals and the output could still have a water content of 5 – 10%. This separation limit is further overcome by using other techniques like coalescers where small water droplets are forced to coalesce with each other to form larger fast settling droplets. If the average droplet size is doubled, separation time decreases by a factor of four. An electrocoalescer takes advantage of the conducting nature of the water droplets and can reduce the water content to under 0.5%. In the presence of an electric field, these conducting water droplets become dipoles whose electric charges, in sufficiently high field strength, can overcome the repulsive surface-surface interactions, resulting in oil film drainage and consecutive coalescence. It follows that a fundamental quantitative understanding of water droplets' dynamics in electric field is very crucial in understanding and improving the operational envelope of this technology. In this work, the behavior of an isolated uncharged conducting drop in external electric field is simulated using CFD. The Navier-Stokes system augmented with Maxwell's stress and Maxwell's equations constitute the governing equations, which are solved using Galerkin finite element analysis. Drop dynamics, deformation, total charge & dipole-dipole interaction of water droplets as a function of applied field strength are discussed.

(108c) Asphaltenes From Tahe and Liaohe Oils: Structural Analysis and Their Interaction with Comb-Type Copolymers

Tao Yang, Jun Xu and **Xuhong Guo**, School of Chemical Engineering, East China University of Science and Technology, Shanghai, China

In this work, asphaltenes were extracted from Liaohe and Tahe crude oils following ASTM D6560-00/IP143. By means of ^1H NMR, XRD and FTIR, their aromaticity, aromatic sheets' size, and the number in a stack cluster were determined. Comb-type poly(maleic anhydride amide–co–α–olefin–co–styrene)s (MASC) were synthesized with different ratio of styrene/α-octadecene, and their effect on the behavior of asphaltenes in model oil consisting 4 wt% hexatriacontane in decane were investigated. The match of aromaticity between MASC and asphaltenes is of critical importance to disperse asphaltenes and thus improve the flow ability of crude oils.

(123a) Scaling in Oil and Gas: Understanding Surface and Bulk Processes

Anne Neville, School of Mechanical Engineering, University of Leeds, Leeds, United Kingdom

Formation of inorganic scale is a particularly troublesome problem in oil and gas flow assurance. Scale forms when the local environment is supersaturated and the kinetics are such that it is possible for crystals to form and grow. To date much emphasis has been on the processes occuring in the bulk solution. This paper discusses the competition between bulk and surface processes in the absence and presence of inhibitors and also discusses means of preventing scale deposition on engineering surfaces.

(123b) The Role of Scale in Flow Assurance

Mason B. Tomson[1], Amy T. Kan[2], Wei Shi[2], Wei Wang[2], Essmaiil Djamali[2], Chao Yan[2], Nan Zhang[2], Haiping Lu[2], Lu Wang[2], Jonathan A. Pennington[2]and Sarah N. Work[3], (1)Civil and Environmental Engineering, Rice University, Houston, TX, (2)Brine Chemistry Consortium, Rice University, Houston, TX, (3)Brine Chemistry Consortium, Rice University, Houston

The ultra-high temperature (150-250°C), pressure (1,000-2,000 bar) and TDS (>300,000 mg/L) in deepwater oil and gas production pose significant challenges to flow assurance and scaling control due to limited knowledge of mineral solubility, kinetics and inhibitor efficiency at these extreme conditions. Brines from multiple wells are collected, mixed in a subsea facility and transported vis subsea flowlines. Often the oil, gas, and brine composition in each well is quite different and consequently will cause significant scaling. The objective of this paper is to present an overview of the on-going research at Rice University to predict and control scale during oil production. Five key areas of research related to flow assurance scale control will be discussed:

1. The challenges of scale prediction at ultra high temperature, pressure, and TDS: Prediction of scale solubility is currently limited by inadequate understanding of the equilibrium constants and activity coefficients associated with common minerals. In some cases, there is also uncertainty about the mineral phases that need to be considered (i.e. silicates, sulfides, etc.). A new apparatus was built to test scale formation and inhibition at high temperatures and pressures up to 250°C, pressure up to 1,500 bar and ionic strength up to 6m in solutions with elevated concentrations of mixed electrolytes (e.g., calcium, magnesium, sulfate and carbonate) representing the maximum range of interferences expected (95%CI) in oil and gas wells. Predictions by a thermodynamic model based on Pitzer's ion interaction theory were updated with these new data to provide reliable predictions of solubilities that are consistent with both experimental and literature data at all conditions tested.

2. Efficient method to study the nucleation kinetics of scale formation and inhibition at these conditions. A rigorous and comprehensive study have been conducted to investigate: (1) the time (minutes to days) dependence of inhibitor thermal (up to 200 °C) degradation, (2) the impact of trace heavy metals, such as nickel and manganese on the degradation of inhibitors, (3) the difference between ageing tests with the presence and absence of core material for adsorption, and (4) the effectiveness of chemical treatment by applying chelating agents to prevent the catalytic degradation caused by trace heavy metals. The results enable a more accurate understanding of thermal degradation kinetics and shed light on potential chemical treatment to enhance inhibitor performance in high temperature environments.

3. Kinetics of nucleation, crystal growth and attachment in the presence and absence of scale inhibitors. In order to assess scaling risk, a better understanding of scale deposition kinetics on steel surface under realistic and complex oil field condition is needed. In this paper, we introduce the development of a novel $CaCO_3$pre-coated (55 μm thick uniform layer of well adhered $CaCO_3$) steel tubing for studies of $CaCO_3$crystal growth kinetics and inhibition kinetics at oilfield condition. This approach provides a relatively stable surface area and eliminates the limits of laboratory batch experiments.

(123c) Emulsion Stabilization, Breaking and Inversion: Advantage or Disaster in Flow Assurance

Jean-Louis Salagerand Ana Forgiarini, Lab FIRP Ingenieria Quimica, Universidad de Los Andes, Merida, Venezuela

@font-face { "Cambria"; }p.MsoNormal, li.MsoNormal, div.MsoNormal { margin: 0in 0in 0.0001pt; font-size: 12pt; ""; }div.Section1 { page: Section1; }

Emulsion type and stablity are directly linked with the physico-chemical formulation, which is a way to characterize the interaction between the surfactant substances at interface with the oil and water phases. Crude oil is generally produced with water, and because the natural surfactants, e.g. asphaltenes and resins, are generally quite lipophilic, the mixture result in a water-in-oil (W/O) emulsion. These emulsions have to be broken in the oil production process, or they have to be inverted to the other type, i.e. oil-in-water (O/W) to transport very viscous crude oils in pipelines. When the conditions are such that gas hydrate can be formed at the interface of water drops, a W/O emulsion could be transformed en an encapsulated drop dispersion. If these solid hydrate covered capsules do not coalesce, no plugging could take place during the transport. This depends on the formulation case which has to stabilize this dispersion. However, such instance is just contrary to the one which is used to break the emulsion at the oil well to separate water. As a consequence, the change in formulation can produce two opposite scenarios. The first one is the separation of water from oil, which is a requirement in the production. The second one is the production of an instability, and even phase inversion, which is going to help coalesce the water, often resulting in a disaster, i.e. the plugging of the pipe by gas hydrate.

Consequently, in the oil production process, the formulation has to be managed in a clever sequence of two opposite steps. In the first one, while the conditions for the formation of gas hydrate are met, the W/O emulsion has to be kept stabilized, even by adding lipophilic surfactants if the natural ones are not sufficient, so that hydrate covered water drops would not coalesce. Then, when the temperature and pressure are appropriate, in the second step, an hydrophilic surfactant of the demulsifier type should be added to break the emulsion and separate water from oil. Alternatively, an O/W emulsion of low viscosity could be used for pipelining. In practice, the formulation is genertally shifted by a combination of adding surfactants and changing temperature.

(123d) An Overview of Emulsion Stabilization Mechanisms in Petroleum-Water Mixtures

Peter Kilpatrick, Chemical and Biomolecular Engineering, University of Notre Dame, Notre Dame, IN

Emulsions of oil in water, but more typically water in oil, are common in the production, transportation, and refining of petroleum and petroleum products. While (1) the adsorption and molecular reorganization of asphaltenic aggregates at the oil-water interface is one of the most common methods of forming these stable emulsions, there are a host of other physical and chemical processes that can also either stabilize emulsion or combine with other processes to stabilize emulsion. Among these are (2) the adsorption of organic acids and their soaps, (3) the adsorption of wax particles, (4) the modification of the rheology of thinning films between emulsion droplets by macromolecules, (5) the adsorption of sub-micron sized inorganic particles such as clays, silica, and iron oxides, and (6) the formation of liquid crystalline films at the oil-water interface. In this talk, we will review the many means of emulsion and thin film stabilization of emulsions in petroleum-water mixtures, as well as the many physical and chemical means of characterizing both the materials that stabilize these emulsions and the means of characterizing the oil-water interface and the emulsion films themselves.

(123e) Effects of Multiphase Flow on Corrosion

Professor Srdjan Nesic, Institute for Corrosion and Multiphase Technology, Ohio University, Athens, OH

The effects of multiphase flow on internal corrosion of mild steel pipelines transporting liquid and gasous hydrocarbons is discussed. The analysis covers research, theoretical, modeling and practical aspects of the corrosion process, with the focus on the main challenges facing us today.

Estimation of Parameters for Simulation of Steady State Foam Flow in Porous Media

Kun Ma, Sibani Lisa Biswal and George J. Hirasaki

Department of Chemical and Biomolecular Engineering, Rice University, Houston, TX 77005, USA

This paper was prepared for presentation at the 2012 AIChE Spring Meeting & 8th Global Congress on Process Safety held in Houston, Texas, USA, April 1-5, 2012.

Summary

In this work, we perform experimental and simulation studies of foam in porous media. Air and surfactant solution (IOS 1518 with a salinity of 1.0% wt NaCl) are co-injected into a silica sand pack with syringe pumps. At a fixed total flow rate, both high-quality and low-quality foam regimes are observed in the experiments. At steady state, the saturation of aqueous phase in porous media is shown to be strongly dependent on foam strength.

We utilize the STARS foam model to fit the experimental results. By conducting superposition of the contour plots of the transition foam quality (f_g^*) and the foam apparent viscosity, one can obtain the optimal reference mobility reduction factor ($fmmob$) and critical aqueous phase saturation ($fmdry$) in the foam model. This approach provides an alternative to fit the foam model in addition to drawing the pressure contours of the experimental data, which may cause inaccuracy of parameter estimation due to the limited number of data points.

With a given surfactant formulation and total injection rate, the presented techniques can be used for prediction of foam apparent viscosity as a function of gas fraction and surfactant concentration, which optimizes the injection strategy for foam enhanced oil recovery (EOR) processes with minimal cost.

Technical background

The foam model using the fractional flow theory (Cheng, Reme et al. 2000), or the STARS^{TM} foam model (Computer Modeling Group 2007), has been widely used to describe foam flow through porous media in the oil and gas industry. These applications include simulation of surfactant-alternating-gas processes (Renkema and Rossen 2007) and simultaneous injection of surfactant solution and gas (Masalmeh, Wei et al. 2011), simulation of oil displacement by foam (Ashoori, van der Heijden et al. 2010; Liu, Andrianov et al. 2011; Zanganeh, Kam et al. 2011), interpretation of experimental data of co-injection of gas and surfactant solution on the 10 m scale (Vassendan, Holt et al. 1999) and in Berea cores (Chalbaud, Moreno et al. 2002), and simulation of foam process at reservoir scale from the Snorre field in Norway (Skauge, Aarra et al. 2002).

The STARS^{TM} foam model is capable to simulate foam flow in both high-quality and low-quality foam regimes. Cheng and coworks (Cheng, Reme et al.

2000) showed how to fit the parameters to steady-state laboratory coreflood data. We briefly reiterate the procedure as follows. (1) Draw vertical ∇p contours through data in the high-quality regime and horizontal contours in through data in the low-quality regime; (2) Pick one representative ∇p contour in the high-quality regime and find $fmdry$ using the equation $\nabla p = u_w \dfrac{\mu_w}{kk_{rw}(fmdry)}$; (3) Determine f_g^* (transition foam quality) by the ratio of flow rates at the transition region of high-quality and low-quality regimes; (4) Estimate $fmmob$ by using $S_w = fmdry$ and applying the fractional flow equation at the transition foam quality f_g^*.

The above modeling process is physically clear, however, it requires the experimental ∇p data to be vertical in the high-quality regime and horizontal in the low-quality regime in order to draw the contours of ∇p, which is not easily obtained in some cases (Alvarez, Rivas et al. 2001; Kim, Dong et al. 2005). Generally, the STARS™ foam model gives good trend of fit to experimental data in both of high-quality and low-quality regimes in these figures, but this model is not able to handle the complex wiggly shape of experimental ∇p contours with a unique set of parameters. Meanwhile, the ∇p contours generated by computer programs are limited to the measured experimental data points, and insufficient data points may cause inaccuracy of the ∇p contours.

Thus, we develop a method to implement the STARS™ foam model without using ∇p contours. Our method is used to model foam flow with a fixed total flow rate of both gas and liquid. Furthermore, we investigate the effect of surfactant concentration on foam strength and verify the function of surfactant in the current foam model in STARS^{TM}. The model can then be used for prediction of foam apparent viscosity as a function of gas fraction and surfactant concentration, in the form of contour plot or 3D plot. This plot will be useful for practical applications to minimize surfactant cost and achieve the required foam strength.

Materials and methods

IOS1518: Internal olefin sulfonate 1518 (19.42% wt active ingredient). This product was manufactured by Stepan (Lot# 18239-032708) with a trade name of Petrostep S-2A.

Silica sand 20/40 is used to pack the porous media in the 1-D glass column. The sand pack has a diameter of 2.58 cm and a length of 27.5 cm. The permeability is 158.0 darcy and the porosity of 36.0% for this system.

IOS1518 with a concentration of 0.1% wt is dissolved in brine and the salinity is fixed at 1.0% wt NaCl. Surfactant solution is co-injected with air to the sand pack at a total superficial velocity of 20 ft/day. A pressure transducer is located at the inflow end, and it is assumed to be measuring pressure drop across the porous media. The outflow end is open to atmosphere. The water saturation in porous media is measured by weighing the sand pack.

Before the experiments, we clean the sand pack by alternately injecting 5 PV (pore volume) DI water and CO_2 (1 psi for 30 min) for 3 cycles. Subsequently, 2

PV of surfactant solution is injected to the porous media to minimize surfactant adsorption in our experiments. A series of experiments starts from 99% air injection displacing surfactant solution and end up with 10% air injection. When a steady state is achieved and the pressure drop reaches a plateau value, the pressure drop is averaged over the period of the steady state and the foam apparent viscosity is calculated through 1-D Darcy's law.

Results and discussions

The effect of foam quality on foam apparent viscosity is shown in Figure 1. We observe two distinctive foam regimes: high quality regimes and low quality regimes (Osterloh and Jante Jr 1992; Alvarez, Rivas et al. 2001). The foam apparent viscosity increases when injected gas fraction increases in the low quality regime, and decreases when injected gas fraction increases in the high quality regime. At the boundary of the two regimes, foam obtains maximal apparent viscosity at a given surfactant concentration and total superficial velocity.

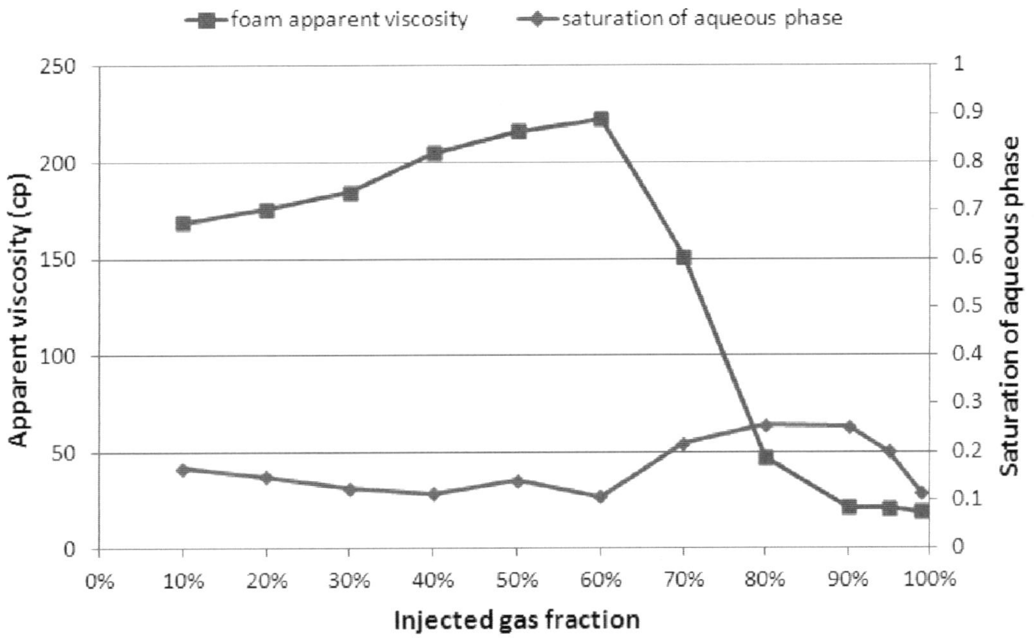

Figure 1. Foam apparent viscosity and saturation of aqueous phase at different injected gas fractions.

The effect of foam quality on the saturation of aqueous phase is also shown in Figure 1. In general, typical fraction flow curves show that higher injected gas fraction results in lower water saturation in the absence of foam. However, it is found in Figure 1 that aqueous phase saturation is more dependent of foam strength than gas fractional flow at high injected gas fractions.

The foam model we use in this study is described as follows (Ashoori, van der Heijden et al. 2010):

$$k_{rgD}^{f} = k_{rg}^{f} / k_{rg}^{f0} = (k_{rg}^{nf} / k_{rg}^{0}) \times FM = (\frac{1 - S_{gr} - S_{w}}{1 - S_{gr} - S_{wc}})^{epg} \times FM$$

where $FM = \dfrac{1}{1 + fmmob \times F_{water} \times F_{surf}}$

$$F_{water} = 0.5 + \frac{\arctan[epdry(S_w - fmdry)]}{\pi}$$

$$F_{surf} = \begin{cases} (\frac{C_{sw,inj} C_{swD}}{fmsurf})^{epsurf} & \text{for } C_{swD} < (fmsurf / C_{sw,inj}) \\ 1 & \text{for } C_{swD} \geq (fmsurf / C_{sw,inj}) \end{cases}$$

Because there are many parameters in the model, it is important to develop a strategy for quickly estimating all of them before running the simulation. To fit the parameters in the model, the first step is to focus on the term F_{water} and lay aside the term F_{surf}. Now, there are only left three parameters to fit: $fmmob$ (reference mobility reduction factor), $fmdry$ (critical aqueous phase saturation) and $epdry$ (a parameter regulating the slope of F_{water} curve near $fmdry$).

If the application focuses on a fixed total superficial velocity, we propose a simple method to fit the parameters. Our process is shown below:

(1) For a fixed surfactant concentration at different foam qualities, identify the transition foam quality $f_g^* = A$ and the maximal foam apparent viscosity $\mu_{foam}^* = B$ from the experimental data.

(2) Draw the contour plot of f_g^* with respect to $fmmob$ and $fmdry$ using the equation $f_g^* = 1 - (1 + \frac{k_{rg}^{nf}(S_w = S_w^*) \times FM(S_w = S_w^*)}{k_{rw}(S_w = S_w^*)} \times \frac{\mu_w}{\mu_g})^{-1}$ based on the fractional flow theory.

(3) Draw the contour plot of μ_{foam} at $f_g = A$ with respect to $fmmob$ and $fmdry$ at steady state.

(4) Make superposition of contour plots of f_g^* in Step (2) and μ_{foam} in Step (3), and identify $fmmob$ and $fmdry$ where $f_g^* = A$ and $\mu_{foam} = B$ cross over.

In this case, $f_g^* = 0.6$ and $\mu_{foam}^* = 222 \quad cp$ are from the experimental data in Figure 1. Contour plots of f_g^* and μ_{foam} are drawn in Figure 2 and Figure 3, respectively. Superposition of these two plots results in $fmmob$ of 11600 and $fmdry$ of 0.34 where $f_g^* = 0.6$ and $\mu_{foam} = 222 \quad cp$ cross over as shown in Figure 4.

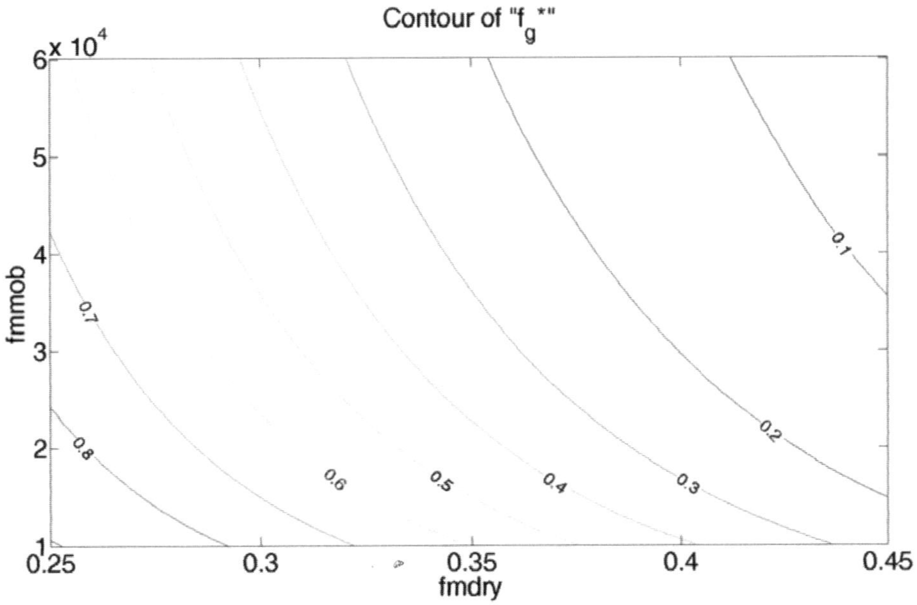

Figure 2. Contour plot of f_g* as a function of *fmmob* and *fmdry*.

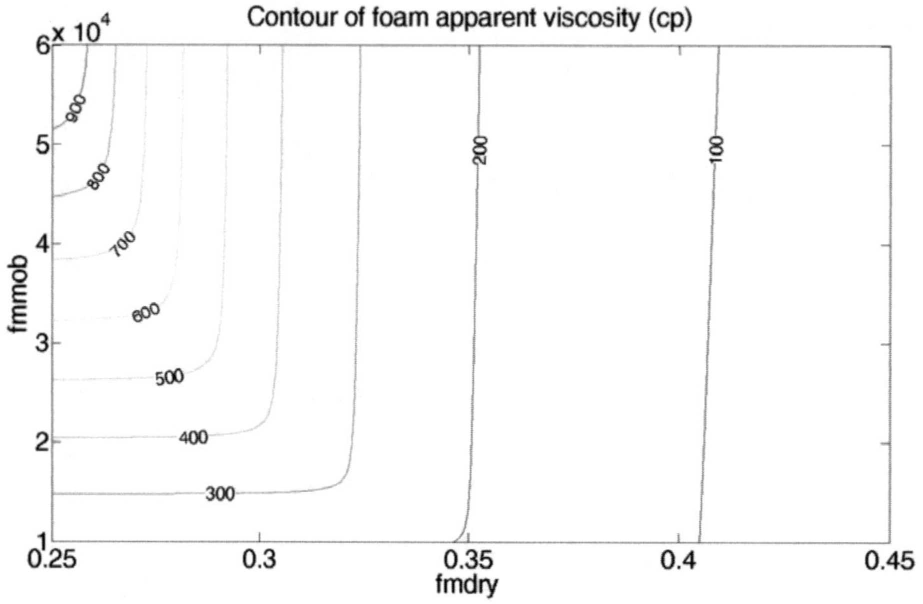

Figure 3. Contour plot of foam apparent viscosity as a function of *fmmob* and *fmdry*. Injected gas fraction is set to be 60% and F_{surf} is set to be 1.

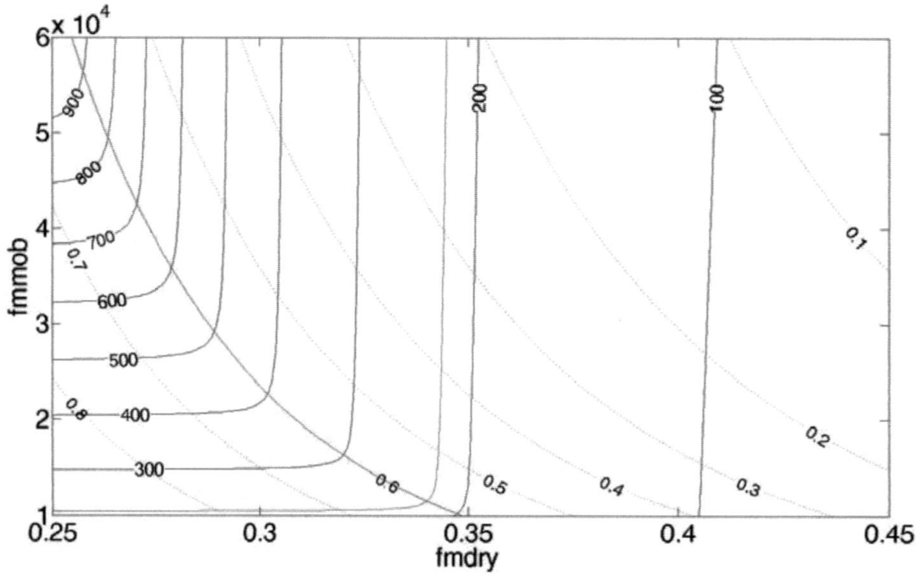

Figure 4. Superposition of contour plots in Figure 2 and Figure 3 to find the position where $f_g^* = 0.6$ and $\mu_{foam} = 222 \ \ cp$ cross over.

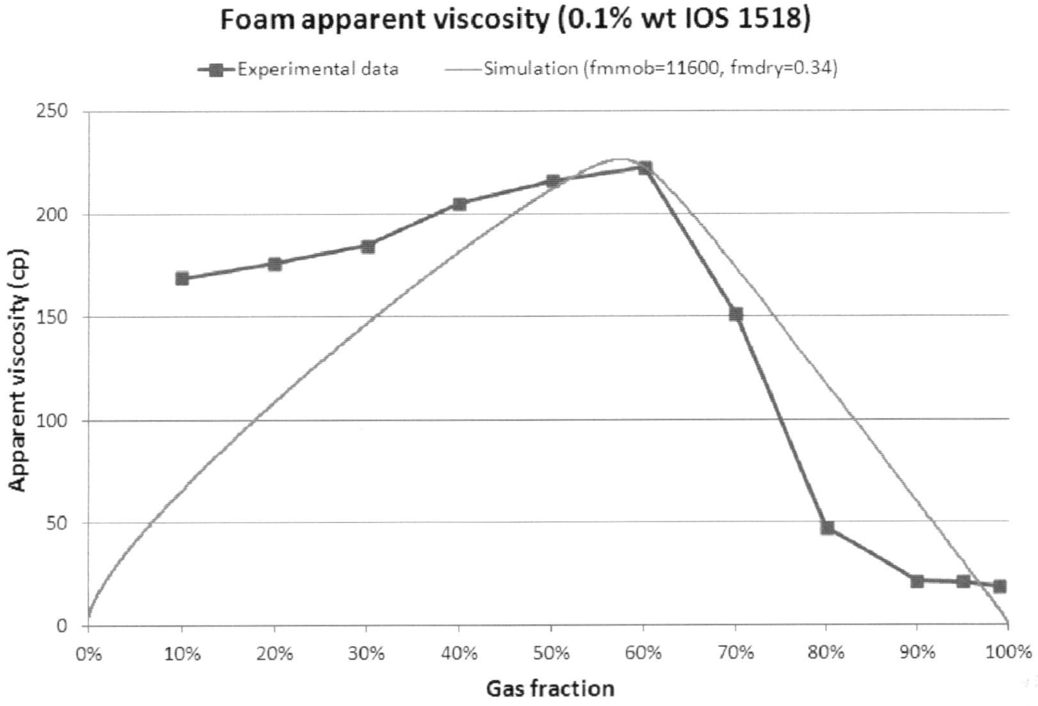

Figure 5. Comparison of foam apparent viscosities between simulation and experiment. $fmmob$ = 11600; $fmdry$ = 0.34.

We use the relative permeability curves in the literature which demonstrated successful foam simulations with the above model (Ashoori, van der Heijden et al. 2010). We also use the value of $epdry$ in the literature (Ashoori, van der Heijden et al. 2010) and will tune it in the future as needed.

The comparison of foam apparent viscosities between simulation and experiment for 0.1% IOS1518 is shown in Figure 5. The simulation shows good match to the experimental data at 50% and 60% gas fractional flow, and the same trend as the experimental data in both low-quality and high-quality regimes. However, deviations are seen in both regimes between simulation and experiment. A possible explanation in the high-quality regime is the requirement of minimum pressure gradient for strong foam generation is not involved in the foam model. In the low-quality regime, the experimental foam viscosity may be dependent on its previous flooding since we decrease injected gas fraction in the experiment. Hysteresis may be existent in the system if the experiment starts from low gas fractional flow to high gas fractional flow.

References

Alvarez, J. M., Rivas, H. J., et al. (2001). Unified Model for Steady-State Foam Behavior at High and Low Foam Qualities. SPE Journal **6**(3).

Ashoori, E., van der Heijden, T. L. M., et al. (2010). Fractional-Flow Theory of Foam Displacements With Oil. SPE Journal **15**(2): 260-273.

Chalbaud, C. A., Moreno, R. A., et al. (2002). Simulating Foam Process for a Venezuelan Pilot Test. SPE Annual Technical Conference and Exhibition. San Antonio, Texas.

Cheng, L., Reme, A. B., et al. (2000). Simulating Foam Processes at High and Low Foam Qualities. SPE/DOE Improved Oil Recovery Symposium. Tulsa, Oklahoma.

Computer Modeling Group (2007). STARS™ User's Guide. Calgary, Alberta, Canada.

Kim, J. S., Dong, Y., et al. (2005). Steady-state flow behavior of CO2 foam. Spe Journal **10**(4): 405-415.

Liu, M., Andrianov, A., et al. (2011). Sweep Efficiency in CO2 Foam Simulations With Oil. SPE EUROPEC/EAGE Annual Conference and Exhibition. Vienna, Austria.

Masalmeh, S. K., Wei, L., et al. (2011). Mobility Control for Gas Injection in Heterogeneous Carbonate reservoirs: Comparison of Foams versus Polymers. SPE Middle East Oil and Gas Show and Conference. Manama, Bahrain.

Osterloh, W. T. and Jante Jr, M. J. (1992). Effects of Gas and Liquid Velocity on Steady-State Foam Flow at High Temperature. SPE/DOE Enhanced Oil Recovery Symposium. Tulsa, Oklahoma.

Renkema, W. J. and Rossen, W. R. (2007). Success of Foam SAG Processes in Heterogeneous Reservoirs. SPE Annual Technical Conference and Exhibition. Anaheim, California, U.S.A.

Skauge, A., Aarra, M. G., et al. (2002). Foam-Assisted WAG: Experience from the Snorre Field. SPE/DOE Improved Oil Recovery Symposium. Tulsa, Oklahoma.

Vassendan, F., Holt, T., et al. (1999). Foam propagation on semi-reservoir scale. SPE Reservoir Evaluation & Engineering **2**(5): 436-441.

Zanganeh, M. N., Kam, S. I., et al. (2011). The Method of Characteristics Applied to Oil Displacement by Foam. SPE Journal **16**(1): 8-23.

Chemical Enhanced Oil Recovery: Novel Surfactants - The Concept of Large-Hydrophobe Alkoxy Carboxylate Surfactants

Upali Weerasooriya [*], Gary Pope, Sriram Solairaj, Jun Lu, Stephanie Adkins, Gayani P. Arachchilage, Do Hoon Kim, Christopher Britton, The University of Texas at Austin
[*] Address correspondence to upali@mail.utexas.edu
(Submitted to the 2012 AIChE Spring Conference)

Abstract

The large hydrophobe concept in surfactants for Chemical Enhanced Oil Recovery (CEOR) is a very recent phenomenon. An elegant method of making a large hydrophobe which can be converted into an effective surfactant molecule is the well known Guerbet reaction. This reaction produces large, branched hydrophobes through the dimerization of linear alcohols. High-performance, cost-effective Enhanced Oil Recovery (EOR) surfactants are produced by carboxylation (carboxymethylation) of large Guerbet alkoxylates. Alkoxy groups such as propylene oxide (PO) and ethylene oxide (EO) can be easily added as extenders to the Guerbet alcohol, followed by carboxylation to make the anionic surfactant. Previously, the use of cost-effective Guerbet alkoxy sulfate surfactants for EOR applications at high temperatures was established by greatly enhancing their thermal stability at high pH (10-11). Alternative thermally and chemically stable anionic surfactant molecules are highly desired, especially for application under conditions where alkali usage is prohibitive. These novel large-hydrophobe carboxylate surfactants meet this need. In addition, the Guerbet alkoxy carboxylate structures can be tailored to fit the specific EOR requirements. These surfactants are stable at elevated temperatures both with and without alkali, and furthermore, they may be used in environments of high salinity and high hardness (high concentration of divalent ions). These carboxylate surfactants have been used in formulations to produce ultra-low interfacial tensions with low microemulsion viscosities for a wide variety of crude oils under a broad range of reservoir conditions. Both sandstone and carbonate corefloods have been conducted with excellent results. Thus, the new carboxylate surfactants provide a breakthrough that greatly enhances the commercial potential of CEOR.

Introduction

The recent developments of new surfactants and the subsequent research advancements have taken CEOR to a new level by greatly broadening the applications of chemical enhanced oil recovery to a much wider range of reservoir conditions in a cost-effective way. A better understanding of the relationship between the surfactant structure and performance improves the process of screening for identifying suitable high-performance surfactants for EOR [1]. Surfactant microemulsion phase behavior scans [2-4] are the most efficient way to screen surfactants and other chemicals for a wide range of crude oils, from light to viscous, and even heavy oils, under a variety of conditions due to their ability to make quick visual observations such as microemulsion formation, viscosity, equilibration time, etc.

Reservoirs with relatively high temperature, high salinity and high hardness conditions, have posed challenges for CEOR, and thus have attracted the attention of many researchers. The key advances are heavily dependent on the novel surfactants described here-in. Surfactants containing large hydrophobes have been identified as being very efficient in achieving ultralow interfacial tension (IFT) for difficult oils [3]. The need for such surfactants to achieve ultra-low IFT while having good tolerance to temperature, salinity and hardness is increasing rapidly, due to ever increasing number of candidate reservoirs with harsh conditions. Previously, it was shown that cost-effective, high-performance surfactants can be produced in the form of Guerbet alkoxy sulfate surfactants (GAS) [5]. Very large hydrophobes can be produced from much smaller linear alcohols using the Guerbet reaction which dimerizes the linear alcohol using base (+Transition Metal) catalysis at high temperatures [6]. High performance Anionic surfactants were then produced by alkoxylation of the Guerbet alcohol with multiple propylene oxide (PO) and ethylene oxide (EO) units, followed by sulfation which is a lower cost alternative to the more complex sulfonation process.

Adkins [5] demonstrated that low cost Guerbet alkoxy sulfates (GAS) made with very large hydrophobes exhibited good performance with difficult oils. Furthermore, the GAS can be made in various hydrophobicities by changing the number of propylene oxide (PO) and ethylene oxide (EO) groups to cater to different reservoir conditions. However, GAS is not chemically stable at high temperature (>50C) unless pH is maintained high (10-11) with the use of alkali. Although use of alkali offers several advantages [7], in many cases, it cannot be used. For example, carbonate formations have anhydrite or gypsum which can result in divalent ion precipitation in the presence of alkali; or hard water is the only available water for injection. Ether sulfonate surfactants such as alkoxy glycidyl sulfonates are tolerant to high temperature. These were tested at high temperatures by Puerto [8]. However, the ether sulfonates are relatively expensive, and none are commercially available at present. Moreover, they usually have short hydrophobes and high hydrophilicity and are not very effective for crude oils with high equivalent alkane carbon number (EACN). Alternative, cost effective, as well as thermally and chemically stable surfactants are very much in need for such conditions.

Alkyl alkoxy carboxylates offer an alternative anionic surfactant class for EOR [9-11]. By including high levels of PO and EO in the molecule, highly versatile alkyl polyoxy propylene polyoxyethylene carboxylates (AEC) can be made [12]. As opposed to alkyl carboxylates which are commonly referred to as soap, AEC exhibit an very high tolerance to water hardness and electrolyte change. Along with high chemical stability to acidic and alkaline pHs [13-15], the AEC thus become a superior alternative to all other anionics.

The carboxylates are produced by the reaction of the terminal hydroxyl group of an alkoxylate with sodium chloroacetate in the presence of sodium hydroxide and can be commercially made at >90% purity [14-16]. Alternatively, the carboxylation reaction can be carried out using chloroacetic acid with two mole equivalents of sodium hydroxide (see schematic in Figure 1). The AEC are high performance surfactants which exhibit good foaming ability, rapid surface wetting, and good detergency, and so they can be used as emulsifying, solubilizing, or dispersing agents [13-15]. Therefore, AEC are highly versatile and can be utilized effectively in many different fields due to advantageous physicochemical properties. In addition, they have good compatibility with representatives of all surfactant classes and ease of handling [14,15]. Most importantly for EOR, the AEC have a high thermal and chemical stability and can be used at both acidic and alkaline pHs, i.e. they are stable to acid and alkali hydrolysis [13,14].

In order to use AEC as an alternative to Guerbet alkoxy sulfate surfactants for chemical EOR, cost-effective, large-hydrophobe AEC must be available. These can be produced by dimerizing linear alcohols to form a Guerbet alcohol, which can then be extended by addition of alkoxy groups such as PO and EO followed by carboxylation (carboxymethylation) to form the final cost-effective anionic surfactant [12]. The Guerbet alkoxy carboxylate (GAC) surfactant structure can be tailored to fit specific EOR needs by varying the number of carbons in the Guerbet alcohol, the number of PO and EO groups. GAC has been successfully tested in the laboratory in phase behavior and core flood experiments with a wide variety of crude oils and reservoir conditions.

Experimental Materials and Procedures

a. Surfactants and Materials

-Anionic Surfactants

Guerbet alkoxy carboxylates were synthesized in the laboratory at The University of Texas. The internal olefin sulfonates (IOS), alcohol propoxy sulfates (APS) and alkyl benzene sulfonates (ABS) were obtained from Stepan Company, Shell Chemical Company, and Huntsman Corporation

-Cosolvents

Triethylene glycol mono butyl ether (TEGBE) and Isobutyl alcohol (IBA) were received from Aldrich Chemicals.

-Polymers

The partially hydrolyzed polyacrylamide polymer Flopaam™ 3330s and 3630s were received from SNF Floerger (Cedex, France).

-Electrolytes and Brines

Sodium chloride, sodium carbonate, calcium chloride, magnesium chloride, and sodium sulfate were obtained from Fisher Chemical company. Specific synthetic brines were made and used based on each specific reservoir brine compositions.

-Crude oil

Several dead crude oils, as well as surrogate oils simulating live crudes, were used in the study and are shown in Table 1 along with their properties. It is well established that reservoir pressure and solution gas can significantly change the microemulsion phase behavior, and thus cannot be ignored. The understanding of equivalent alkane carbon number (EACN) concept is very useful in modeling the live oil [17-22]. It is much slower, difficult and expensive to perform phase behavior and core flood experiments under high pressure. Therefore, the surrogate oil, a mixture of dead crude and a less volatile, low-EACN hydrocarbon to match the live oil EACN, were used for the experiments at ambient pressure to account for the effect of solution gas on phase behavior.

b. Phase Behavior Experiments

Phase behavior tests are studied as a quick way to identify effective surfactant formulations for specific oils at the reservoir temperature and brine conditions. The detailed procedure for phase behavior experiments was reported in several papers [2-4]. The surfactant mixtures with oil and brine, after mixing, were carefully observed over an extended period of time. The formulations that quickly form low viscosity microemulsions and display the ultra-low IFT with both oil and aqueous phase were selected for further evaluation. The aqueous stability of the mixtures of surfactant and brine was tested at both room temperature and reservoir temperature. The aqueous solutions were observed to ensure that no cloudiness and/or phase separation occurs up to the desired injection salinity. The selected formulation was further evaluated in a core flood experiment in order to demonstrate its efficacy.

c. Aqueous Solubility Experiments

The aqueous solubility of each chemical formulation was tested by adding the surfactant solution prepared for the phase behavior experiments to 20 mL glass ampules. Typically a scan (either salt or sodium carbonate) that mirrors the phase behavior scan is produced. In some cases polymer (usually at concentrations of about 1000-3000 ppm) is added to the aqueous solubility ampules. The ampules are blanketed with Argon and sealed using the propane-oxygen torch. The ampules are mixed using the Vortex Genie 2 until a homogenous solution is created. Observations of the aqueous solutions are recorded first at room temperature and then the ampules are equilibrated at reservoir temperature using an oven. The aqueous solutions continue to be monitored after reaching the desired temperature.

d. Stability Experiments

The hydrolytic stability at elevated temperatures of the alkoxy sulfate surfactants were tested by studying surfactant solutions over time at 85-100 °C [5]. The thermal and chemical stability of the GAEC surfactants were studied by comparing the aqueous solubility and phase behavior results over time at elevated temperatures.

e. Core floods

The chemical flood experiments were designed with favorable salinity gradients to maximize robustness of the floods [23-25]. Phase behavior tests identified the surfactant formulations and the salinities at which the surfactant slugs and polymer drives were injected. Differential pressure transducers were used to measure the pressure drop across the entire core, as well as individual sections of the core. The effluent from the core was collected at a set time intervalin fraction collectors in glass test tubes for analysis. There are many papers describing the detailed procedure for core flood experiments for many years and will not be explained here [2-4]. The properties of the cores used for core floods are listed in Table 2, and all the core floods in this paper are summarized in Table 3. Table 4 lists the brine compositions used in each coreflood.

Results and Discussions

a. Phase Behavior and Stability of GEC

Phase behavior experiments were designed to study the behavior of the mixtures of oil, brine, and chemicals (surfactant formulation including co-solvent plus additional salt or alkali) at reservoir temperature. Ideally, after a few days of mixing, a middle-phase microemulsion will form that has a low viscosity and an ultra-low interfacial tension with both the excess aqueous and oil phases. The GAC surfactants have been used under a wide variety of conditions of temperature, salinity, divalent ion concentration, pH, and crude oil to develop effective surfactant formulations. Table 5 presents several formulations with long-term aqueous solubility.

Figure 2 shows a photograph of the microemulsion phase behavior pipettes with a formulation consisting of 0.5% C32-7PO-32EO carboxylate and 0.5% C19-23 IOS at 100 °C with a WOR of 1 using Oil #5 after more than 290 days. The elevated temperature of this experiment made the need for a co-solvent unnecessary. The phase behavior scan consists of brine #1 (composition in Table 6) with additional Sodium chloride added to produce a salinity gradient (increasing salinity from left to right). The Winsor microemulsion type is labeled in the photograph where the microemulsion goes from Type I to Type III to Type II as the salinity of the pipettes is increased. The interface is marked where it is difficult to discern from the photograph. The corresponding oil (red) and water (blue) solubilization ratios are plotted in Figure 3 as a function of salinity after 610 days. The optimal salinity occurs at 38,000 ppm with an optimal solubilization ratio of 13 cc/cc. A solubilization ratio of 10 cc/cc (the minimum value required of good phase behavior) corresponds to an IFT of 0.003 dynes/cm, which is sufficiently low for EOR applications. In addition to the ultra-low IFT that was observed with this phase behavior experiment, a low viscosity of the microemulsion phase was also observed. For comparison with the solubilization plot at 610 days of equilibration, the solubilization ratio curves after only 137 days at reservoir temperature are included in Figure 4 for this surfactant formulation. The chemical and thermal stability of the surfactants in this phase behavior experiment are thus confirmed.

A similar formulation consisting of 0.5% C28-25PO-25EO carboxylate and 0.5% C15-18 IOS was developed for use with Oil #6 at 100 °C with varying concentrations of brine #2 to produce the salinity scan. Figure 5 presents a photograph of the phase behavior pipette near the optimal salinity and in the Type III region where each phase is labeled after equilibrating for over 170 days for a WOR of 1. Figure 6 presents the solubilization ratio curves for this formulation after 170 days of equilibration. The optimal salinity is 60,000 ppm with a solubilization ratio of 12 cc/cc. The aqueous solution is clear until a salinity of 61,000 ppm is reached. The clear solution at a salinity of 57,000 ppm is shown in the photograph of Figure 5. For reference, the solubilization ratio curves are also plotted after only 70 days in Figure 7. Neither the phase behavior nor the aqueous stability has changed after 170 days again demonstrating the stability of the surfactants under conditions of high temperature, salinity, and divalent ion concentration. This stability would be impossible to obtain with an alkoxy sulfate surfactant due to severe hydrolysis, especially at a neutral pH. Further evidence of the stability of the carboxylate surfactants is also shown by the long term aqueous solubility of the formulations presented in Table 5 in hard brines.

The activity diagram is very important for designing the ASP flood for active oils. It shows the effects of produced soaps on phase behavior with changing oil concentration, which is critical in the design of the salinity gradient. For active oils with alkali present, a slightly negative slope of the activity region is more desired. An example of this favorable negative slope is illustrated using oil #1 without alkali presented (Fig. 8). In the activity diagram, the red data indicates Type III, blue Type II, and green Type I. The formulation tested for this oil at neutral pH and at 30 °C was 0.25% C28-25PO-15EO Carboxylate, 0.15% C15-18 IOS, 0.10% C20-24 IOS. From the activity diagram, we can see that even though the experiments were conducted at neutral pH condition, this oil still showed some activity. This suggested that there could be some mobilization of soap from oil even without alkali being present to induce in-situ soap generation. This implies that GAC may help mobilize some organic (naphthenic) acids from the oil to the interface, thereby reinforcing surface activity. Another example, oil #2, is moderately active at a temperature of 30 °C. The formulation was developed using 0.25% C24-25PO-18EO carboxylate, 0.25% C13-13PO sulfate, 0.25% C12 ABS, and 0.25% TEGBE. The activity diagram for this formulation is shown in Figure 9. Again, it can be seen that there is a slight shift in optimum salinity from 44,000 ppm to 54,000 with changes in oil concentration from 50% to 10%, which indicates there may be some interaction between naphthenic acid in this oil and the GAC surfactant. One possible hypothesis would be that the interaction between the carboxylate surfactant and the naphthenic acids is much stronger than that between the naphthenic acids and sulfate or sulfonate surfactants because of the similarities of the functional groups (ie. the sulfate/sulfonate surfactants do not engage the naphthenic acids from the oil under neutral pHs). Further studies are needed to elucidate the underlying mechanism of this interesting phenomenon.

b. Corefloods results

Recent advances in CEOR technology have broadened the scope of application to an ever increasing number of reservoirs. Although viscous oils tend to be more difficult to recover than light oils, they have many advantages and are more attractive when effective formulations with sufficient mobility control are applied. Yang [26] gave an insightful discussion on CEOR processes for viscous oils. Viscous oils usually have higher organic acid contents that generate in-situ soaps with alkali which could benefit ASP flooding very significantly. However, in many situations, the usage of alkali is not feasible due to divalent ion precipitation and consequently only non-alkaline (SP) formulations can be used to meet the requirements. Crude oil # 3 with 75 cp viscosity is such an example. The SP formulation developed for this oil consists of 0.5% C28-35PO-10EO carboxylate, 0.5% C20-24 IOS, 1% IBA. Figure 11 gives a solubilization ratio over 50 at optimum salinity of ~32,000 ppm. An SP coreflood #1 was then designed using the results of phase behavior. A 0.6 PV (PV*C=30) surfactant slug was injected at the optimum salinity of 32,000 ppm followed by a polymer drive of 22,000 ppm salinity. The final oil recovery was 90.1% of water flood residual oil saturation and the final oil saturation after

SP flood was 0.041 (Figure 12). The total surfactant retention was measured to be 0.28 mg surfactant per gram of rock.

A significant amount of oil reserves remain in carbonate reservoirs, many of which are at high temperature and high salinity. Such harsh conditions greatly challenge the application of CEOR. Levitt [27] described an ASP flood in a high temperature, high salinity carbonate. However, alkali cannot be used in many instances where anhydrite is present, high cation exchange capacity exists at increased pH, or soft injection water is not available. Recent developments of GAC have enabled CEOR to be expanded to these reservoirs.

Many reservoirs containing considerable amount of divalent cations and high salinities in brines offer formidable challenges to the CEOR process. The divalent cations, such as Ca^{2+} and Mg^{2+}, consume alkali and may precipitate surfactant and polymer, leading to injection problems as well as affecting phase behavior,thus making the surfactant selection difficult. When the softening of the hard injection water is not an option, chelating agents such as tetrasodium ethylene-diaminetetraacetate (EDTA-Na_4) can be used to sequester divalent cations, reported by Yang [26]. However, EDTA-Na_4 is relatively expensive, and the required amount to complex divalent cations is much higher than the theoretical value of 9 to 1 weight ratio as the temperature increases. GAC surfactants are tolerant to divalent cations, and by changing the number of PO and EO groups, they can accommodate the high salinity and hardness without adversely affecting the solubilization ratio.

The Core flood #2 with oil #5 was designed based on the results of phase behavior in Figure 13. The core used was of Silurian dolomite and is very vuggy and heterogeneous based on visual observations. A tracer test of the clean core is shown in Fig.14. Early breakthrough (~0.45 PV) and the long tail (nearly 3 PV) indicated the extent of heterogeneity of the core. Since this was the very first test of GAC in a core, a higher than normal amount (PV*C=50) of surfactant slug was injected into the core, followed by the polymer drive. The final residual oil recovery after waterflood was 90.5%, with S_{orc} of 0.044, as shown in Figure 15. The core flood effluent samples were analyzed by HPLC to measured surfactant retention. The total surfactant retention was 0.33 mg surfactant per gram of rock. The pressure drop during the chemical flood was ~1.5 psi/ft at 2 ft/day frontal velocity (Fig. 16). Such low pressure drop is very favorable and close to the field pressure gradient.

A successful example with high salinity and hardness was an SP formulation developed for oil #4 containing 1% C28-25PO-55EO carboxylate, 0.5% C15-18 IOS, 0.5% C19-23 IOS, 1% TEGBE at 83°C (Fig. 17). The formulation was tested in an Estaillade core (Coreflood #3). The formation brine and the injection slug salinities are 230,000 ppm and 67,000 ppm, respectively (Table 4). Identifying an effective surfactant formulation for such high salinity and hardness is difficult while meeting low IFT and aqueous stability requirements. The GAC surfactant composition still performed very well and yielded good results. Injecting a 0.2 PV (PV*C=40) SP slug, the oil recovery was 92% and final oil saturation was 0.016 (Fig.18). Surfactant retention was 0.3 mg/g. The results demonstrate that GAC surfactants are excellent candidates for CEOR even for high temperature, high salinity and hardness carbonate reservoirs.

Conclusions

Linear alcohols can be dimerized to produce large hydrophobe structures via the cost-effective Guerbet reaction. The addition of alkoxy groups (such as PO and/or EO) to these Guerbet alcohols followed by carboxymethalyation (carboxylation) creates novel high-performance Guerbet alkoxy carboxylate surfactants. These surfactants differ from the previously presented Guerbet alkoxy sulfate surfactants in terms of their thermal and chemical stability at both acidic and alkaline pHs. These novel GAC surfactants are highly desired for application under challenging conditions. This new class of surfactants proved its effectiveness in phase behavior and core flood experiments. Mixtures of surfactants containing GAC produced ultra-low IFT and high recovery of crude with a variety of oils under various harsh conditions. The application of the GAC surfactants to mobilize viscous oil has also been successful in the laboratory. The successful core flood examples showed that the GAC surfactant formulations can tolerate extremely high temperature, high hardness and salinity of brine and produce high oil recovery. Thus, the advent of this novel class of cost-effective thermally and chemically stable GAC surfactants greatly broadens the scope of chemical EOR.

Acknowledgements

The authors would like to thank the industrial affiliates of the chemical enhanced oil recovery project at the University of Texas at Austin for the financial support of this research.

References

1. Solairaj, S. "New Method of Predicting Optimum Surfactant Structure for EOR" M.S. Thesis, University of Texas at Austin, Austin Texas, December 2006.
2. Levitt, D.B., Jackson, A.C., Heinson, C., Britton, L.N., Malik, T., Dwarakanath, V., and Pope, G.A. "Identification and Evaluation of High Performance EOR Surfactants," SPE 100089, presented at SPE/DOE IOR Symposium, Tulsa, OK, April 2006.
3. Zhao, P., Jackson, A.C., Britton, C., Kim, D.H., Britton, L.N., Levitt, D.B., and Pope, G.A., "Development of High-Performance Surfactants for Difficult Oils," SPE 113432, presented at SPE/DOE IOR Symposium, Tulsa, OK, April 2008.
4. Flaaten, A.K., Nguyen, Q.P., Zhang, J., Mohammadi, H., and Pope, G.A. "ASP Chemical Flooding without the Need of Soft Water," SPE 116754, presented at SPE IOR Symposium, Tulsa, OK, April 2008.
5. Adkins, S., et al., A New Process for Manufacturing and Stabilizing High-Performance EOR Surfactants at Low Cost for High-Temperature, High Salinity Oil Reservoirs in SPE Improved Oil Recovery Symposium 2010, Society of Petroleum Engineers: Tulsa Oklahoma.
6. Jr., A.J.O.L., Guerbet Chemistry. Journal of Surfactants and Detergents, 2001. 4(3): p. 311-315.

7. Jackson, A.C. “Experimental Study of the Benefits of Sodium Carbonate on Surfactants for Enhanced Oil Recovery” M.S. Thesis, University of Texas at Austin, Austin Texas, December 2006.
8. Puerto, M, Hirasaki, G. J., Miller, C. A., and Barnes, J. R., "Surfactant Systems for EOR in High-Temperature, High-Salinity Environments", SPE 129675 presented at SPE IOR Symposium, Tulsa, OK, 25-28 April, 2010.
9. Shaw, J.E., Carboxylate Surfactant Systems Exhibiting Phase Behavior Suitable for Enhanced Oil Recovery. Journal of the American Oil Chemists' Society, 1984. 61(8): p. 1395-1399.
10. Li, G.-Z., et al., What is the Criterion for Selecting Alkaline/Surfactant/Polymer Flooding Formulation: Phase Behavior or Journal of Dispersion Science and Technology, 2000. 21(3): p. 305-314.
11. Li, G.Z., et al., Design and Application of an Alkaline-Surfactant-Polymer Flooding System in Field Pilot Test. Journal of Dispersion Science and Technology, 2005. 26: p. 709-7=17.
12. Abe, M., et al., Phase Behavior of Branched Tail Ethoxylated Carboxylate Surfactant/Electrolyte/Alkane Systems. Journal of Dispersion Science and Technology, 1987. 8(2): p. 157-72.
13. Rosen, M.J., Surfactants and Interfacial Phenomena2004, Hoboken, NJ, USA: Wiley.
14. Ansgar Behler, et al., Industrial Surfactant Syntheses2001: Taylor & Francis Group LLC.
15. DanChem Technologies, I. NEODOX Alcohol Ethoxycarboxylate. April 2011]; Product Information]. Available from: www.danchemtechnology.com/neodox.html.
16. Marlowet Ether Carboxylic Acids. April 2011; Product Description. Available from: www.innovadex.eu/Lubricants/Detail/3857/110895/Marlowet--Ether-Carboxylic-Acids.
17. Cayias, J.L., Schechter, R.S., Wade, W.H., “Modeling Crude Oils for Low Interfacial Tension”, SPE 5813, presented at SPE-AIME IOR Symposium, Tulsa, OK, 22-24 March 1976.
18. Salager, J.L., Bourrel, M.U., Schechter, R.S., and Wade, W.H. “Mixing Rules for Optimum Phase-Behavior Formulations of Surfactant/Oil/Water Systems”, SPEJ 19(5), 271-278, 1979.
19. Glinsmann, G.R., “Surfactant Flooding with Microemulsions Formed In-situ – Effect of Oil Characteristics”, SPE 8326, presented at SPE-AIME ATCE Symposium, Dallas, TX, 1979.
20. Puerto, M.C., and Reed, R.L. “A Three Parameter Representation of Surfactant/Oil/Brine Interactions”, SPE Journal 23(4), 669-682, 1983.
21. Roshanfekr, M., Johns, R. T., Pope, G.A., Britton, L., Linnemeyer, H., Britton, C., Vyssotski, A. “Effect of Pressure, Temperature and Solution Gas from Surfactant Polymer Floods”, SPE 125095 presented at SPE ATCE, New Orleans, LA, 4-7 October, 2009.
22. Roshanfekr, M. “Effect of Pressure and Methane on Microemulsion Phase Behavior and its Impact on Surfactant-Polymer Flood Oil Recovery”, PhD Dissertation, The University of Texas at Austin, December 2010.
23. Glover, C.J., Puerto, M.C., Maerker, J.M., and Sandvik, E.L., "Surfactant Phase Behavior and Retention in Porous Media", SPEJ, June, 183-193, 1979.
24. Pope, G.A., Wang, B. and Tsuar, K. “A Sensitivity Study of Micellar/Polymer Flooding,” SPEJ, December, 357-368, 1979.
25. Hirasaki, G.J., van Domselaar, H.R., and Nelson, R.C., "Evaluation of the Salinity Gradient Concept in Surfactant Flooding," SPEJ, June, 486-500, 1983.
26. Yang, H., Britton, C., Liyanage, P.J., Solairaj, S., Kim, D., Nguyen, Q., Weerasooriya, U., and Pope, G.A. “Low-cost, High-performance Chemicals for Enhanced Oil Recovery”, SPE 129978, presented at the SPE IOR Symposium, Tulsa, OK, April 2010.
27. Levitt, D.B., Dufour, S., Pope, G. A., Morel, D., and Gauer, P., "Design of an ASP Flood in a High-Temperature, High-Salinity, Low-Permeability Carbonate", IPTC 14915 presented at IPTC, Bangkok, Thailand, 7-9 February, 2012.

Table 1—Oil Properties

Oil #	Temperature (°C)	API	Total Acid Number (mg KOH/g oil)	Surrogate Oil Viscosity (cp)
1	30	19	2.4	74
2	30	31	0.7	21
3	46	21	2	75
4	83	36.5	0	1.8
5	100	34	0.05	0.5
6	100	22	0.15	2.1

Table 2—Core Properties

Coreflood #	Core name	Type	PV (ml)	Length (cm)	Diameter (cm)	Porosity	Brine Permeability (md)
1	Berea	Sandstone	114.2	29.5	5.0	0.195	283
2	Silurian	Dolomite	98	30.2	5.0	0.17	478
3	Estaillade	Limestone	80.3	29.7	3.8	0.24	144

Table 3—Coreflood Summary

Coreflood #	1	2	3
Oil #	3	5	4
Initial salinity (ppm)	4,392	67,135	230,800
Surfactant Slug			
Surfactant concentration (wt%)	0.5	2	2
PV injected	0.6	0.25	0.2
PV×C (%)	30	50	40
Polymer	FP 3630s	FP 3330s	FP 3330s
Polymer concentration (ppm)	4200	4500	5300
Viscosity (cp)	90	15	15
Salinity (ppm)	32,000	40,000	67,000
Velocity (ft/day)	1.2	2	1
Polymer Drive			
Polymer concentration (ppm)	4000	4500	4700
Viscosity (cp)	85	18	15
Salinity (ppm)	22,000	25,000	41,300
Velocity (ft/day)	1.2	2	1
Coreflood Results			
Recovery (%)	90.1	90.5	92
Final Residual Oil Saturation, S_{orc}	0.041	0.044	0.016
Surfactant Retention (mg/g)	0.28	0.33	0.3

Table 4—Brine Compositions

Coreflood #	Brine	Na^+	Ca^{2+}	Mg^{2+}	K^+	SO_4^{2-}	Cl^-	HCO_3^-	TDS
1	Initial brine	1660	26	11	28	0	2667	0	4392
	Slug	12472	26	11	28	0	19355	0	31892
	Polymer drive	8737	26	11	28	0	13590	0	22392
2	Initial brine	19249	4360	938	0	650	41524	0	67135
	Slug	12810	455	1477	0	3003	22540	84	40369
	Polymer drive	7320	260	844	0	1716	12880	48	23068
3	Initial brine	65560	18400	3300	0	250	143240	0	230800
	Slug	20400	840	2900	0	5320	37600	0	67060
	Polymer drive	12560	515	1800	0	3275	23140	0	41300

Table 5—Formulations of Long-term Aqueous solubility

Surfactant Formulation	Temperature (°C)	Aqueous Salinity (ppm)	Clarity Time (days)
0.5% C28-25PO-25EO Carboxlate, 0.5% C15-18 IOS	100	58,000	>261
0.5% C28-25PO-45EO Carboxylate, 0.5% C15-18 IOS	100	46,000	>470
0.7% C28-25PO-55EO Carboxylate, 0.3% C12 ABS	100	31,000	>422
0.5% C28-25PO-25EO Carboxylate, 0.5% C15-18 IOS, 0.5% TEGBE	105	56,300	>292
0.5% C28-25PO-25EO Carboxylate, 0.4% C15-18 IOS,	105	30,800	>286

0.1% C19-23 IOS,
0.5% TEGBE

Table 6—Brine Compositions for Oil # 5, 6, and 4 Phase Behavior Experiments

Ions	Brine #1 (ppm)	Brine #2 (ppm)	Brine #3 (ppm)
Na^+	18,300	12,200	41,450
K^+	0	0	0
Ca^{2+}	650	500	3,900
Mg^{2+}	2,110	1,350	150
Cl^-	32,200	21,100	71,000
SO_4^{2-}	4,290	3,250	500
HCO_3^-	120	0	0
TDS	57,670	38,400	117,000

▶

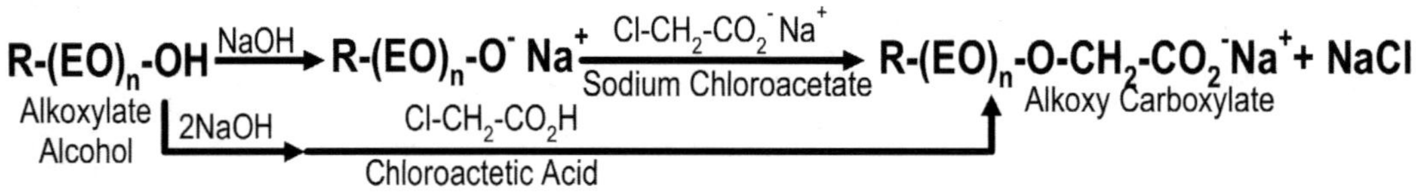

Figure 1. Schematic of the reaction to form the alkoxy carboxylate from the corresponding alcohol alkoxylate where R is the carbon chain plus any number of PO.

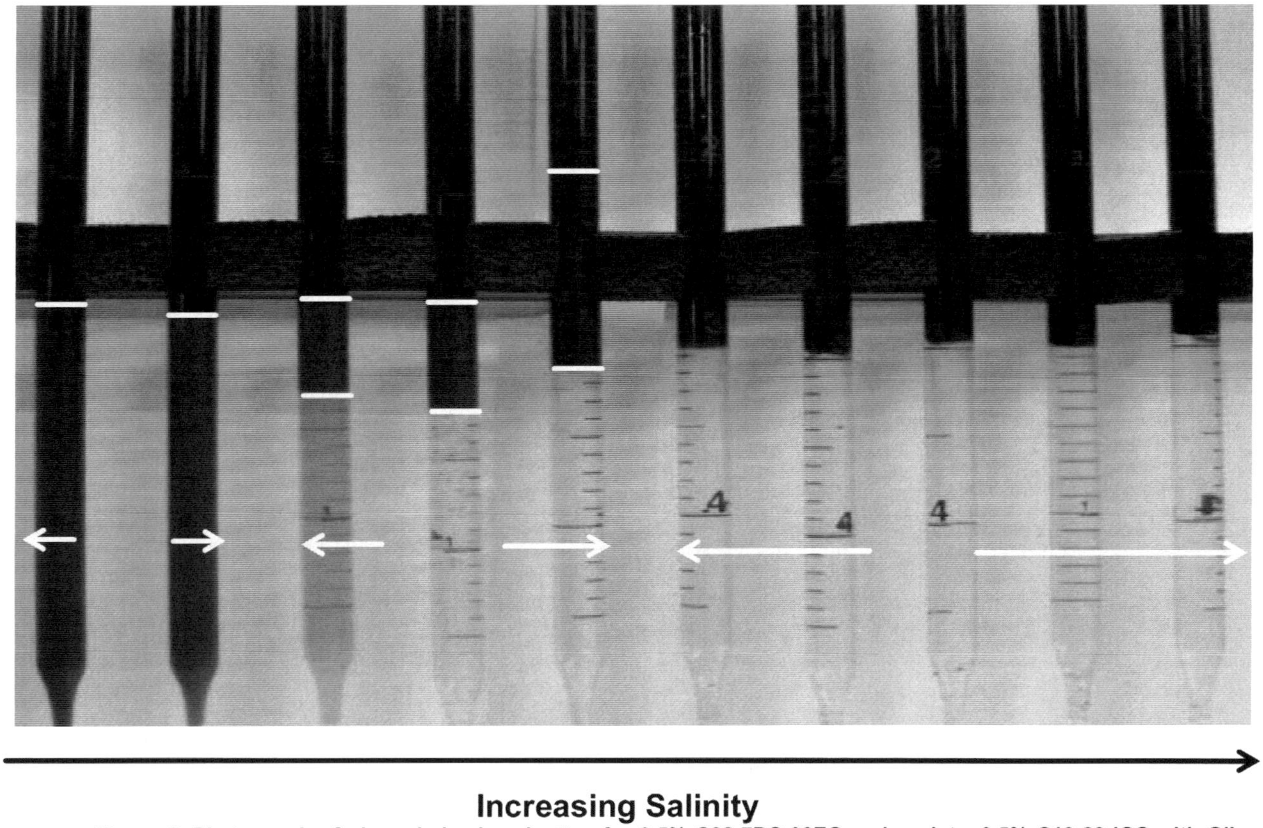

Figure 2. Photograph of phase behavior pipettes for 0.5% C32-7PO-32EO carboxylate, 0.5% C19-23 IOS with Oil 56 at 100 °C for over 290 days.

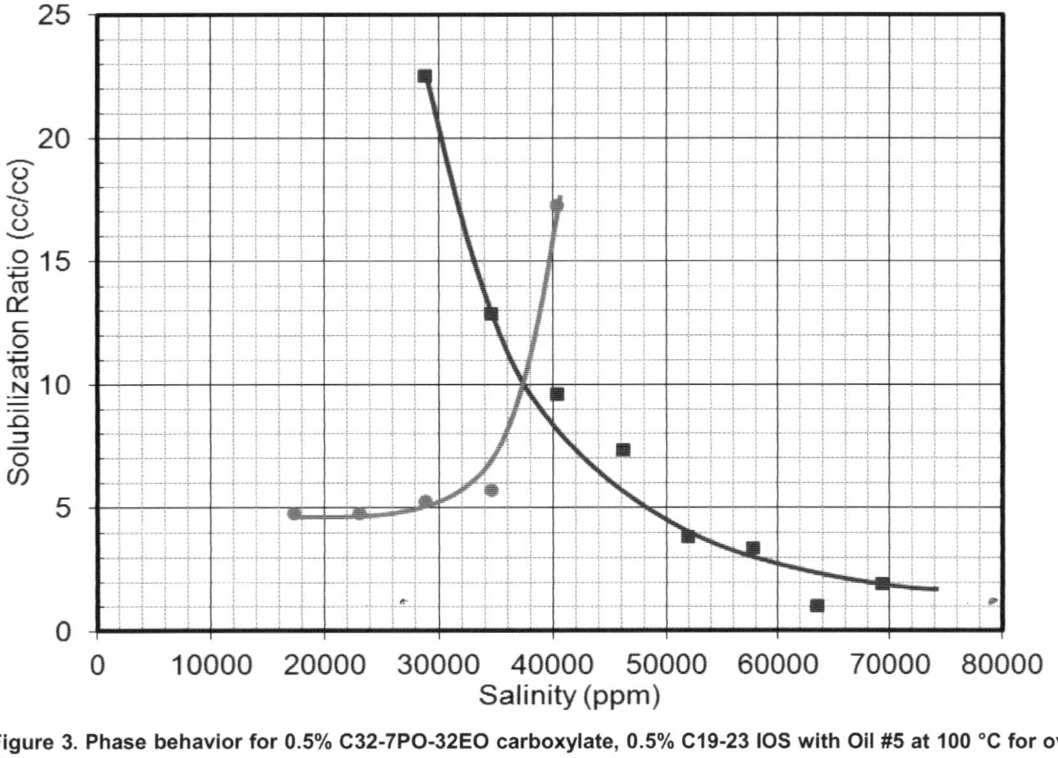

Figure 3. Phase behavior for 0.5% C32-7PO-32EO carboxylate, 0.5% C19-23 IOS with Oil #5 at 100 °C for over 610 days.

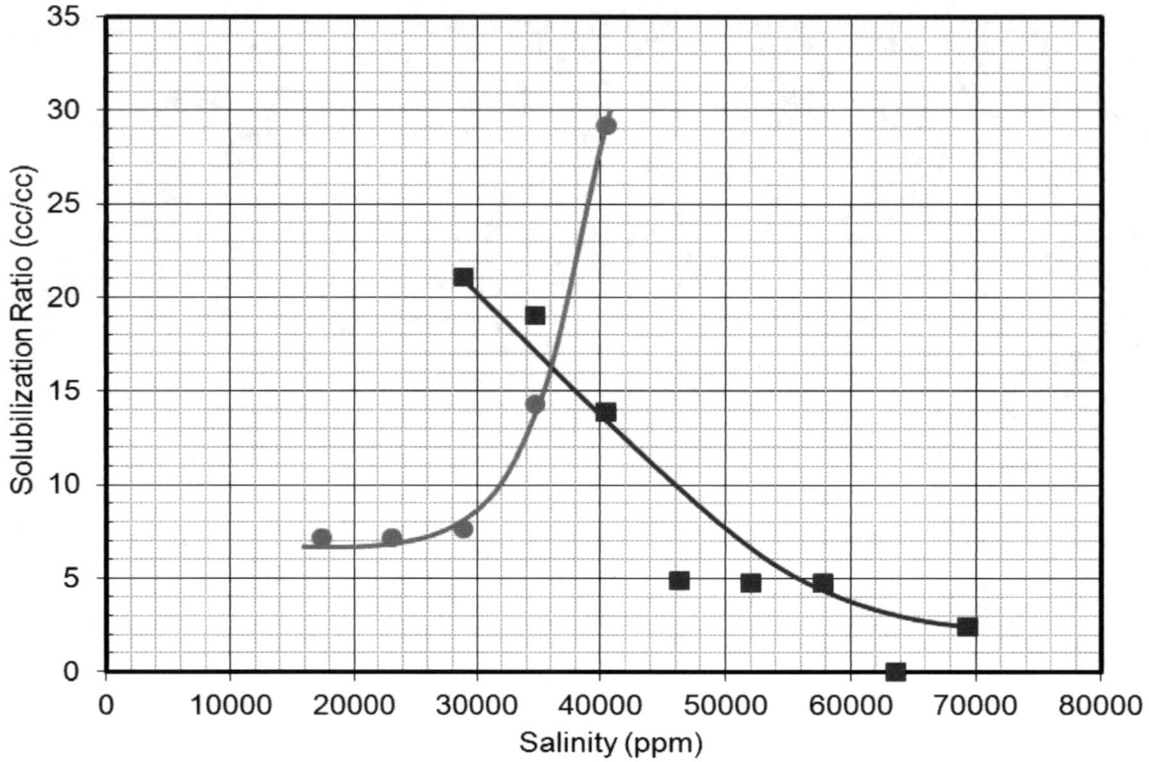

Figure 4. Phase behavior for 0.5% C32-7PO-32EO carboxylate, 0.5% C19-23 IOS with Oil #5 at 100 °C for over 137 days.

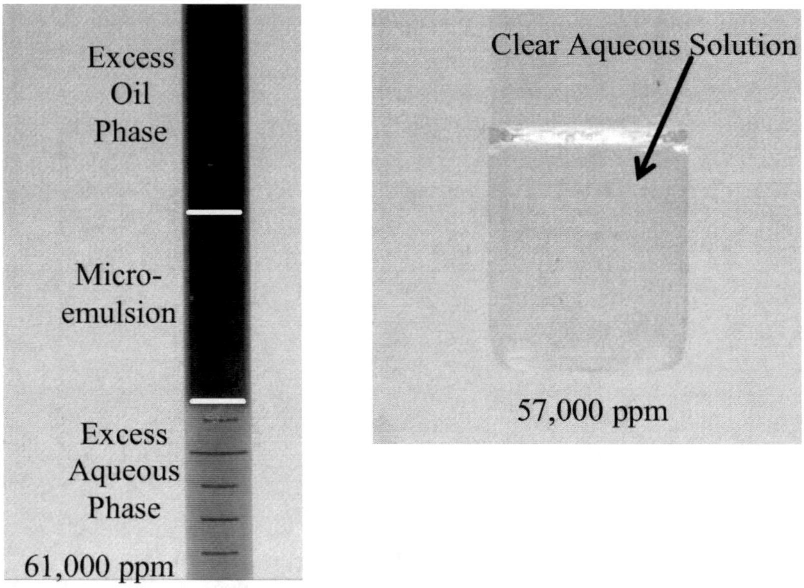

Figure 5. Photograph of the phase behavior pipette for 0.5% C28-25PO-25EO carboxylate and 0.5% C15-18 IOS with Oil #6 at 100 °C at 61,000 ppm and the aqueous solubility sample at 57,000 ppm.

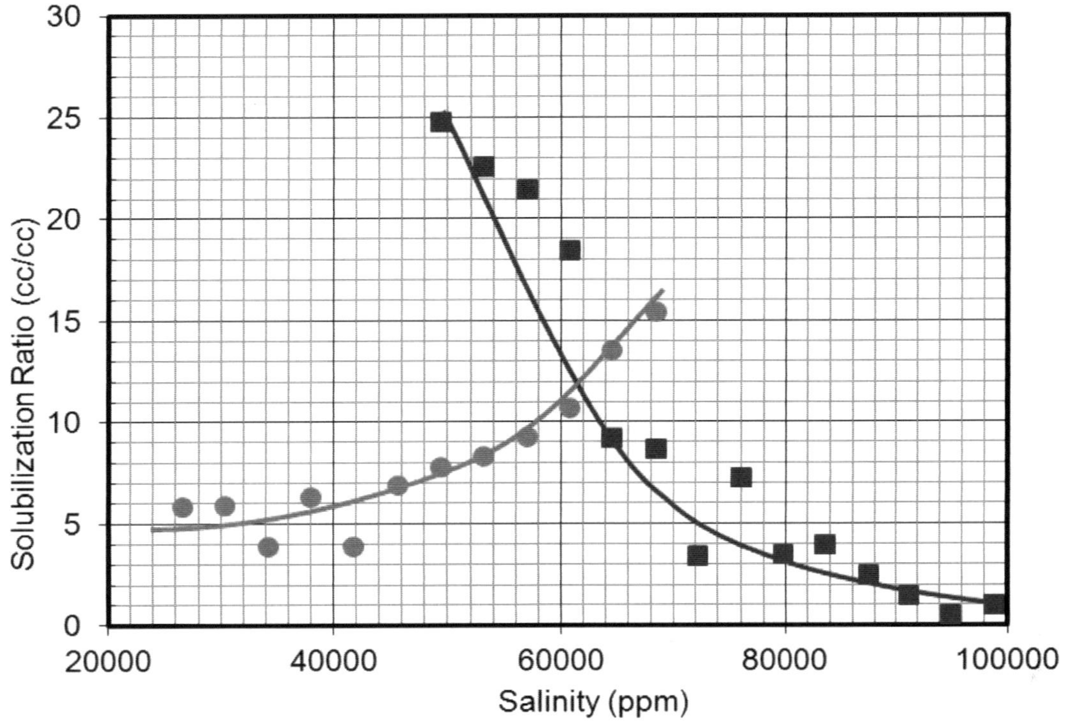

Figure 6. Phase behavior for 0.5% C28-25PO-25EO carboxylate and 0.5% C15-18 IOS with Oil #6 at 100 °C after 170 days.

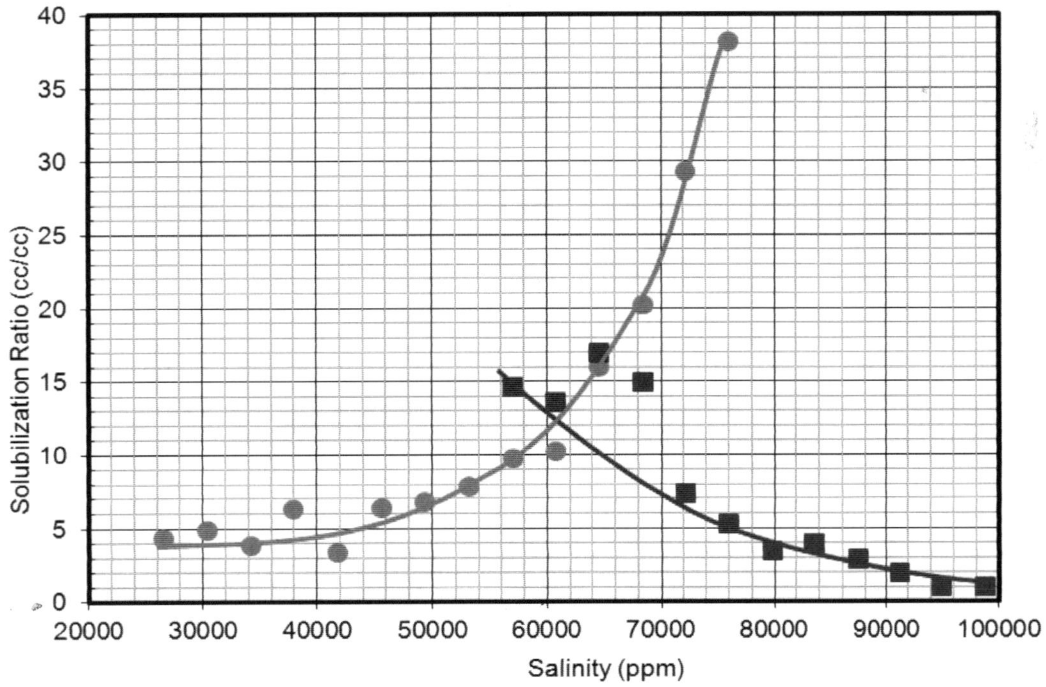

Figure 7. Phase behavior for 0.5% C28-25PO-25EO carboxylate and 0.5% C15-18 IOS with Oil #6 at 100 °C after 70 days.

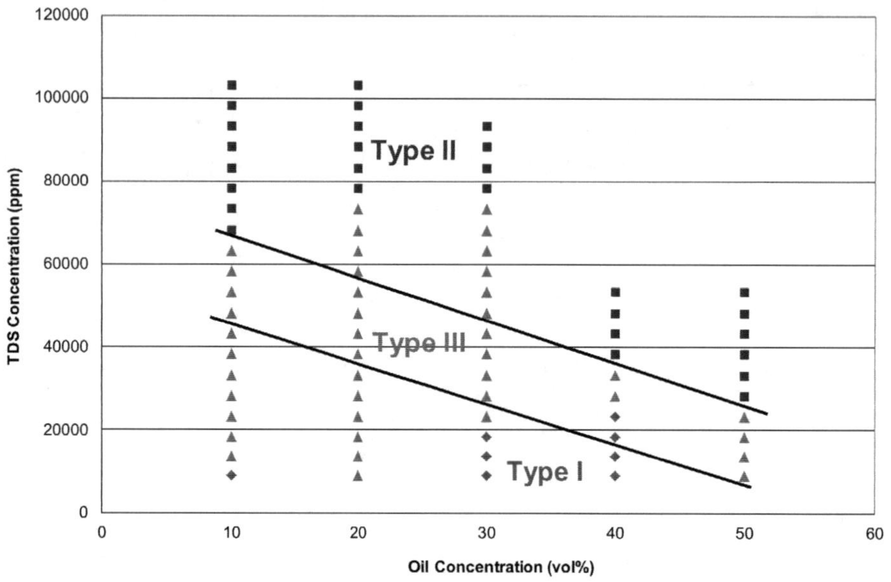

Figure 8—Activity diagram of oil #1 at 30 °C.

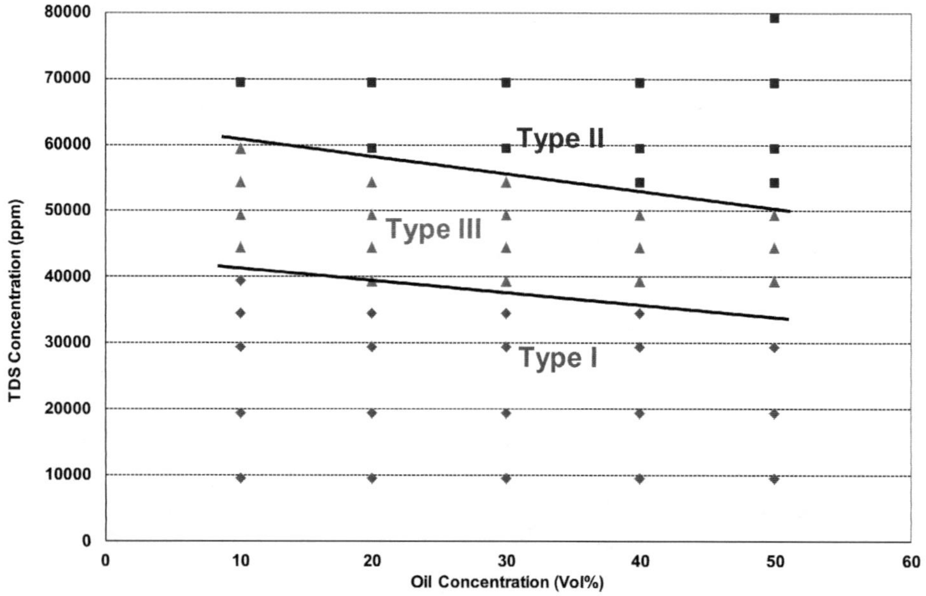

Figure 9—Activity diagram of oil #2 at 30 °C.

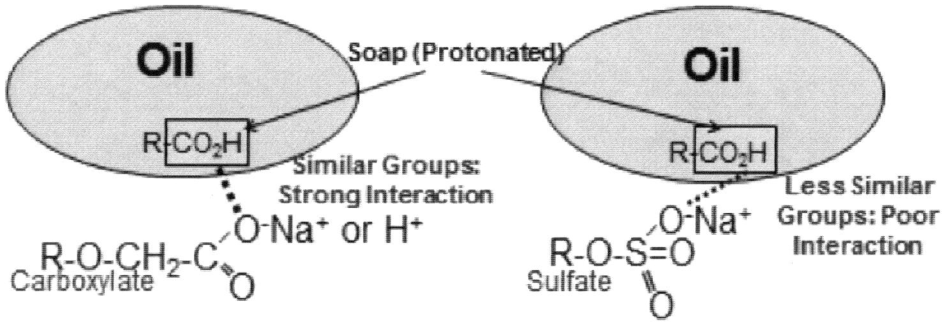

Fig. 10—Schematic of the postulated interaction between carboxylate group and the naphthenic acids in oil.

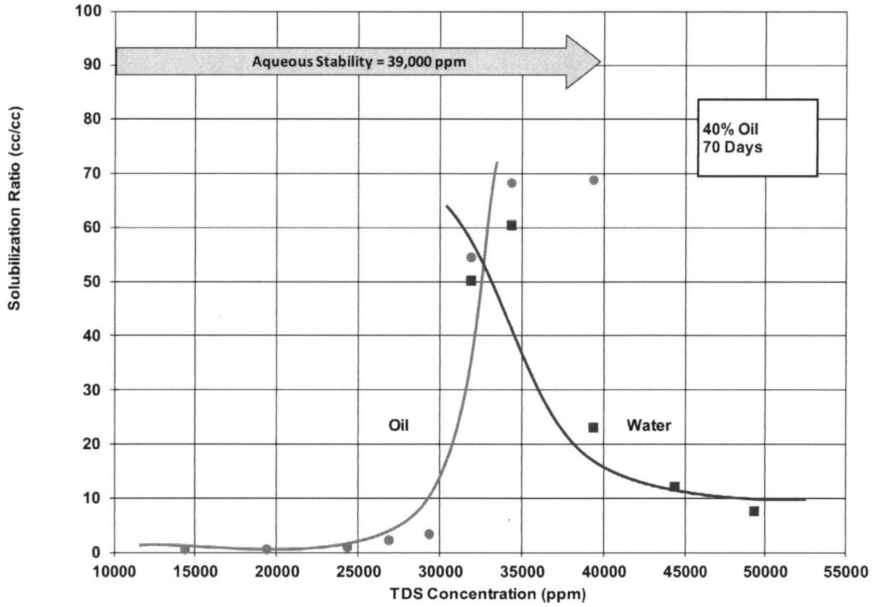

Figure 11—Phase behavior of 0.5 wt% C28-35PO-10EO Carboxylate, 0.5 wt% C20-24 IOS for oil #3 at 46 °C.

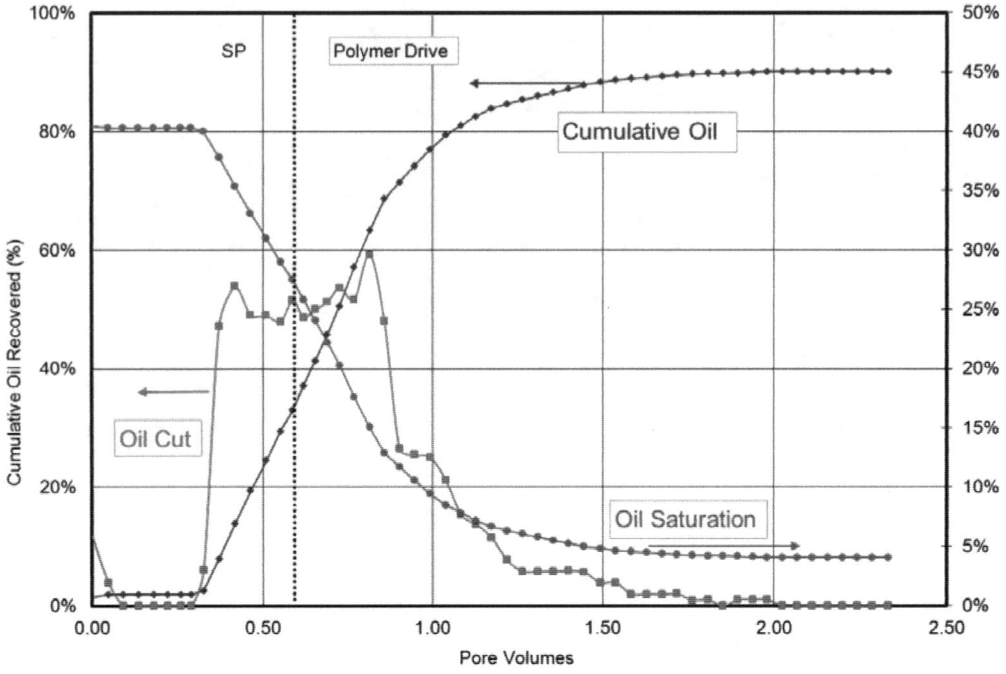

Fig. 12—Oil recovery from core flood 1 for oil #3 in Berea sandstone at 46 °C.

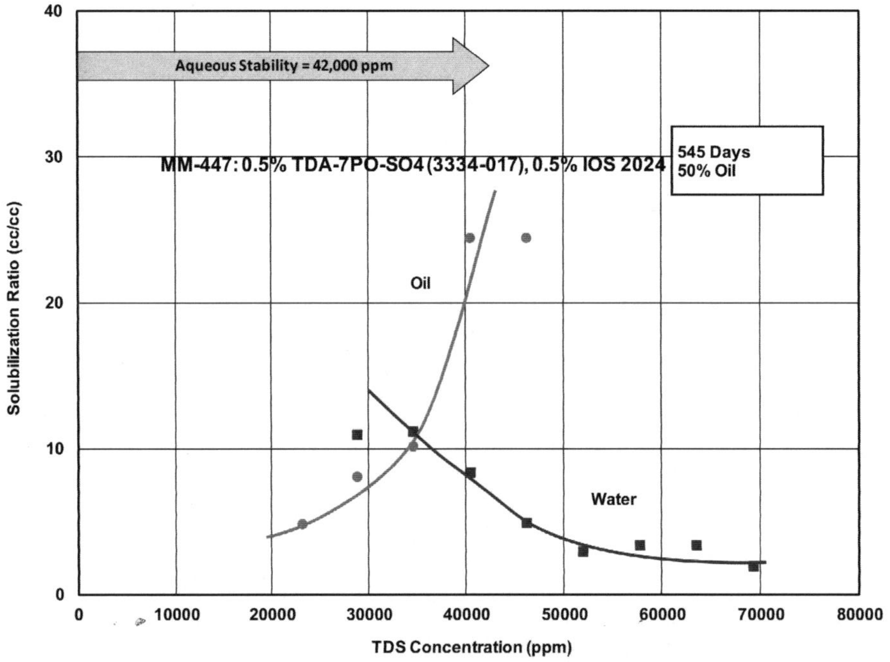

Figure 13—Phase behavior of 1.0 wt% C32-7PO-32EO Carboxylate, 1.0 wt% C19-23 IOS for oil #5 at 100 °C.

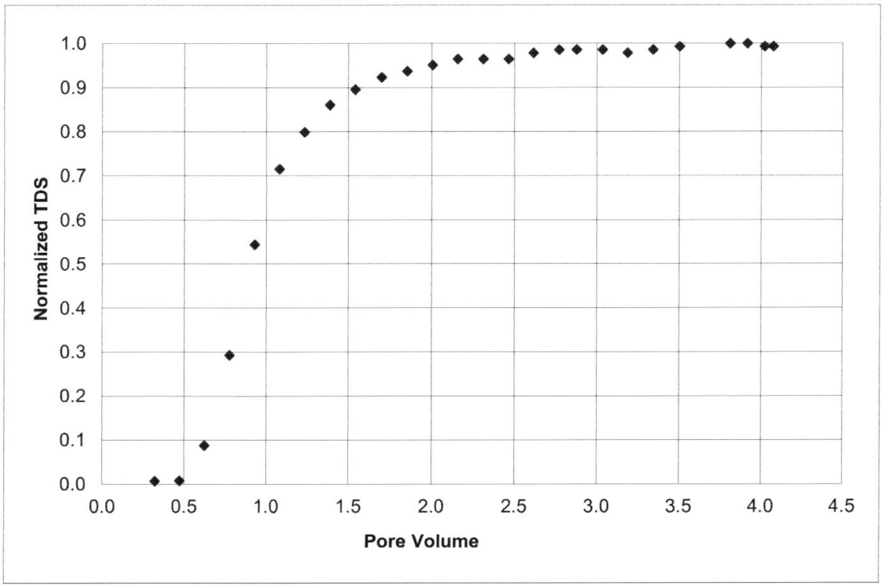

Fig. 14—Tracer test in Silurian dolomite.

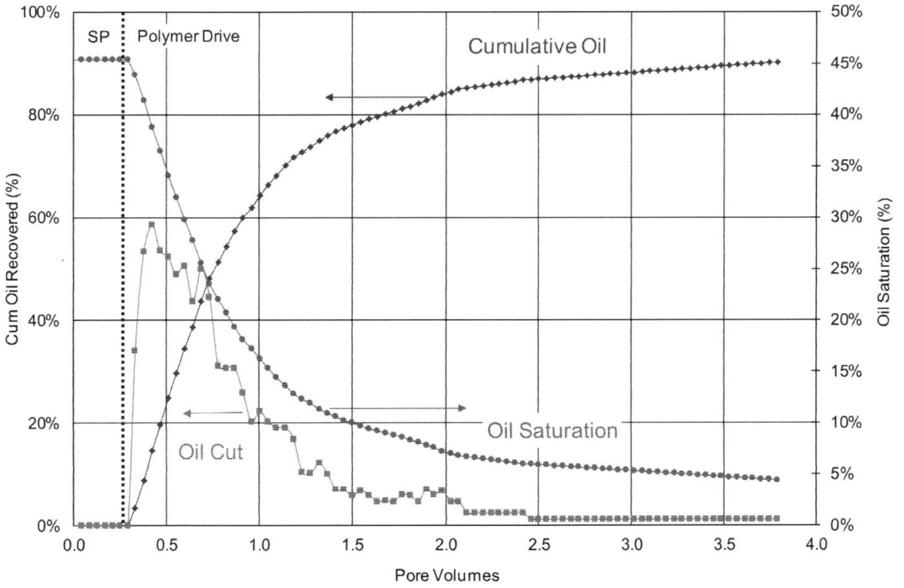

Fig. 15—Oil recovery from Core Flood 2 for Crude #5 in Silurian dolomite at 100 °C.

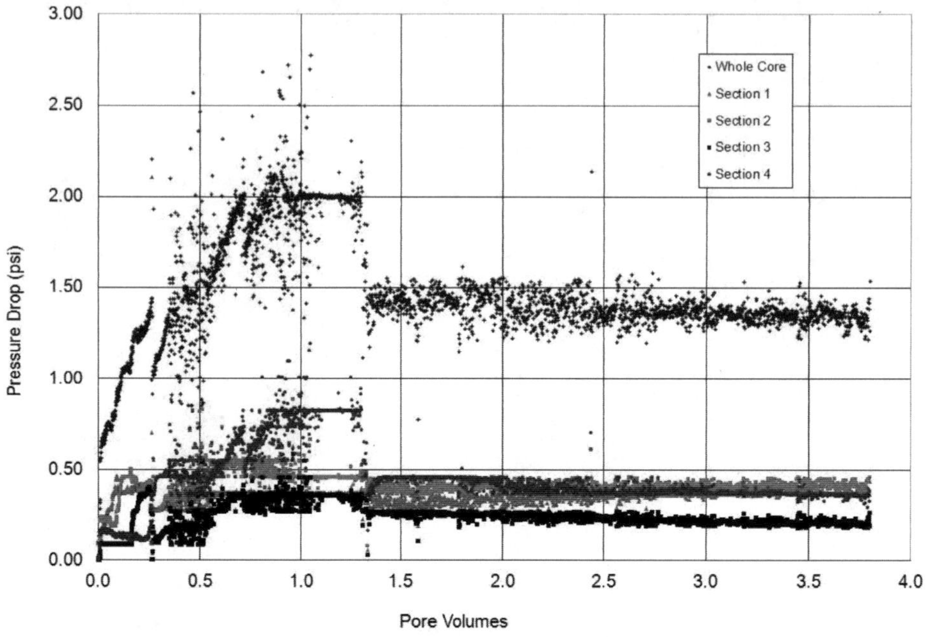

Fig. 16—Pressure drop for Core Flood 2 at 100 °C.

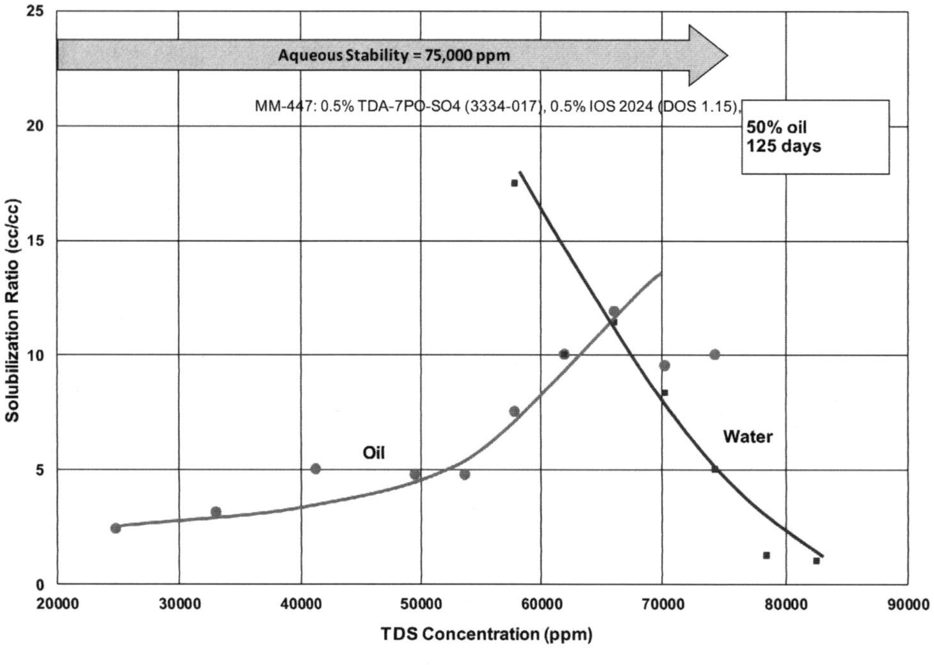

Figure 17—Phase behavior of 1.0 wt% C28-25PO-55EO caboxylate, 0.5 wt% C15-18 IOS, 0.5 wt% C19-23 IOS, 1.0 wt% TEGBE for oil #4 at 83 °C.

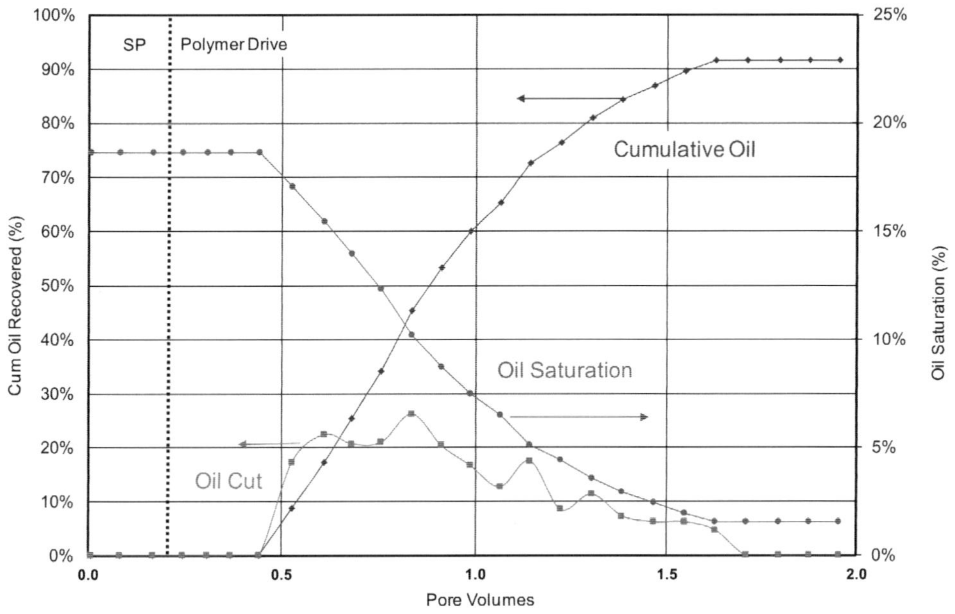

Fig. 18—Oil recovery from Core Flood 3 for oil #4 in Estaillade limestone at 83 °C.

(126c) A New Acid System for Matrix Stimulation of Deep Oil and Gas Wells

Hisham A. Nasr-El-Din, Petroleum Engineering, Texas A& M University, College Station, TX

Matrix acidizing is used in carbonate formations to create flow channels from the formation to the well bore, whereas in sandstone formations the goal is to dissolve materials that impair the well production. However, the use of acids in deep wells has some major drawbacks including high reaction rate and corrosion to well tubulars.

A new stimulation fluid was developed that can be used as a replacement for or in combination with acid treatments in deep wells. A polyacid whose structure allows for acidification is described. The chelate-based fluid effectively dissolves CaCO3 and it is less corrosive to the equipment.

This paper discusses the reaction of the new GLDA chelate with calcite and will compare its performance with other available chelates, including EDTA, HEDTA, NTA and EDG. GLDA dissolves calcite over a wide pH interval, although it is less effective than HEDTA at > pH 5. A unique property of GLDA is its high solubility - solutions exceeding 40 wt% can be achieved at a ~ pH 2 where HEDTA solubility is limited to ~ 10 wt%. A mole of GLDA acid with a natural ~ pH 1.5 is capable of dissolving up to two moles of CaCO3. Throughout the pH range GLDA appears to be as thermally stable as HEDTA.

As an additive to HCl, GLDA is as effective as HEDTA in preventing precipitation of moderate levels of Fe in spent acids. At high Fe concentration GLDA is slightly less effective than HEDTA on a molar basis – but to deal with high Fe GLDA may be better as significantly higher concentrations of it are possible in various acids. In 28 wt% HCl, HEDTA has limited solubility while GLDA's solubility exceeds 40 wt%.

From an environmental standpoint GLDA, is readily biodegradable. As a replacement for HCl acid, GLDA is significantly safer and is less corrosive. GLDA was successfully applied to stimulate a deep gas well, and the results were above expevtations.

Chemical EOR by Wettability Alteration

G. Sharma, R. Gupta, K. Mohan, and K.K. Mohanty, University of Texas at Austin
mohanty@mail.utexas.edu

(Submitted to the 2012 AIChE Spring Conference)

Abstract

Waterflooding is not very efficient in non-fractured or fractured oil-wet reservoirs. The ionic composition of the brine and addition of surfactants can change the wettability of the reservoir rocks from oil-wet to intermediate-wet. More than 50% OOIP (original oil in place) can be recovered from fractured carbonates by the use of a dilute (~0.1 wt%) surfactant solution. This process is sensitivity to temperature and fracture spacing. In secondary flooding of non-fractured carbonate rocks, wettability alteration agents enhances oil recovery over that of just waterflooding.

Introduction

Many carbonate reservoirs are oil-wet and fractured[1-2], and waterflooding gives poor oil recovery[3]. One of the methods to recover oil from oil-wet fractured carbonate reservoirs is to alter the wettability of these reservoirs. Previous work in the last ten years has shown that surfactants[4-11] and electrolytes[12] can be used to alter the wettability of originally oil-wet carbonate rocks. However, a systematic approach to screen surfactants for wettability alteration is lacking.

Most of the studies in the surfactant flooding have been aimed at obtaining ultra low interfacial tension to increase capillary number and thus decrease the residual oil saturation to recover extra oil. The surfactant combinations which give low IFT may or may not give wettability alteration. Levitt et al.[13] showed that changing the ratio of propoxy sulfates, internal olefin sulfonates (IOS) and alfa olefin sulfonate surfactants can optimize the surfactant slug for various salinity conditions. Zhang et al.[14] worked with low surfactant (0.2 wt %) concentration of branched propoxylated sulfate along with IOS to tune the optimal salinity of the system. Zhao[15] observed that IOS is a good candidate for oils having high viscosity and high wax content. They further discussed that the carbon number of the surfactant can be tailored to optimize IOS for the given reservoir conditions. Maerker and Gale[16] showed that the optimal surfactant can be tailored by adjusting the hydrophilic - lipophilic balance using different degrees of ethoxylation (EO) and propoxylation (PO) in a surfactant molecule. The phase behavior of the ethoxylated nonyl phenols is studied as a function of alkane carbon number (ACN) and EO content by Hayes et al[17]. They found that for each oil composition there exists an optimal EO number which gives the least IFT. Addition of EO groups gives more temperature and salinity tolerance to a surfactant. Alkyl/aryl sulfonates forms calcium and magnesium sulfonates that would precipitate or partition into the oil phase[14]. Other factors like carbon chain length and branching, size and type of hydrophilic head group, number of EO/PO groups, and addition of co-surfactants can also change hydrophilic - lipophilic balance, which in turn changes the optimal salinity. Healy et al[18] conducted a similar parametric study using anionic surfactants. Hirasaki and Miller[19] emphasized the use of alcohol or branched hydrophobes to avoid formation of viscous liquid crystalline microemulsion phases during surfactant flooding experiments. Austad and others[6,7,11] have identified cationic surfactants which alter the wettability of chalk to water-wet and increase oil recovery.

Buckley et al.[20] showed that interactions between polar oil components and solid, surface precipitation of asphaltene type molecules from oil, acid/base and ion binding between ion sites on solid/brine and oil/brine or a combination of them could be responsible for wetting of mineral surfaces by a crude oil. Hamouda et al.[21] worked with a system of n-decane/fatty acid/water on calcite plates at elevated temperatures. They showed that the wettability (contact angle) follows the same trend as partition coefficient of fatty acid and IFT which in turn depends on pH and thus on the zeta potential of calcite.

This work studies the screening of surfactant system in a systematic manner for oil-wet fractured reservoirs. The surfactants are screened for structural stability for each reservoir condition. This step is followed by oil-brine phase behavior studies and wettability studies. The surfactants are screened which give wettability alteration from oil-wet to water-wet conditions for low salinity-low temperature, low salinity-high temperature, high salinity-low temperature, and high salinity-high temperature. The performance of screened surfactants is demonstrated by imbibition experiments of oil-wet carbonate rocks. Secondary waterfloods have also been conducted with or without wettability altering agents.

Results and Discussion

The results of imbibition experiments for low salinity-low temperature system are shown in Figure 1. The experiments were conducted at 25 °C at optimal salinity. The spontaneous imbibition volumes for Alf 35 and Alf 38 solutions are about the same and higher than those for the other two cases. About 50% of the oil is produced in about 100 days of spontaneous imbibition with Alf 35 and Alf 38. The final recovery is about 62% OOIP. These two surfactants are good wettability altering agents for the calcite plate. The production for Alf 68 is about 35% in 100 days. Alf 68 induces spontaneous imbibition, but to a lower extent compared to the first two cases. There was very little imbibition with the NaCl brine alone. This indicates that the core was oil-wet at the start of spontaneous imbibition tests. Either low IFT or wettability alteration is required to recover oil.

Figure 2 shows the results of some high performing surfactants for high temperature-low salinity system. Recovery above 50 % OOIP is observed in 30 days for the cases with Alf 68, MX and 15-S at 90 °C and 15-S at 70 °C with tight oil-wet carbonate rocks (~15 md). About 65 % OOIP was recovered for MX at 90 °C and 15-S at 70 °C. These two surfactants are effective for oil-wet carbonate reservoirs with low salinity (< 3 wt %) and high temperature (> 70 °C). About 55 % OOIP is recovered with Alf 68 and 15-S at 90 °C. MX was also found to perform well at 50 °C. It recovered about 34 % OOIP in 30 days and 60 % OOIP in 260 days (not shown in the plot). Thus, MX is a suitable surfactant for low salinity reservoirs with temperature above 50 °C. The imbibition experiment result with S70 at 90 °C for high salinity-high temperature system is shown in Figure 3. About 70 % of OOIP is recovered in 30 days in a ~15 md oil-wet carbonate rock.

Imbibition was also carried with 0.2 wt % a nonionic surfactant solution in a brine at 50 °C and 90 °C, respectively (in a 4 md core). 9 % and 19 % of OOIP was recovered in just 14 days at 50

°C and 90 °C, respectively. Figure 4 shows the picture of the core. The pictures of cores both at 50 and 90 °C shows oil drops around and top of the core (Figure 4). It clearly shows that wettability is altered to water-wet or intermediate-wet conditions. This leads to positive capillary pressure, and thus oil is recovered by counter-current imbibition. As time progresses the oil is mostly recovered from the top portion (mainly top cross section) of the core leading to co-current imbibition supported by gravitational forces. Though wettability is altered by the surfactant creating positive capillary pressure, it also reduces the IFT which reduces the magnitude of the capillary pressure. In this case, oil recovery is a gravity driven process.

Secondary surfactant floods were conducted to determine if the wettability-altering (WA) surfactant E can improve the recovery rate while maintaining high ultimate recovery. Surfactant brine (0.2 wt% surfactant E in 36,792 ppm salinity brine) was injected in both the floods at 3 ml/hr for 4 PV. The results of the floods are given in Table 3. Figure 5 shows a comparison in cumulative oil recovery using a wettability altering surfactant to that of a typical waterflood on similar rock under similar conditions. For WA-1, breakthrough occurs at about 0.4 PV. The oil recovery is 82% in 3 PV and 90.3% in 4 PV injected. For WA-2, injected water broke through at about 0.4 PV; the oil recovery was 86% at 3 PV and 94.8% at 4 PV injected. The two water floods (WF-1 and WF-2) in similar cores are compared with the surfactant floods (WA-1 and WA-2) in Figure 5. Oil recovery is about 62% in the water floods at 3 PV injection, whereas it is about 85% OOIP in the secondary surfactant floods. An extra 23% oil was recovered in 3 PV by changing the wettability from oil-wet to intermediate-wet. The oil recovery increases due to the wettability alteration.

Conclusions

Surfactants have been identified which can alter the wettability of oil-wet carbonate rocks to water-wet conditions. Anionic and nonionic surfactants are screened for high temperature (~90 °C) and high salinity (~8 wt %) systems containing significant concentration of magnesium and calcium ions. Sodium carbonate was found to be a good option for the brine with low divalent ions, while sodium metaborate was found to be an option for brine containing these ions. The optimal salinity can be tuned by either changing surfactant concentration or using a mixture of surfactants. Above a given salinity a surfactant cannot alter the wettability. Surfactants have been identified that can change the contact angle of an initially oil-wet calcite plate for all four types of reservoirs: low temperature-low salinity (<50 °C, < 3 wt %), high temperature-low salinity (70-90 °C, < 3 wt %), low temperature-high salinity (<50 °C, 4-8 wt %) and high temperature-high salinity. The contact angle was found to decease with the increase in temperature for all surfactants. About 35-70 % OOIP was recovered with most of the screened surfactant in all four regimes. The oil recovery was found to be a gravity-driven process. Rate of oil recovery increases with an increase in the extent of wettability alteration towards water-wet condition, temperature, and fracture density. Increase in IFT increases the oil recovery rate if the wettability can be changed to water-wet conditions and vice versa if the rock remains oil-wet.

Acknowledgements

This work was partially funded by a grant from the ACS Petroleum Research Fund..

Reference

1. Roehl, P.O. and Choquette, P.W.: *Carbonate Petroleum Reservoirs,* Springer-Verlag, New York (1985).
2. Chillenger, G.V. and Yen, T.F.: "Some Notes On Wettability and Relative Permeability Of Carbonate Rocks. II," *Energy and Sources*, **7**, 67-75 (1983).
3. Manrique, E. J., Muci, V. E., and Gurfinkel, M. E.: "EOR Field Experiences in Carbonate Reservoirs in the United States," SPE 100063, proceedings of Symposium on Improved Oil Recovery, Tulsa, 22–26 April, 2006.
4. Seethepalli, A., Adibhatla, B., and Mohanty, K.K.: "Physicochemical Interactions During Surfactant Flooding of Fractured Carbonate Reservoirs," *SPEJ*, **9**(4), 411-418 (December, 2004).
5. Adibhatla, B., and Mohanty, K.K.: "Oil Recovery from Fractured Carbonates by Surfactant-Aided Gravity Drainage: Laboratory Experiments and Mechanistic Simulations", SPE 99773, *SPEREE*, 11, 119-130 (Feb., 2008).
6. Austad, T. and Milter, J.: "Spontaneous Imbibition of Water Into Low Permeable Chalk at Different Wettabilities Using Surfactants," paper SPE 37236, proceedings of SPE International Symposium on Oilfield Chemistry, Houston, 18–21 February, 1997.
7. Standnes, D.C. and Austad, T.: "Wettability Alteration in Chalk 2. Mechanism for Wettability Alteration from Oil-Wet to Water-Wet Using Surfactants," *JPSE*, **28,** 123 (2000).
8. Hirasaki, G. and Zhang, D.L.: "Surface Chemistry of Oil Recovery from Fractured, Oil-Wet, Carbonate Formation," *SPEJ*, **9**(2), 151-162 (June 2004).
9. Xie, X., Weiss, W.W., Tong, Z., and Morrow, N.R.: "Improved Oil Recovery from Carbonate Reservoirs by Chemical Stimulation", SPE 89424, proceedings of the Symposium on Improved Oil Recovery, Tulsa, 17-21 April, 2004.
10. Spinler, E.A., D.R. Zornes, D.P. Tobola, A. Moradi-Araghi : "Enhancement of Oil Recovery Using a Low Concentration of Surfactant to Improve Spontaneous and Forced Imbibition of Chalk," SPE 59290, proceedings of the SPE/DOE IOR Symposium, Tulsa, 3-5 April, 2000.
11. Wu, Y., Shuler, P., Blanco, M., Tang, Y. and Goddard, W. A. : "A Study of Wetting Behavior and Surfactant EOR in Carbonates with Model Components", SPE 99612, proceedings of symposium on Improved Oil Recovery, Tulsa, 22–26 April, 2006.
12. Tweheyo, M. T. , Zhang, P. and Austad, T. : "The Effect of Temperature and Potential Determining Ions Present in Seawater on Oil Recovery From Fractured Carbonates," SPE 99438, proceedings of the Symposium on Improved Oil Recovery, Tulsa, 22-26 April, 2006.
13. Levitt, D.B, Jackson, A.C., Heinson, C., Britton, L.N., The Malik, T., Dwarakanath, V., Pope, G.A.: "Identification and Evaluation of High-Performance EOR Surfactants," *SPE Reservoir Evaluation & Engineerin,* **12**, 243-253 (April 2009).
14. Zhang, D.L , Liu, S., Yan, W., Puerto, M. and Hirasaki, G. J.: "Favorable Attributes of Alkali-Surfactant-Polymer Flooding," paper SPE 99744, proceedings of the Symposium on Improved Oil Recovery, Tulsa, 22-26 April, 2006.
15. Zhao, P., Jackson, A.C., Britton, C, Kim, D.H., Britton, L.N., Levitt, D.B., and Pope, G.A.: "Development of High-Performance Surfactants for Difficult Oils," SPE 113432, proceedings of the Symposium on Improved Oil Recovery, Tulsa, 19-23 April, 2008.
16. Maerker, J.M. and Gale, W.W.: "Surfactant Flood Process Design for Loudon," *SPE Reservoir Engineering*, **7(1),** 36-44 (February, 1992).
17. Hayes, M.E., Bourrel, M., El-Emary, M.M., Schechter, R.S., and Wade, W. H.: "Interfacial Tension and Behavior of Nonionic Surfactants," **19**, *SPEJ*, 349-356 (1979).

18. Healy, R.N., Reed R.L., and Stenmark, D.G.: "Multiphase Microemulsion Systems," *SPEJ*, **261**, 147-60 (June 1976).
19. Hirasaki, G. J., and Miller, C.A.: "Recent Advances in Surfactant EOR," IPTC 115386, proceedings of IPTC, Kuala Lumpur, 3-5 December 2008.
20. Buckley, J. S., Liu, Y. and Monsterleet, S.: "Mechanisms of Wetting Alteration by Crude Oils," *SPEJ*, **3(1),** 54-61 (March, 1998).
21. Hamouda, A. A. and Gomari, K.A.R. : " Influence of Temperature on Wettability Alteration of Carbonate Reservoir", paper SPE 99848, IIOR Symposium, 22-26 April, 2006

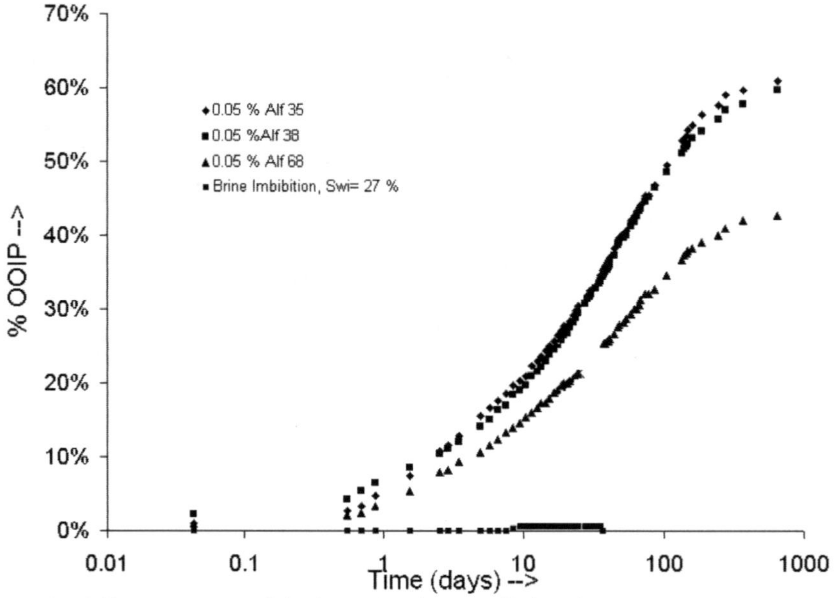

Figure 1: Oil recovery with time for low salinity-low temperature system

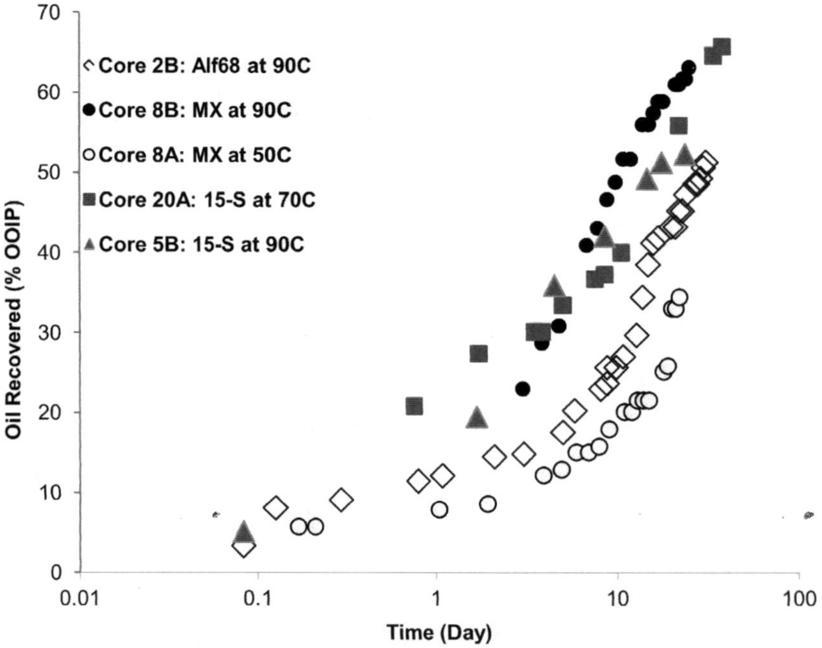

Figure 2: Oil recovery with time for low salinity-high temperature system

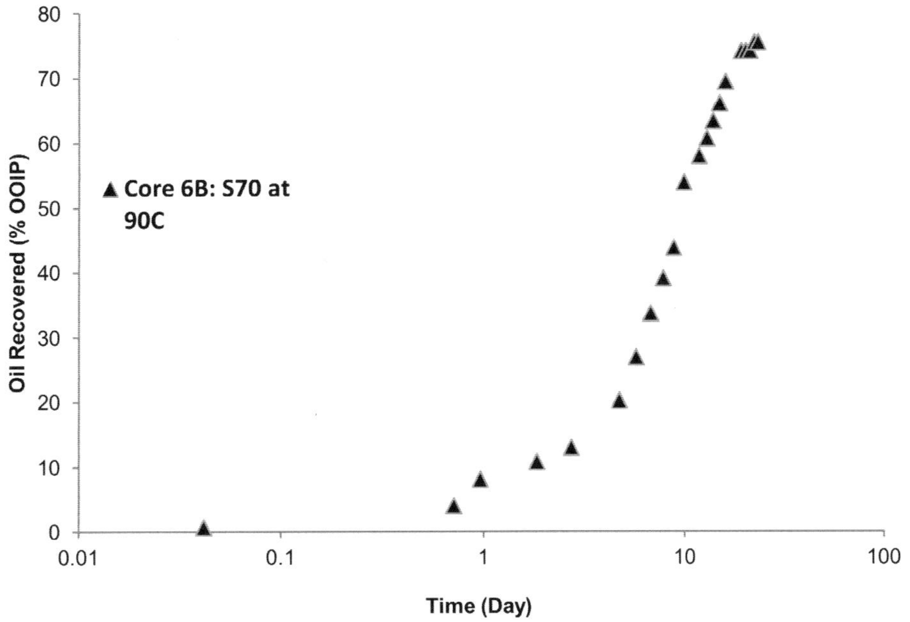

Figure 3: Oil recovery with time for high salinity-high temperature system

Figure 4: Core 1D at 50 °C after 5 days

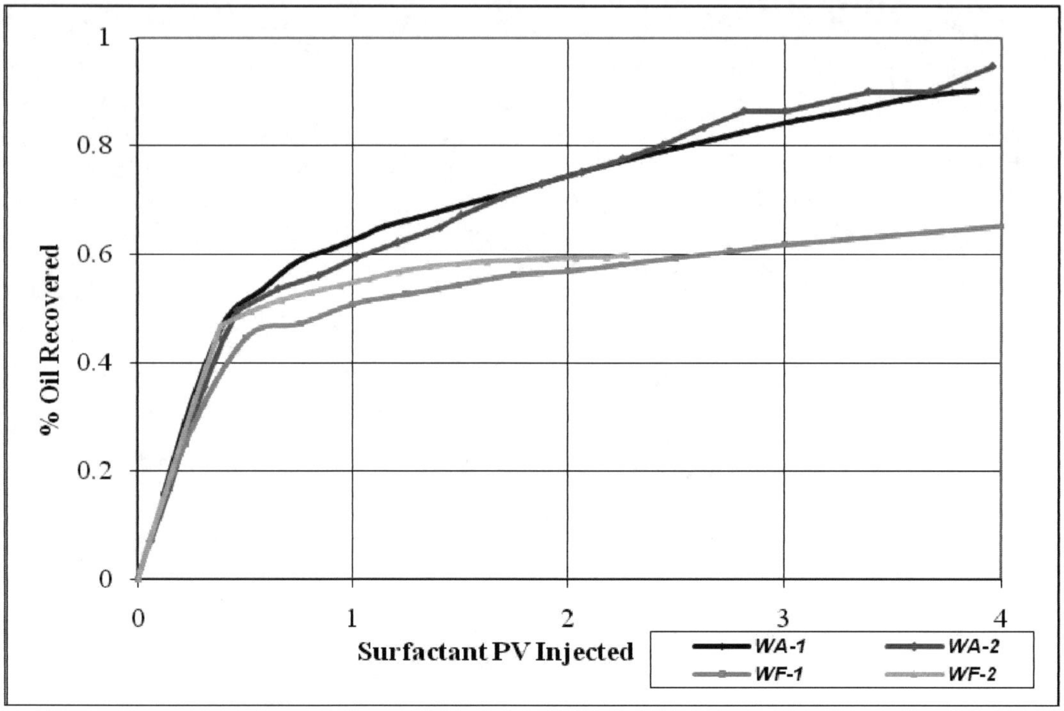

Figure 5. Comparison of oil recovery during secondary surfactant flood using wettability altering surfactant to that of typical water floods

Polyelectrolyte Nanoparticles for Flow Assurance

Stephen J. Johnson & Jenn-Tai Liang*
University of Kansas, Lawrence, KS, USA
**Address correspondence to sjohn@ku.edu*

Abstract

Nanoparticles consisting of self-assembling complexes of polyanions and polycations are able to entrap a variety of charged species, including many oil and gas chemicals. Entrapment in these polyelectrolye complexes (PEC) is through a combination of electrostatic and steric interactions. Entrapped material is protected from the hostile subterranean environment, and its release delayed while it is transported to target locations underground. The size and surface charge of the nanoparticles can be tuned to favor propagation or retention in porous media. This is a flexible technology with great promise for the upstream oil and gas industry. We review our work on applying it to two aspects of flow assurance: (i) delayed gelation of polymers used in water shut-off and conformance control by entrapping crosslinking agents such as Cr(III), and (ii) hydraulic fracturing fluid cleanup using PEC-entrapped enzymes.

Introduction

Chemicals used to improve oil and gas recovery are used in large quantities to account for loss through adsorption and reduced performance due to extremes of pressure, temperature, salinity and pH in the subterranean environment. By entrapping such chemicals in PEC nanoparticles with carefully selected properties, chemicals may be transported efficiently to the intended point of application, and their activities may be delayed, enhanced or prolonged. We have developed nanoparticle systems for a range of applications and have published proofs-of-concept for delayed gelation of HPAM polymers for in-depth profile modification,[1;2] and improved clean-up of guar-borate gels used as viscosity enhancers during hydraulic fracturing operations.[3-5] Desirable features of such particles include the formation of stable, homogeneous, colloidal suspensions able to penetrate into geological formations (particle size in the sub-micrometer range), controllable surface charge (indicated by ζ), stability at reservoir salinities, temperatures and pressures, and predictable release rates of the active components.

PEC nanoparticles were originally developed in the pharmaceutical industry and have been used to deliver a variety of therapeutic compounds in a delayed-release format. They are assembled by mixing solutions of polycations and polyanions, along with a charged active component, which becomes incorporated into the particles and is retained by electrostatic and steric forces (Figure 1). In some cases, the active component may be one of the polyelectrolytes. Properties of the particles such as particle size, zeta potential (ζ), entrapment efficiency and rate of release can be tuned by altering parameters such as molecular weight, branching, functionality, charge density and relative proportions of the polyelectrolytes, as well as manipulating environmental factors such as temperature, pH and salinity.

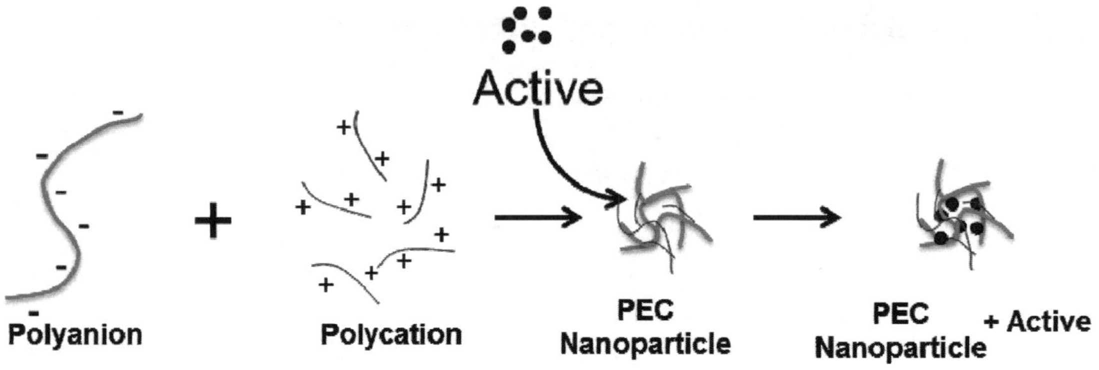

Figure 1. PEC preparation by mixing polyanions, polycations and an active component

In-Depth Profile Modification

Channeling of injected water and chemicals during waterflooding can result in poor volumetric sweep efficiency, high oil-water ratios and excessive recycling of injected fluids. Polymer gels have been used extensively to improve sweep efficiency by selectively blocking high conductivity channels so that subsequent waterflooding is diverted through previously unswept, lower permeability formations (Figure 2).

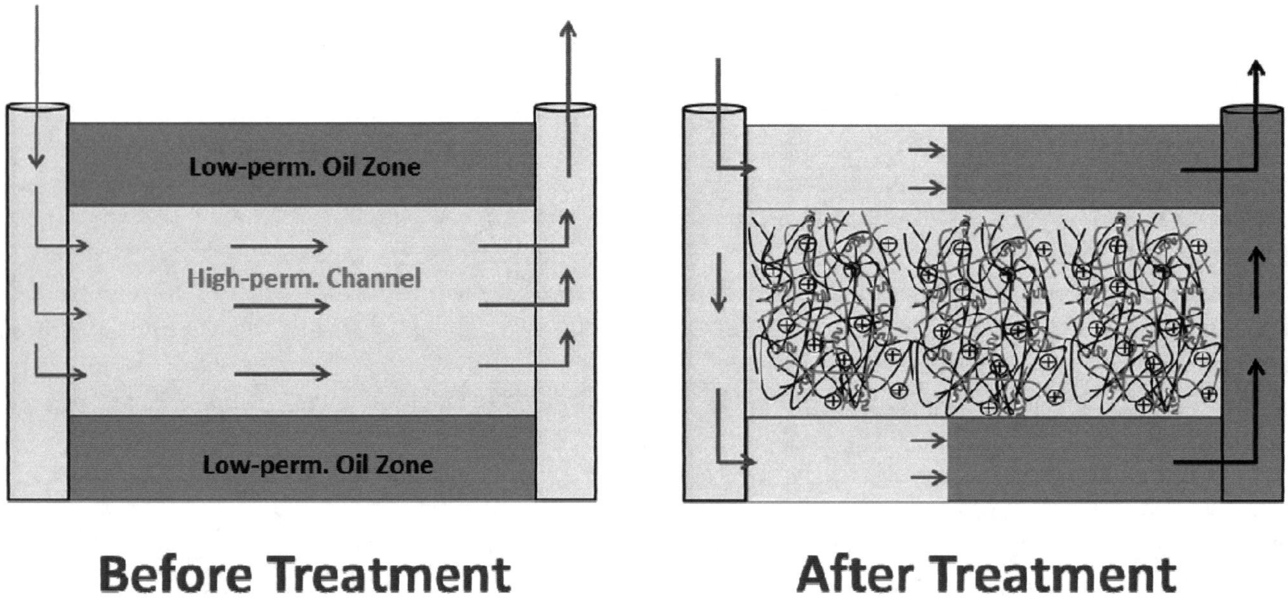

Figure 2. Gelant injection to block high permeability channel and divert subsequent water flood through previously unswept zone

One of the most commonly used gelant systems in this application is partially-hydrolyzed polyacrylamide (HPAM) crosslinked with trivalent chromium. HPAM-chromium gels are very effective at reducing permeability but they suffer from a short gelation time, which limits the volume of gelant that can be pumped into a well before gelation occurs and it is no longer possible to propagate the fluid. The use of ligands such as chromium acetate or propionate can act to delay gelation by a few hours, but even so the system is most effective in the near wellbore region where there are fractures near to, or intersecting, the well.

We have shown that if the chromium is supplied entrapped in PEC nanoparticles (Table 1) comprising, for instance, polyethylenimine and dextran sulfate, the bulk gelation time, as indicated by viscosity of otherwise identical compositions of HPAM and Cr(III), may be greatly extended (Figure 3) and that the colloidal stability of these PECs is maintained in representative synthetic field brine and seawater.[1;2]

Table 1. Typical properties of PEI/DS PEC nanoparticles containing Cr(III)

Diameter, nm	ζ, mV	Entrapment Efficiency, %
100-200	+25	>85

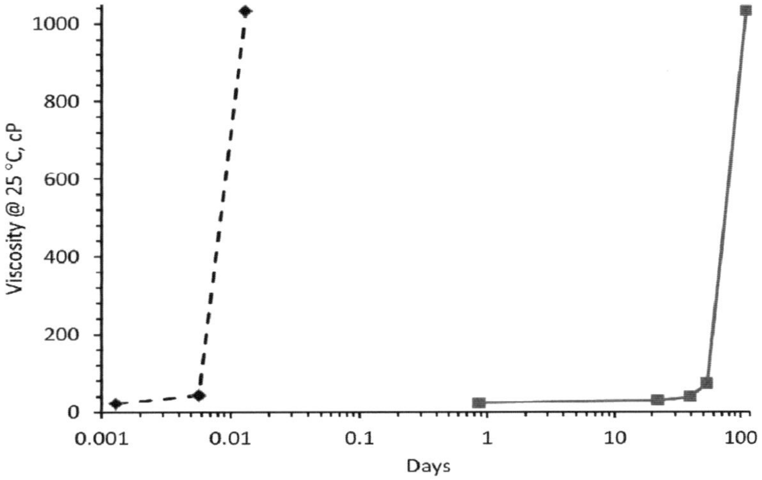

Figure 3. Viscosity of gelants comprising 5000 ppm HPAM + 100 ppm Cr where Cr was supplied as $CrCl_3$ solution (--◆--) or entrapped in PEC nanoparticles (—■—), incubated at 65°C to show delayed release of Cr crosslinker.

By varying the PEC formulation, choice of HPAM, and concentrations of the various components of the gelant system it has been possible to control the magnitude of the delay in gelation from hours to months over a wide range of temperatures and in several representative brines (unpublished data).

Hydraulic Fracturing

Hydraulic fracturing is a common and increasingly used method to improve productivity of oil and gas wells. Water-based fracturing fluid is injected at high pressure to form fractures leading to the well bores, thus greatly increasing the effective surface area of the well. A typical hydraulic fracturing fluid contains proppants, such as sand, to maintain the fracture aperture after the fracturing pressure has been removed; polymer or gel to increase the viscosity to better transmit pressure and efficiently carry the proppant; and chemicals such as oxidizers or enzyme to "break" the gel and return it to water-like viscosity so that it can be produced from the well to make way for hydrocarbon production.

Existing breakers are subject to some limitations. If used at too high a concentration, the gel may be degraded early, causing poor fracturing performance, ineffective placement of proppants, and potentially excessive invasion of the matrix by polymer. Early breaking is prevented by using encapsulated breakers that are released when the fracture closes but this may result in inhomogeneity in distribution of the breaker. If the gel is broken incompletely, well productivity will also be poor. Finally, enzymes are generally preferred to oxidizing breakers because they are less hazardous and more compatible with other fluids, tubing and pumping equipment, but their activity is dependent on temperature and pH.

In conventional hydraulic fracturing operations, guar and guar derivatives such as hydroxypropyl guar are commonly used, often crosslinked with borate to form a highly viscous gel that recovers from exposure to the shear encountered during injection (Figure 4).

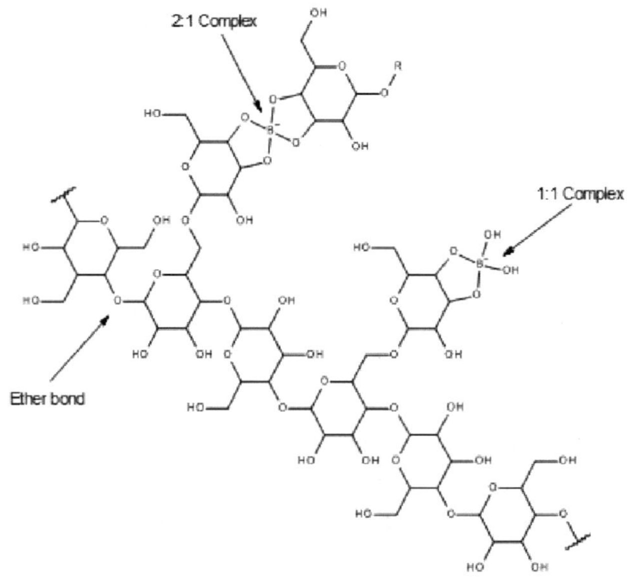

Figure 4. Structure of guar-borate gel, showing borate crosslinking and ether bonds broken by pectinase. – R indicates a second guar molecule

Stable nanoparticle formulations were developed using polyethylenimine and dextran sulfate to entrap reagent grade pectinase (Table 2) and a commercially applied enzymatic gel breaker (not shown).

Table 2. Properties of pectinase-containing PEC nanoparticles for gel breaking

Diameter, nm	ζ, mV	Entrapment Efficiency, %
435	+36	>75

Through shear loop experiments, it was found that these particles were insensitive to the shear regime typically encountered during an hydraulic fracturing injection. Breaking of a 4000 ppm guar + 1000 ppm borate gel incubated with 0.1 wt% pectinase as supplied or entrapped in PEC nanoparticles at 25°C showed delayed release of enzyme, with breaking occurring approximately 12 hours later with the PEC-entrapped enzyme (Figure 5).[3] This is an attractive time-frame for commercial hydraulic fracturing activities since it allows injection with minimal loss of viscosity, followed by breaking with a relatively short shut-in time to allow prompt return of the well to production.

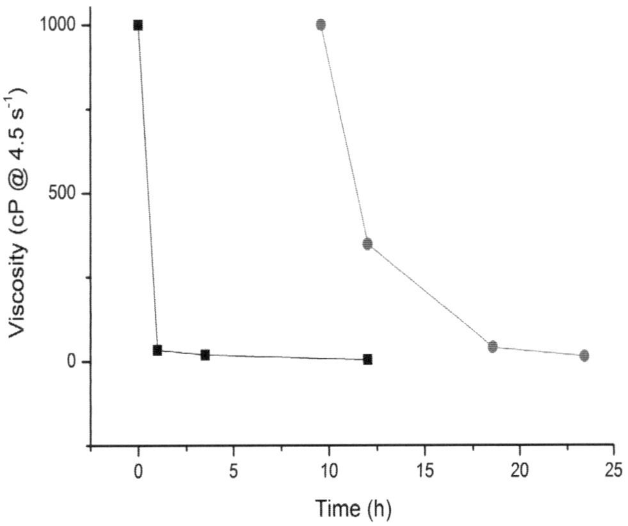

Figure 5. Viscosity of 4000 ppm guar + 1000 ppm borate gel with 0.1 wt% pectinase as supplied (—■—) or entrapped in PEC nanoparticles (—●—) at 25°C to show delayed release of enzyme.

In addition to the expected delay, it was found that entrapment in PEC nanoparticles confers some protection to the enzyme from high pH and temperature. Figure 6 demonstrates this effect.

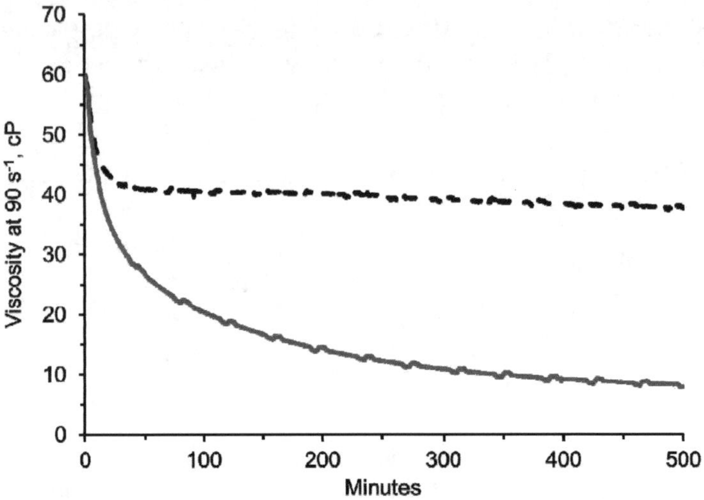

Figure 6. Viscosity of 5000 ppm guar with 0.1 wt% pectinase as supplied (---) or entrapped in PEC nanoparticles (—) at 40°C and pH 9.1 to show protection of enzymes from temperature and pH.

The pectinase used in this experiment has temperature and pH optima of 25°C and 7.0, respectively. At 40°C and pH 9.1 it is rapidly denatured and becomes inactive, as shown by the initial drop in viscosity of a guar solution incubated with unentrapped enzyme, and subsequent stabilization of the polymer solution at about 40 cP. When the same concentration of enzyme is supplied as PEI/DS PEC nanoparticles, the viscosity reduction is much more pronounced, and continues for a longer time, reaching a value of <10 cP after 500 minutes. We speculate that the microenvironment within the nanoparticles, with multiple charged moieties and pH buffered by the PEI, acts to stabilize the enzyme despite the pH and temperature experienced in the bulk solution.[4;5]

The improved clean-up of guar-based hydraulic fracturing fluids has been proven as a concept. Future work will include conductivity cell experiments, and optimization of the PEC formulation to introduce other polyelectrolytes and temperature- and pH-tolerant enzymes under conditions that better represent the wellbore and reservoir conditions.

Other Potential Applications

Further targets of this general approach include other flow assurance applications such as near-wellbore surfactant placement for condensate remediation, enhanced retention and delayed release of scale inhibitors; and EOR efforts such as in-depth nutrient delivery for MEOR and bioremediation, alkaline-surfactant-polymer (ASP) flooding, and in-depth placement surfactants for wettability alteration.

Acknowledgments

The authors would like to acknowledge our coauthors on the work described in this review: Dr. Mary Cordova, Dr. Min Cheng, Dr. Julieta Trejo, Professor G. Paul Willhite, Professor Cory Berkland, Ms. Masi Veisi, Dr. Reza Barati, Dr. Stan McCool and Professor Don W. Green. Funding for this work was provided by ConocoPhillips Company and the Tertiary Oil Recovery Project at the University of Kansas.

References

1. Cordova, M., Cheng, M., Trejo, J., Johnson, S.J., Willhite, G.P., Liang, J.-T., and Berkland, C., "Delayed HPAM gelation via transient sequestration of chromium in polyelectrolyte complex nanoparticles". *Macromolecules*, 2008. **41** (12): 4398-4404.
2. Johnson, S.J., Trejo, J., Veisi, M., Willhite, G.P., Liang, J.-T., and Berkland, C., "Effects of divalent cations, seawater and formation brine on positively charged polyethylenimine/dextran sulfate/Cr(III) polyelectrolyte complexes and HPAM/Cr(III) gelation". *Journal of Applied Polymer Science*, 2010. **115** (2): 1008-1014.
3. Barati, R., Johnson, S.J., McCool, S., Green, D.W., Willhite, G.P., and Liang, J.-T., "Fracturing fluid cleanup by controlled release of enzymes from polyelectrolyte complex nanoparticles". *Journal of Applied Polymer Science*, 2011. **121** (3): 1292-1298.
4. Barati, R., Johnson, S.J., McCool, S., Green, D.W., Willhite, G.P., and Liang, J.-T., "Polyelectrolyte complex nanoparticles for protection and delayed release of enzymes in alkaline pH and at elevated temperature during hydraulic fracturing of oil wells ". *Journal of Applied Polymer Science*, 2012. **IN PRESS**.
5. Johnson, S., Barati, R., McCool, S., Green, D.W., Willhite, G.P., and Liang, J.-T., *Polyelectrolyte compex nanoparticles to entrap enzymes for hydraulic fracturing fluid cleanup*, in *American Chemical Society Annual Meeting*. 2011. American Chemical Society: Denver, CO. 168-172.

Consideration in the Production of Liquids from Shales

Milind D. Deo, Department of Chemical Engineering, University of Utah, Salt Lake City, Utah 84112

Abstract

The world energy demand is being increasingly met by exploitation of unconventional resources. Unconventional gas production accounts for about a third of the total natural gas production in the United States for example. Large shale gas discoveries in the United States have made possible an abundant supply of clean natural gas for decades. Technically recoverable U.S. shale gas resources are now more than double the volume of previous shale gas resources and exceed 800 TCF. Production of liquids from shale has also seen a significant increase in recent times. The technologies for the production of oil from oil shale have not advanced as quickly – but several pilot operations are ongoing. In this paper, we review the unconventional resource base in the United States and the evolution of production technologies for fluids production from shale.

It is now certain that specific aspects of the shale geology and reservoir characteristics are important in identifying and producing fluids from shales. Geochemical and thermodynamic factors that result in the presence of producible liquids from a given resource should be noted. The primary geologic and reservoir characteristics of the major fluids producing plays will be described followed by an analysis of the controlling features.

Liquids are being produced in large quantities from Bakken and Eagle Ford and to a smaller extent in the Nibrara, Woodford and the liquid leg of the Barnett. In this paper, we explore the different production behaviors of these plays in light of their preliminary geologic characterization. Fluid models were created to model bubble or dew point transitions, phase viscosities and relative permeabilities. Reservoir simulations of selected fields show conditions under which production rates and available field history can be matched. We also discuss sensitivity of important parameters on liquid rates from the various plays. The study highlights the distinct geologic flavor of each of the liquid plays and how it impacts production of liquids.

Introduction

Shale gas plays have emerged as most promising for providing sustained natural gas supplies for years to come. Figure 1 shows shale gas basins in 32 countries. Shale gas plays offer energy security to a number of countries with geopolitical implications by providing vast amounts of sources of local energy.

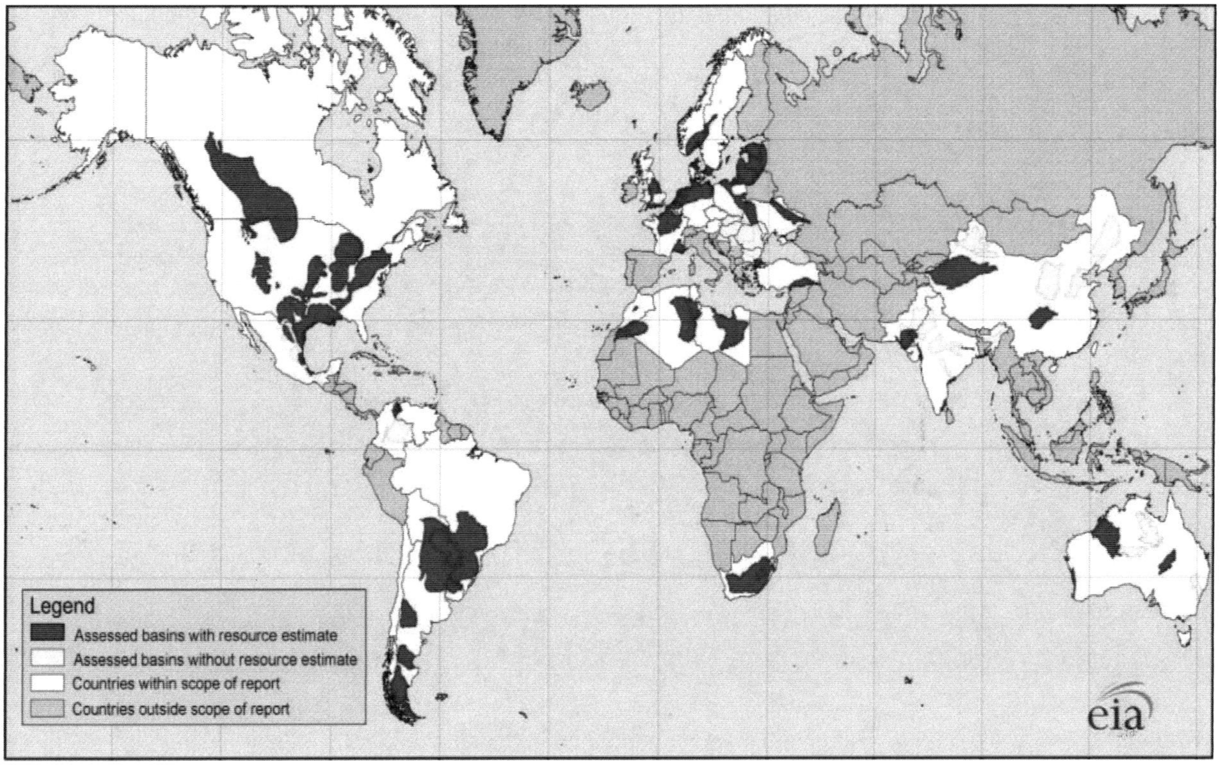

Figure 1: Shale reservoirs in 32 different countries (source – Energy Information Administration, www.eia.doe.gov)

The United States is at the forefront of shale gas production technology development. A vast number of shale gas plays have been identified and a number of these plays have either been developed or being developed. The most significant shale gas plays are the Barnett, Haynesville, Fayetteville and Marcellus. Eagle Ford in Texas and the Bakken in North Dakota have emerged as significant oil plays in shaly reservoirs. These are shown in Figure 2. Barnett led the production as shale gas reservoirs began contributing a significant fraction to the overall gas production in the United States. About a quarter of the total gas production from the United States (25%) was from shale gas reservoirs in 2010. As the Barnett is leveling off, Haynesville and Marcellus appear to be the most promising in terms of production potential (Figure 3). Bakken, Eagle Ford and Niobrara are the key liquid (oil and condensate) plays in the near term.

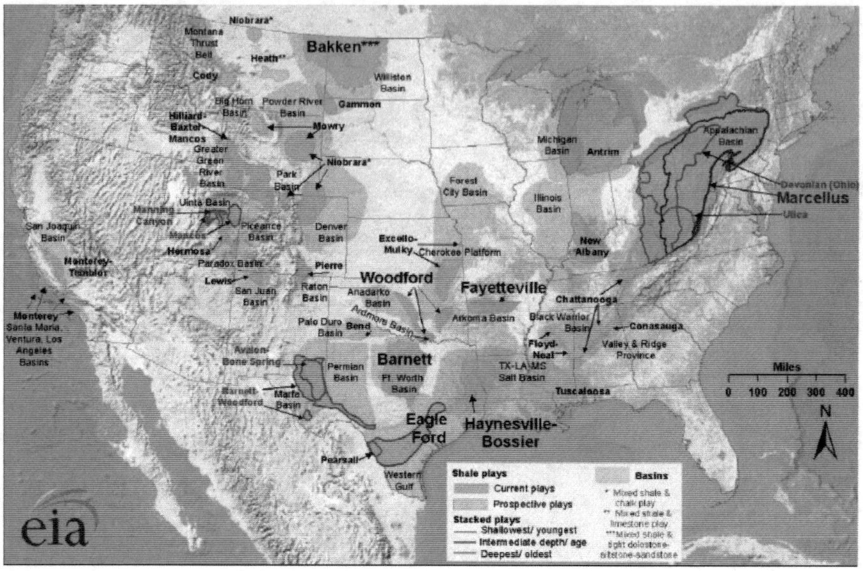

Figure 2: The major shale plays in the United States. Eagle Ford, Bakken and Niobrara have emerged as main oil production plays from shales. (Source: Energy Information Administration – www.eia.doe.gov).

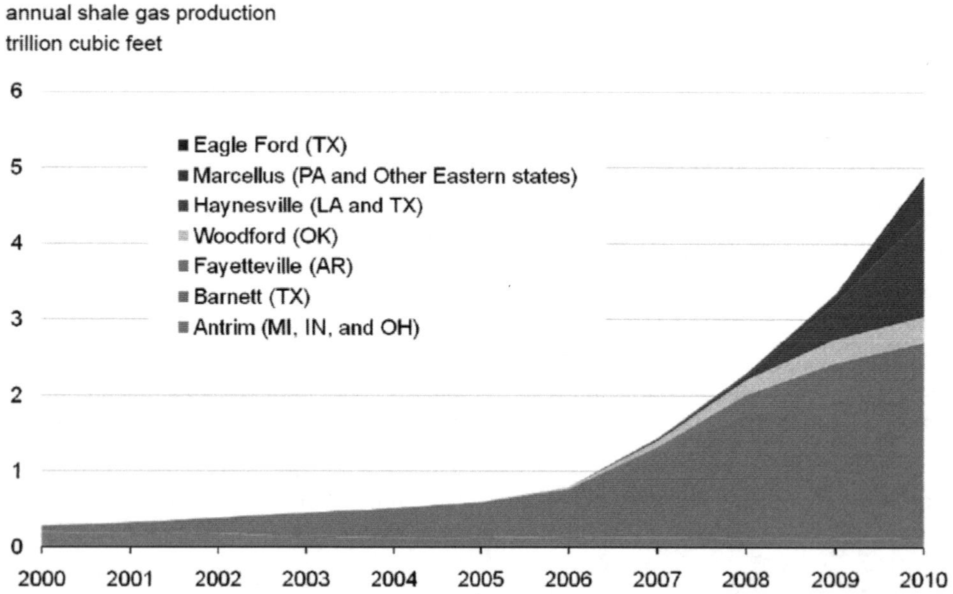

Figure 3: This figure shows the breakdown of gas production from the various shale gas plays. Barnett led the production increase through 2008 while Haynesville and Marcellus appear most promising going forward.

Horizontal well technology with multiple staged hydraulic fractures has made large volumes of gas production possible from shale reservoirs. The contrast between the numbers of horizontal and vertical well drilling in the Barnett is clearly seen in Figure 4. The production from these newly drilled horizontal wells was the main reason for the shale gas reservoirs providing a significant portion of the total gas production in the United States (Figure 5).

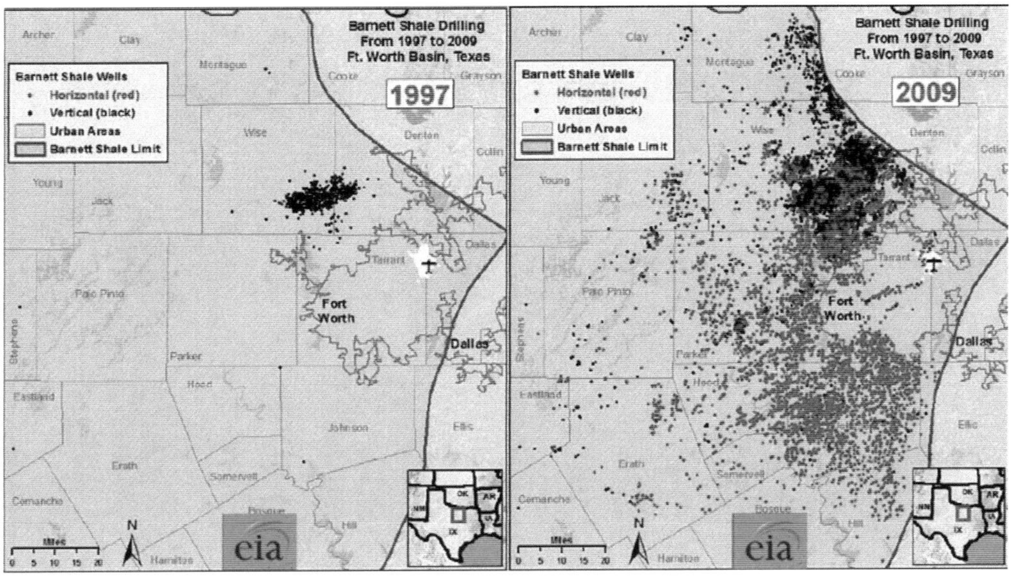

Figure 4: There are now over 13,000 wells in the Barnett. The preponderance of horizontal well drilling is clearly seen in this figure as drilling expanded from 1997 to 2009.

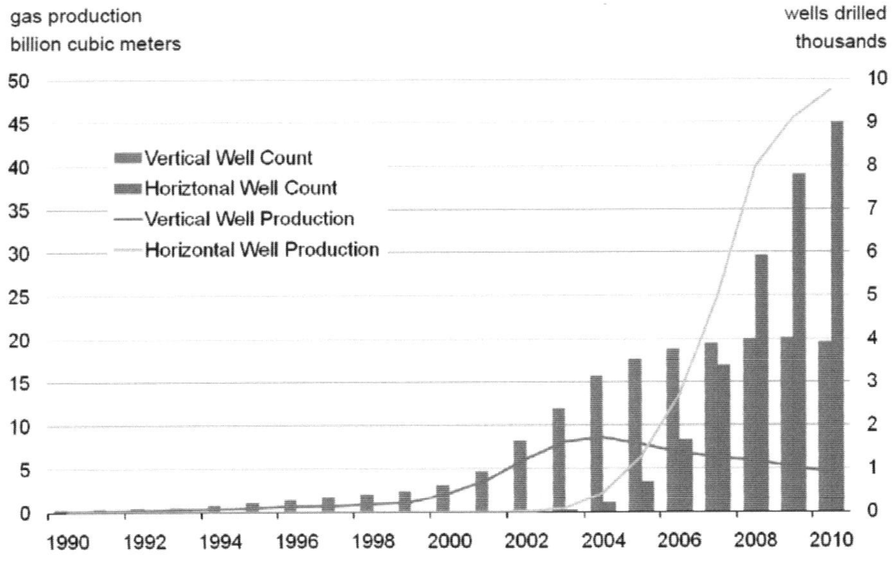

Figure 5: Production from the vertical and horizontal wells in shale plays. (Source: Energy Information Administration – www.eia.doe.gov).

Considerations in Producing Liquids

Companies are targeting oily plays in shale. Eagle Ford formation in Texas is indicative of this trend. The oil production from Eagle Ford is shown in Figure 6. The current production has reached almost 70,000 barrels/day. The rate of increase in just two years is astounding.

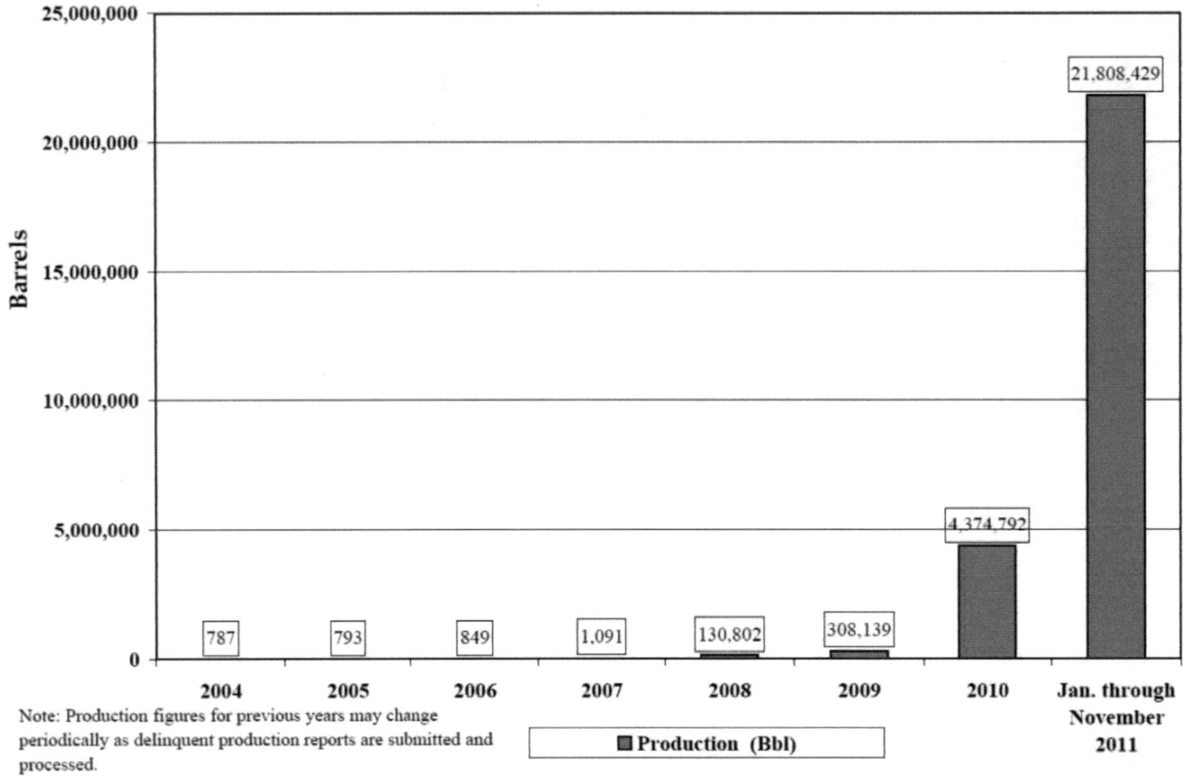

Figure 6: The growth in the production of liquids from the Eagle Ford shale in Texas. Data from the Texas Railroad Commission.

Different Liquid Plays and Geologic Considerations

Tight oil plays are complex than the gas counterparts due to fluids consideration and relative permeabilities. Every shale is different, and is characterized by complex mineralogy. The best acreage has high resistivity, adequate thermal maturity and enough reservoir energy to mobilize the oil. Petrophysically, 'sweet spots' have a higher brittleness index (ratio of Young's Modulus to Poisson's Ratio) and lower plasticity (ductility), higher effective porosity, and higher TOC-feet. The roles of fractures are still being debated, but many industry reports state that presence of natural fractures is required to make the well successful.

The Upper and Lower Bakken organic black shale members serve as both source rock and reservoirs. The Middle Bakken is a conventional, but tight, clastic and carbonate reservoir. The upper Three Forks can have similar characteristics as the Middle Bakken and be a target as well. Some early Bakken development was selectively successful in the black shales where natural fractures were prevalent.

The Eagle Ford Shale is a shale primarily in terms of particle size only, and is better described as a mixed siliclastic and carbonate unit. It is a world-class organic source rock, known for sourcing the overlying Austin Chalk. In addition to changing depth and thickness, there is also dramatic change in mineralogy, TOC, thermal maturity, pressure (and overpressure), and fluid type. Depending on location, the Eagle Ford produces dry gas, wet gas, condensate, and oil. As compared to other shale gas plays, the Eagle Ford has a high carbonate content and lower quartz content, increasing brittleness and 'fracability'. The Eagle Ford exhibits a high positive correlation between porosity and TOC.

Development challenges for the Eagle Ford are primarily driven by geologic variability including mineralogy, TOC, thermal maturity, pore pressure, and amount of natural fracturing. Other issues include variable choice of hydraulic fracture fluid type based on clay mineralogy. Vertical wells (or vertical portions of the laterals) provide thorough understanding of these geologic and reservoir parameters facilitating lateral well development. Seismic data analysis and interpretation are important, especially in geo-steering, and using the attributes of curvature, coherence, dip azimuth, and pre-stack anisotropy.

Due to its complexity, exploration of the Niobrara has been slow and the play is still in its infancy when compared to the Bakken or Eagle Ford. Niobrara wells will have on average lower rate of return on wells than the Bakken and Eagle Ford due to higher variability as well as lower recoveries offsetting the lower costs. The Niobrara remains largely untested outside of Wattenberg and Silo even after the advent of hydraulic fracturing.

A number of factors have influenced many poor well results in the Niobrara.

- Drilling in low resistivity pay (high water saturation)
- Drilling in regions without natural fractures and with low reservoir energy (characterized by low reservoir pressure and low-to-no natural gas)
- Drilling in regions where the Niobrara may not have reached thermal maturity necessary for oil generation
- Non-optimal wellbore placement (not staying in zone or non-optimal well orientation) or completion design (possibly drilled without the benefit of seismic)

Sample Reservoir Simulations

Simulations were conducted using a block reservoir with dimensions of 2000ft X750 ft X 200 ft located 12000 ft below the surface with one horizontal well (middle of the reservoir along X-axis) and one vertical fracture (parallel to XY plane) of width 0.05 ft. Fracture permeabilities are kept constant.

The initial reservoir pressure and initial water saturation were 5300 psi and 16% respectively. Initially, reservoir has single phase hydrocarbon with 84% saturation. Phase diagrams for the black oils and condensates were constructed using the compositional information available. The phase diagrams for the oil is shown in Figure 7 for reference.

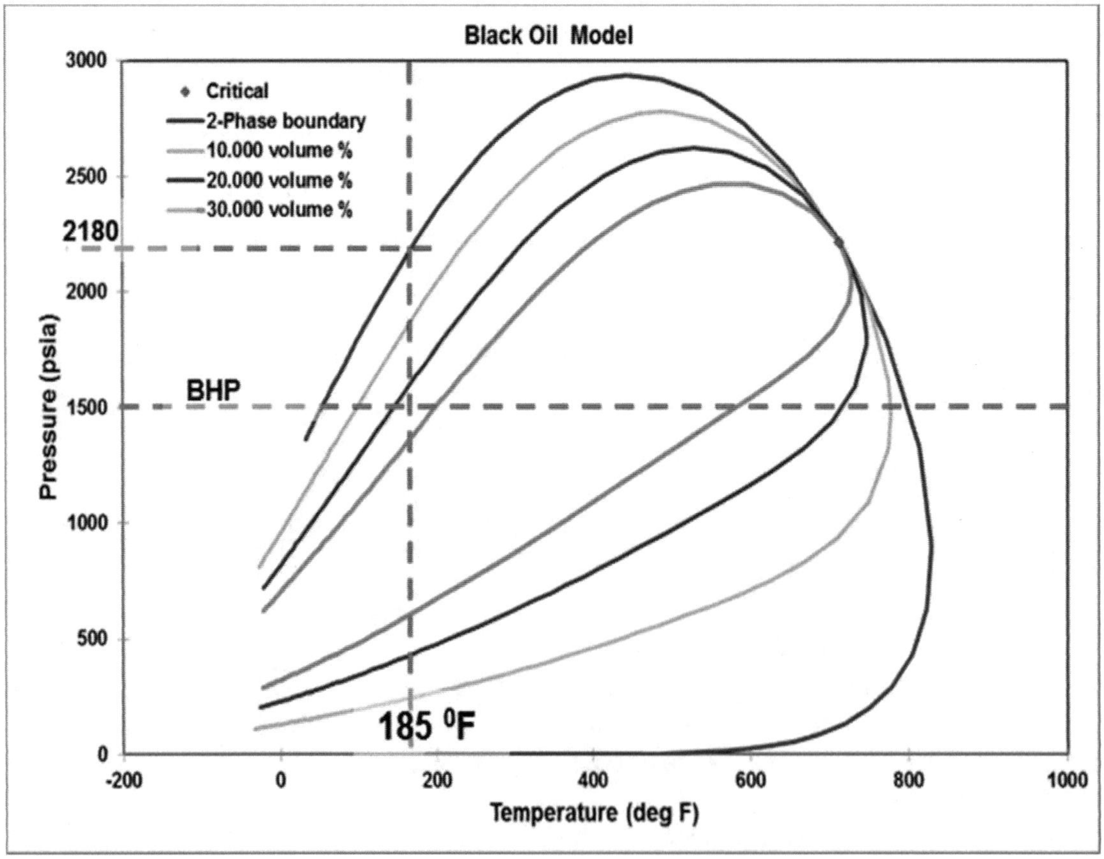

Figure 7: Phase diagram for the oil constructed using compositional information.

The produced gas oil ratios at various matrix permeabilities are shown in Figure 8.

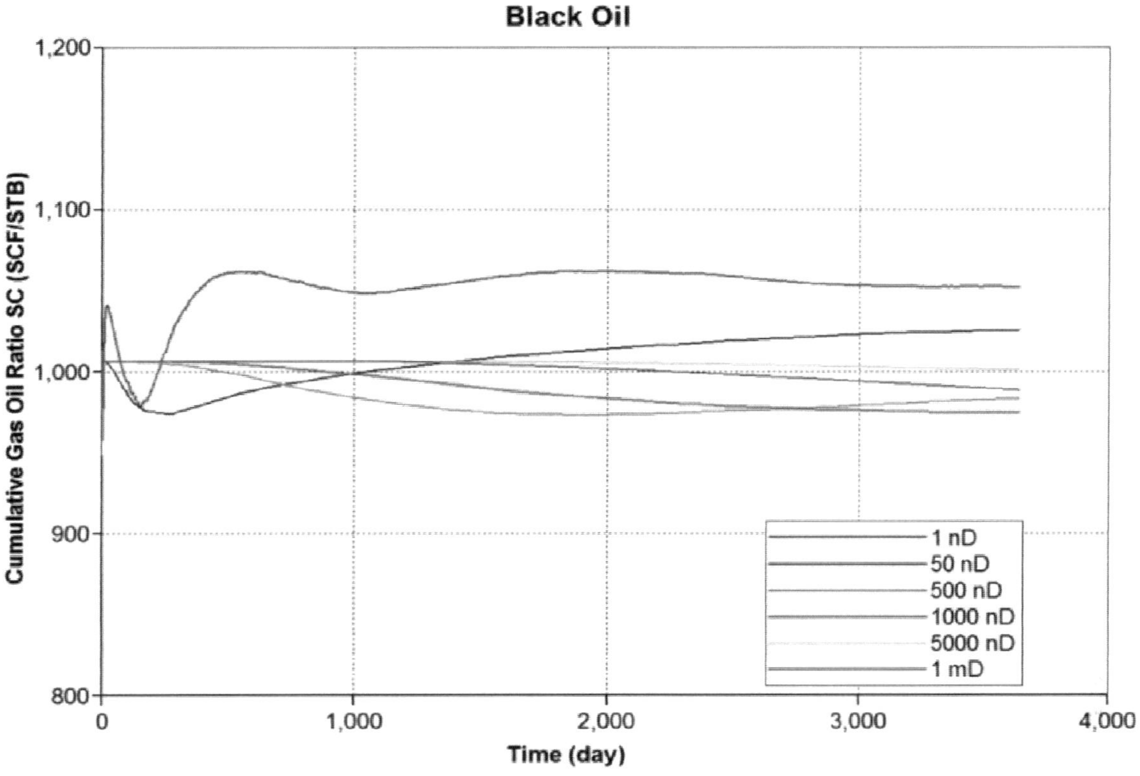

Figure 8: Gas oil ratios at various matrix permeabilities.

GOR decreases as matrix permeability increases until reservoir acts like an infinite boundary reservoir. When pressure front reaches the boundary, the GOR starts increasing unless majority of reservoir is under bubble point pressure. Similar trend is observed in case of gas condensate reservoir.

Conclusions

Production of liquids from shale has been increasing exponentially in just the past couple of years. The promising plays are Bakken, Niobrara, Eagle Ford, Woodford and Mowry. Bakken is not considered a true shale. All of these plays have their own unique characteristics. Understanding production of liquids from shale requires combining the compositional and phase behavior information of the fluids with multiphase flow considerations.

References

Energy Information Administration: www.eia.doe.gov

(129e) Multi-Scale Discussions on Mass Transport in Gas Shale Reservoirs

I. Yucel Akkutlu, University of Oklahoma, Norman, OK

Fluid mechanics of natural gas in organic-rich shale involves nano-scale phenomena which could lead to potential non-Darcian effects during gas production. In general, these are low-Reynolds number and non-continuum effects and, more importantly, pore-wall dominated multi-scale effects. In this study we introduce a new lattice Boltzmann method to investigate these effects numerically in simple pore geometries. The standard method has been developed in the 1980s to overcome the weaknesses of lattice gas cellular automata and has emerged recently as a powerful tool to solve fluid dynamics problems, in particular in the areas of micro- and nano-fluidics. The new approach takes into account molecular-level interactions using adsorptive/cohesive forces among the fluid particles and defining a Langmuir-slip boundary condition at the organic pore walls. The model allows us to partition mass transport by the walls into two components: slippage of free gas molecules and hopping (or surface transport) of the adsorbed gas molecules. Using the standard two- and three dimensional lattices, low-Reynolds number gas dynamics is simulated in a one-hundred nanometer model organic capillary and later in a bundle of smaller size organic nanotubes. The results point to the existence of a critical Knudsen number value for the onset of laminar gas flow under typical shale gas reservoir pressure and temperature conditions. Beyond this number the predicted velocity profile shows that the mechanisms of slippage and surface transport could lead to molecular streaming by the pore walls which enhances the gas transport in the organic nanopores. The work is important for development of new-generation shale gas reservoir flow simulators and it can be used in the laboratories for organic-rich shale characterization.

(129f) Supercritical Fluid Interaction Between Kerogen and Gas in Shale

James Holste, Texas A& M University, College Station, TX and Kenneth R. Hall, Chemical Engineering, Texas A& M University, College Station, TX

Recently, there has been widespread activity to develop natural gas resources available in various shale formations throughout the United States and beyond. The resources available in these formations represent a long-term source of abundant natural gas.

There is a significant difference between the reservoir matrix materials in shale formations and those in sandstone or carbonate reservoirs in that the shale matrix contains substantial amounts of heavy organic materials such as kerogen. Some of these organic materials have the potential for dissolution into the gas phase, especially in at higher temperatures and pressures. The concentrations of heavier hydrocarbons in the gas phase are small, but, as large quantities of gas are produced, the accumulated amount of heavy hydrocarbon can become significant, possibly leading to flow assurance challenges. The solubility effects may also be exacerbated by the use of light hydrocarbon fluids such as propane in the fracturing process.

The flow assurance challenges arise because the solubility of heavy hydrocarbons in methane gas is strongly influenced by the gas pressure. As the pressure of the gas decreases as it flows through the reservoir matrix and then through the production tubing to the surface, the decreased solubility may lead to precipitation of solids or viscous liquids that would impede flow through the reservoir or production tubing. The effect is similar to the precipitation of asphaltenes and waxes that provide flow assurance challenges in the production of more conventional reservoirs, especially in deep-water production systems.

This paper uses phase-equilibrium thermodynamics to explore the potential impact in the shale-gas production environment. There is little information available about the details of the shale matrix, which is composed of kerogen and associated geological decomposition products. In addition, the composition will differ for each reservoir. As a result, although this paper will draw upon knowledge gained through previous applications of supercritical fluid extraction processes, a detailed and accurate representation of the shale-gas behavior also will require experimental results that currently are not available in the public domain.

(129g) Methodology for Characterization of Fluid Properties

Matt Honarporand Nagi Nagarajan, Hess Corporation, Houston, TX

Early characterization of fluid properties in unconventional reservoirs is critical to lateral well placement, completion and stimulation. Unfortunately, determination of PVT properties based on standard downhole fluid sampling in pilot hole is not practical because of the extremely low permeability nature of the formation. Surface fluid sampling after stimulation and well completion is not suitable for determination of fluid properties even after several weeks of production because of the contamination from fracturing fluid and absence of stable GOR.

One approach is to preserve cores in Canisters at the wellsite for chromatographic analysis and HC extraction from full diameter cores since the center of the core may hold fluids at virgin reservoir conditions for an extended period of time. Integration of these data with mud gases, C1-C6+ collected during drilling and coring provide the opportunity to predict fluid properties of unconventional reservoirs in a timely manner that help decision making about completion and stimulation. In this presentation I will provide an example fluid characterization from Eagle Ford play in TX.

(148a) Hydrate Blockage Formation

Jefferson Creek, PAC, Chevron ETEC, Katy, TX

Abstract: Hydrate Blockage Formation

A primary focus of Flow Assurance Engineers during project development is the management of gas hydrates. Traditionally is was realized that no blockage could form if no hydrates formed. The past decade as seen intensive research to discover ways of allowing hydrate formation yet safely managing the flow in the project production system with maximum uptime and safety. Early attempts focused on "low dosage" chemical additives called AA's or antiagglomerate chemicals. These chemicals could accelerate the formation of hydrates while simultaneously limiting the total growth and dispersing the hydrates in the hydrocarbon liquid phase.

More recently researchers have sought methods to physically manage hydrate formation in a more holistic manner. This requires some detailed knowledge of how blockages are formed. This presentation will discuss the mechanisms of blockage formation and how this knowledge enhance the ability of production operators to successfully produce fluids with "natural" inhibition.

(148b) Field Results for Slurry Flow

Michael Eaton[1], **Jason W. Lachance**[1], Don P. Shatto[2], Larry D. Talley[1] and Doug J. Turner[2], (1)ExxonMobil Upstream Research, Houston, TX, (2)ExxonMobil Development Company, Houston, TX

The formation of non-agglomerating hydrate slurries has been reported by several laboratories. Pumpable hydrate slurries have been generated successfully in re-circulating loops. However, a slow hydrate film growth mechanism prevented sustained slurry flow in a recent field trial. This paper reports the observations of hydrate film formation in a once-through 3/8-inch lab apparatus and its relationship to observations in a once-through 4-inch field facility. The origin of this film formation will be discussed.

(148c) Do We Have New Solutions to the Old Problem of Gas Hydrates?

Bahman Tohidi, Ross Anderson, Antonin Chapoy, Jinhai Yang and Rod Burgass, Institute of Petroleum Engineering, Heriot-Watt University, Edinburgh, United Kingdom

Gas hydrates are crystalline compounds formed as a result of combination of suitable size gas molecules and water under suitable pressure and temperature conditions. The resemble ice but unlike ice can form at temperatures well above ice formation. Their formation in oil and gas pipelines could result in serious operational problems and safety concerns. The conventional techniques in avoiding gas hydrates problems are, dehydration, insulation and/or heating, injection of thermodynamic or low dosage hydrate inhibitors. In this communication we discuss some new techniques for preventing gas hydrate problems that could improve the reliability of hydrate prevention techniques and/or reduce the associated costs. The new solutions include; hydrate safety margin monitoring, hydrate early hydrate detection systems, latest results and techniques for evaluating low dosage hydrate inhibitors, and cold flow. The latest development in thermodynamic modelling will also be presented.

(148d) Developing a Comprehensive Transient Hydrate Formation Model: From Laboratory to Field Applications

Amadeu K. Sum, Chemical & Biological Engineering, Colorado School of Mines, Golden, CO

Gas hydrates pose the major flow assurance problem in the production and transportation of oil and gas. Managing the formation of gas hydrates is central for safe and continuous operation. In this talk, we will provide an overview of the Center for Hydrate Research and its efforts toward better understanding the formation and agglomeration of hydrates in multiphase flow. The talk will discuss the projects in the Center aimed at quantifying how hydrates can be managed by first understanding the fundamental processes for nucleation, growth, agglomeration, deposition, and plugging, knowledge in these areas that have been accumulated over decades of research. While still a work in progress, significant advances have been made in describing the hydrate formation in oil-dominated, water-dominated, and gas-dominated systems. One of the end goals of our effort is to develop the knowledge and tools to manage hydrates, as opposed to avoidance, in the production and transportation of oil and gas.

(148f) Hydrates in the Ocean--Beneath, Around, and Above Production Equipment

Gregory Hatton, Shell Oil Company, Houston, TX

Hydrates amass beneath and around production equipment, and can form in hydrocarbon-seawater jets/plums. The source of the hydrocarbons in these hydrates may be natural seepages, small production system leaks, or large leaks. In this presentation, some of i) the formation parameters for these hydrates and ii) the impacts on normal production operations and hydrocarbon-spill capture systems will be discussed.

(154a) Understanding and Modeling Hydrate Formation in High Water Producing Oil Pipelines

Sanjeev Joshi, Colorado School of Mines, Golden, CO

As oil/gas subsea fields mature, the amount of water produced increases, which results in an increased risk of hydrate plug formation in the flowlines. It is important to understand the mechanism of hydrate plug formation in high water cut systems in order to manage hydrate plug prevention. In this work, we will present an extensive series of hydrate formation and dissociation experiments in a four-inch diameter flowloop at the ExxonMobil research facility at Friendswood, TX. The effect of mixture velocity (1 to 2.5 m/s) and liquid loading (50 to 90%) on hydrate plug formation was studied for 100% water cut (no oil present) systems and 85% water cut systems, with methane as the gas hydrate former. The pressure drop across the pump due to flow did not increase until a certain concentration of hydrates, defined as $F_{transition}$, was reached. The 5-50 times rise in pressure drop was explained by a combination of hydrate deposit at the pipe wall (reduction in the flow cross-section) and high frictional losses due to formation of hydrate bed (high concentration hydrate-water mixture). $F_{transition}$ was found to be unaffected by liquid loading and the presence of salt in water while it increased with increasing flow velocity. A conceptual picture for hydrate growth and plugging mechanism for 100% water cut systems was proposed, which involved a transition from homogeneous to heterogeneous suspension flow of hydrate particles in water which leads to increased interaction between continuously growing hydrate particles leading to the formation of a hydrate bed/deposit in the pipe.

Design and Optimize Field Architecture for Deepwater Production Wells Based on Flow Assurance Criteria

Sridhar Reddy Appanapalli and Manjunath Terwad
Department of Mechanical Engineering, University of Houston, Houston, Texas, USA

Abstract

As Oil Exploration and Production (E&P) companies are venturing further out into the sea and deeper to tap into pockets of oil and natural gas around the world, the Front End Engineering and Design (FEED) analysis of deepwater field development has become very critical due to the high risks in Capital Expenditure (CAPEX) and Operating Expenditure (OPEX). The field architecture for most deepwater developments has been dictated by flow assurance constraints and is a critical part in the design and operation of offshore oil and gas systems.

In this case study, a flow assurance analysis using PIPESIM was performed for a field development and is used to determine the ideal field architecture and reconcile the operational conditions against the field requirements. The field has five deepwater wells that are located at depth greater than 3000ft and the corresponding reservoirs were assumed to have same fluid composition. The flow assurance analysis was used to determine the field architecture including co-mingling of wells, flowline routes and FPSO location. The other parameters that were calculated are production tubing, flowline and riser line sizing along with topside arrival pressure and temperature values. Flowline and riser insulation requirements to maintain production temperature above wax appearance temperature and gas lift injection requirements to maintain topside arrival pressure were also determined. In addition, an erosion analysis was performed to verify the production fluid velocity does not exceed the erosion velocity limit per API 14E. The possible flow assurance issues that could arise from this design were deduced and necessary strategies to overcome were suggested.

Introduction

One of the major challenges in the offshore industry is to develop cost-effective management strategy for the flow assurance issues, both design and operating procedures. This paper presents the flow assurance modeling and analysis made on the hypothetical deepwater field with design solutions to manage the flow assurance challenges. Flow Assurance has emerged as an engineering discipline with the emergence of deepwater field developments. The severity of consequences of many flow assurance issues have increased with increased water depth and many traditional shallow water operating procedures cannot always be applied.

The most significant characteristics of deepwater are low temperature and a dramatic increase in the hydrostatic head pressure. This results in energy loss which is primarily due to Joule Thompson cooling and Potential energy loss. The deepwater condition is also ideal for the formation of the solid deposits such as hydrate and wax, which increases the risk of flow degradation or even kills the flow.

Typically field layouts consist of topside facilities located near to the center of field and the production and injection wells are spread around at relatively short distances. To maximize the economic gain and also to reduce the CAPEX, subsea tie-backing of marginal satellite hydrocarbon fields to the nearby topside facility is done so that existing production infrastructure can be used rather building a new one. A tie-back system magnifies the impact of the in-field lines on flow assurance management of the field. The importance of the heat losses to the environment is increased and even greater focus is paid to the thermal performance. It is indeed demonstrated that a pipeline with infinite insulating performance will still experience temperature decrease in the in-field flowline, due to elevation changes.

Based on the flow assurance modeling and analysis results, operational strategies were developed for the scenario of steady state flow and for different water cuts to consider the effects on production with aging of the fields.

Field Architecture

The five production wells (P1 through P5) in the subsea field as shown in Figure 1 are in water depths ranging from 1100 to 1250 meters. The field will be developed as subsea tiebacks to an FPSO with offset distances less than 2 km and is positioned approximately in the middle of the field. The reservoirs are at around 2120 meters below the sea bed.

THE GRAPHIC ON THIS PAGE WAS NOT AVAILABLE ON THE E-MEDIA VERSION OF THIS PAPER.

Figure – 1: Subsea field layout with location of wells and FPSO

The system comprises of flowlines and risers with capability for one-way pigging. There were multiple solutions available to connect the wells to the FPSO. Some options considered were, one to connect all 5 wells with five risers, two to connect 4 wells to one riser & 1 well to another riser and option three to connect 3 wells to one riser and 2 wells to another riser. In this study, authors choose to perform the analysis on option three by connecting Wells P1, P2 and P4 together to one riser and wells P3 and P5 together to the other riser and the risers tied back to the FPSO.

Flow Assurance strategy

System line sizing were optimized based on available pressure constraints from reservoir to topside separator conditions using downhole gas-lift. The flowlines and riser internal diameters were designed at 10 inch for the entire system and all the wells were designed to flow through a wellbore tubing diameter of 7 inches. It must be noted that all five wells had same casing profile for the wellbore tubing. Table 1 shows the total distance for each component of the tie-back system.

Table – 1: Length of each component and total length of each tie-back system

Distance (m)	Tubing	Flowline			Riser	Total (m)	Total (ft)
P1->P2->P4->FPSO	3420	632	566	632	1175	**6425**	**21079**
P5->P3->FPSO	3420	825	632	-	1175	**6052**	**19855**

Peak erosional velocity for the entire system was checked against the acceptable erosional velocity limits. Erosional velocity limits were calculated using API 14E equation for downhole tubing, flowlines and risers using a C-factor of 200.

Wax management is primarily done by providing thermal insulation to the flowlines and riser. The subsea production system has been designed with thermal insulation to achieve flowing temperatures at the topside to be over 80°F. Most of the production life will flow with no wax issues except during late life in the three-well tie-back system may lead to some wall deposition. It is recommended to reduce the rate of deposition by injecting wax dispersant chemicals into the flowline. Another way is to use one-way pigging option to further manage wax deposition and cleaning the pipelines.

Additionally, Wax deposits and Hydrates may form during transient activities such as cold start-up period and system shut-in period. Hence the system design has incorporated thermal insulation on the flowlines and riser to slow cool down period without operator intervention. If wax and hydrates were to form during start up period, it is necessary to remove all water using pigs to push dry or circulate hot oil around the system. During a shut down period there may be wax deposited on the walls of the flowlines. The wax is assumed to re-melt as flowline and riser heat up upon production restart. If Hydrates were to form it can be remediated by depressurizing the flowlines to a point below where hydrates are stable at sea bed temperature. Methanol or Kinetic hydrate inhibitors can be injected to control hydrates formation which will also ensure the flowing temperatures to remain above the cloud point temperature to keep wax deposition in control.

Input data and Assumptions made

Input data:

- Reservoir and Fluid data
- Design data for the production wells

- Ambient sear water temperature profile
- Well tubing profile
- Production wells sear water depth
- Production profile or early life, mid life, and late life
- Thermal properties
- Gas-lift system parameters

Assumptions:

- Same well tubing profile for all production wells.
- Vertical well completion type.
- Water cut – 10%, 40% and 90% for Early, mid and late life respectively.
- Manifold or chokes were not used in the analysis. Instead adder component will be used to input production flowrate from other wells.
- Assumption is made that flow is always upstream and there will be no negative flow from high pressure wells into the low pressure wells.
- Downhole equipment like Gas-lift will be used inside the well tubing profile.
- Used steel pipe with roughness ratio of 0.0018" for entire pipeline system.
- No slugging analysis performed.

Modeling and Analysis Approach/Methodology

The design and analysis of the oil and gas production wells from reservoir to topside was modeled using PIPESIM software, a steady-state, multiphase flow simulator and hence transient temperature conditions during system start up and shut-in could not be calculated. In this study, multiphase flow was modeled from the reservoir to the wellhead and combined with flowlines and riser to topside. In-built modeling algorithms were utilized for gas lift and erosion modeling.

Modeling

The basic steps listed below were utilized to build the model.

- Units - Engineering
- Fluid data - Black Oil
    - Calibration data (no calibration used)
- Components in the model
    - Well components (vertical completion well, tubing)
    - Pipeline component (Flowline, Riser, and Multiplier/Adder)
    - Field equipment (gas-lift)
- Heat transfer options (Coated – 2 Btu/hr-ft2-F for wellbore tubing, Insulated – 0.2 Btu/hr-ft2-F for flowlines and risers)
- Flow correlation for multiphase flow (Vertical- Orkiszewski, Horizontal - Beggs and Brill Revised)

Analysis Approach

PIPESIM model was built for one production well with wellbore tubing and the other well(s) were connected thru the flowlines using an adder to input production flowrate for a given well. The flowline elevations for each well were taken into account per the subsea field layout. The model was repeated for different stages of production life.

Iterations were initially run for early life production rates for both well tie-back systems. While optimizing the flow, simulation was run to check for topside arrival pressure & temperature, well head pressure & temperature and pipeline erosion along the total length. To optimize the flow to topside, the ID's of tubing, flowline and riser were modified until the flow was able to reach the topside facility. After running several iterations it was noted that with changing tubing, flowline and riser diameters ranging from 5" to 14" the flow would not reach to the topside. It was then determined to use downhole gas-lift equipment at the bottom of the wellbore tubing to boost the production flowrate. The iterations were rerun with range of downhole gas-lift flowrates for varied diameters on the tubing, flowlines and riser until both tie-backs were optimized for top side pressure & temperature, wellhead pressure & temperature, erosion limits and wax formation temperatures.

Table 2 shows the diameters of each component of the tie-back system and their overall lengths that meets the design requirements.

Table – 2: Tubing, Flowline and Riser lengths and ID

	Tubing	Flowline					Riser
		P1 to P2	P2 to P4	P4 to FPSO	P5 to P3	P3 to FPSO	
Length (m)	3420	632	566	632	825	632	1175
ID (in)	7	10	10	10	10	10	10

While repeating the models for later stages of production life like mid and late life iterations, the ID's of tubing, flowline and riser were unchanged. The only variables that changed were the production flowrates for the respective wells and gas lift flowrates to meet the topside constraints.

Note 1: With the option 3 chosen for analysis, different paths were simulated for optimum results. Flow path P3 to P5 to FPSO and flow path P4 to P2 to P1 to FPSO were also analyzed. It was noted that topside temperature would fall below the wax appearance temperature for both tie-back systems during late life production.

Note 2: Analysis was also performed with coated pipe option for all components including wellbore tubing, flowlines and riser. It was noted that the arrival temperatures would fall well below wax appearance temperature even at the early life production rates. Arrival pressure would also fall below the topside pressure requirement with the downhole gas lift exceeding the design limits.

Analysis Results

Multiphase flow modeling has been performed from reservoir to the topside separator to size the production flowlines and risers for the entire field. Oil production rate from each well during earlylife, midlife and latelife were analyzed for desired topside pressure, temperature, wax appearance and erosion limits. Average well flow rates range from 12400 (early) ~ 1560 (late) BOPD. Peak rate in well P5 is up to 15000 BOPD. The flow path was determined to have low energy wells and hence downhole gas-lift was added at the bottom of the wellbore tubing to obtain desired flow rates. Analysis was performed for early life, mid life and late life of the production stages with water cut 10%, 40% and 90% respectively.

The simulation results indicate the topside pressure and temperature were met for the entire field during all stages of production life except for the late life of three-well profile where the arrival temperature falls short by 10°F to wax appearance temperature. The temperature remains above WAT at the wellhead and gradually fall down to 70°F before it reaches topside. It must be noted that since the production rates are lower at the later stages and the watercut percentage remains higher, wax deposits along the flowline wall will be negligible and controllable either by using pigging or chemical injection into the flowlines if needed. Figure 2 explains the pressure and temperature profile for the late life case with 90% water cut.

The erosion study performed on all the production wells met well within the allowable erosion limits and are shown in the charts below. The erosion velocity ratio graph in Figure 3 for the late life case with 90% water cut shows that as long as the ratio of fluid mean velocity to the erosion limit velocity falls below one, the erosion in the pipeline system is in control.

Gas lift amount needed for the entire system to boost flowrate and maintain the production rates were monitored and were kept within the design limits. However, it must be noted that for each tie-back the PIPESIM model had utilized only one well tubing profile modeled with adders for production rates from other wells, it was unclear if the gas-lift amount utilized in the simulation would be considered only to the well with tubing details or it must be included to the adders. In any case if the gas-lift was considered to be combined for all the wells in each tie-back, the total gas-lift amount would still fall within the maximum allowable per design requirement. For example, in early life production stage the two-well tie-back uses 15.7 mmscf/d and with one adder the total would be (2*15.7) 31.4 mmscf/d and the three-well tie-back uses 10.6 mmscf/d and with two adders the total would be (3*10.6) 31.8 mmscf/d. Together the combined gas-lift rate will be 63.2 mmscf/d which is within the max allowable of 65 mmscf/d.

Finally, efforts were made to ensure the wellhead pressure and temperature was met within the design criteria during the entire production life. Results are tabulated in Tables 3 and 4 for the two tie-back systems.

All the analysis were performed using insulated pipes with heat transfer co-efficient (U-value) of 0.2 Btu/hr-ft2-F for flowlines and riser except the wellbore tubing was modeled as coated pipe with 2 Btu/hr-ft2-F.

Table – 3: Results for production life of two-well tie-back system

P5->P3->FPSO	WC (%)	GL (mmscf/d)	TP (psia)	TT – Insulated (°F)	WHP (psia)	WHT (°F)	FMV (ft/s)	EVR
Earlylife	10	15.7	216	86	853	100	51	0.53
Midlife	40	4.1	217	88	961	99	11	0.14
Latelife	90	8.4	216	83	1071	94	18	0.14

Table – 4: Results for production life of three-well tie-back system

P1->P2->P4->FPSO	WC (%)	GL (mmscf/d)	TP (psia)	TT – Insulated (°F)	WHP (psia)	WHT (°F)	FMV (ft/s)	EVR
Earlylife	10	10.6	216	85	953	98	67	0.73
Midlife	40	1.7	217	82	894	92	10	0.14
Latelife	90	4.6	219	70	754	80	23	0.18

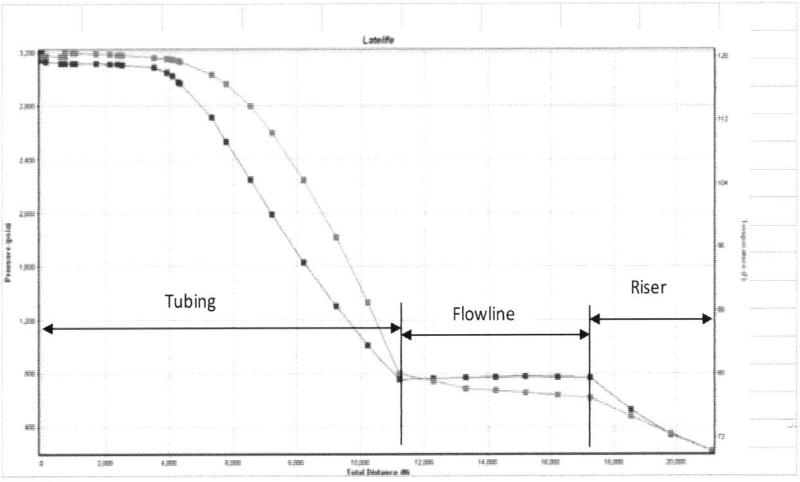

Figure- 2: Pressure & Temperature vs Total distance. (WC- 90%, Arrival temperature 70°F)

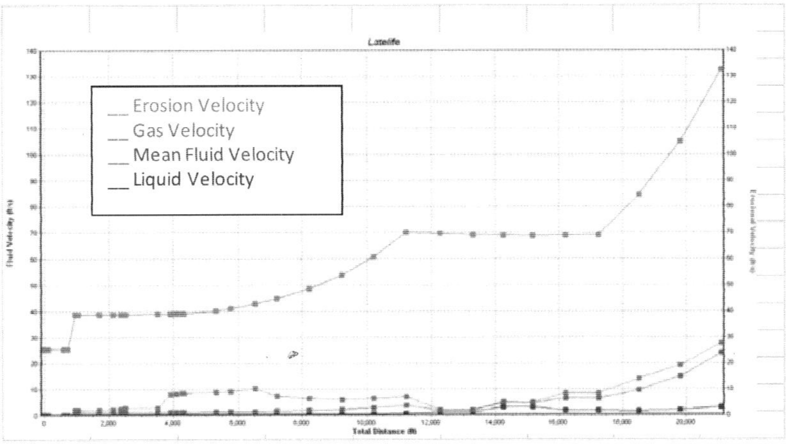

Figure – 3: Erosion analysis for the Total distance. (WC- 90%)

Conclusions

- Flow assurance is central to the design of subsea systems.
- A production system has been modeled for the given location of the reservoirs at different bathymetry.
- System line sizes are optimized based on available pressure and temperature constraints (reservoir and topside facilities) using downhole gas-lift.
- The production wells are commingled in the increasing order of well head pressures so as to avoid the choking of the pipelines in the network.
- The model is developed using series connections to FPSO which will result in lower total length of the piping (and hence cost), but a severe limitation is the shutting down of the entire loop whenever there is a problem in anyone of the lines.
- Joule-Thompson heating effect is observed at the starting depths of the tubing and this is due to sudden drop of pressure and thus sudden increase in temperature, after that the pressure and temperature drop occurs predictably.
- The fluid mean velocity, gas and liquid velocities remain well below the erosional velocity limit in any life. The erosional velocity ratio lies well below the value of one and hence problems due to erosion will never be a concern.
- Wax deposition is controlled primarily by thermal insulation and the temperature may never fall below the specified WAT limit.

Recommendations

- It is recommended that a slugging analysis be performed to determine the worst case liquid slug size. Based on these results the topsides equipment can be sized to handle a worst case liquid slug.
- The use of a corrosion inhibitor injection is recommended if corrosion of the steel tubing or flowline/riser piping is a concern based on the contaminants existing within the produced hydrocarbons.
- If insulation is not sufficient then the use of a low dose hydrate inhibitor injection is recommended to mitigate hydrate formation especially during shut-down conditions.
- It is recommended that if required, the rate of wax deposition can be reduced by injecting wax dispersant chemicals into the flowline. Another solution recommend is to use one-way pigging option to manage wax deposition, if required and to clean the pipelines.

References

1. MECE 6397- Flow Assurance Class Notes, University of Houston, Spring 2011, Phaneendra Kondapi and Janardhan Davalath.
2. PIPESIM Production System Analysis Software by Schlumberger.
3. Pipe Flow 1, Single-phase Flow Assurance, Ove Bratland, 2009.
4. Pipe Flow 2, Multi-phase Flow Assurance Ove Bratland, 2010.

(154c) Lithologic Heterogeneity and Focused Fluid Flow Governing Gas Hydrate Distribution in Marine Sediments

Sayantan Chatterjee, Chemical Engineering, Rice University, Houston, TX

Solid gas hydrates can form when low molecular weight gas molecules and water combine at relatively high gas concentrations, high pressures, low temperatures, and low salinity conditions. These conditions are present along many continental margins (as well as in permafrost environments) where hydrocarbon gases, especially CH4, accumulate in sediment pore space within a finite depth interval known as the gas hydrate stability zone (GHSZ). While the amount and distribution of marine gas hydrates remain uncertain, they may constitute a potential energy source, a deep-water geohazard, and an important component of the global carbon cycle.

Previous numerical models have coupled mass, momentum and energy transport equations to obtain steady-state solutions for gas hydrate distribution in marine sediments. These homogeneous 1-D models provide first-order insights on hydrate occurrence, but do not capture the complexity and heterogeneity observed in natural gas hydrate systems. Numerous studies have shown that gas hydrate and free gas accumulation in marine sediments is highly heterogeneous at the m-scale, which complicates views on their formation. Fracture network systems and dipping permeable layers are common heterogeneities in natural gas hydrate settings; these localize flow, resulting in accumulation of localized, concentrated hydrate deposits. To incorporate these heterogeneous features, lateral fluid flow and to broaden our knowledge of primary controls of hydrate concentration and distribution, we extend our previous 1-D model to two dimensions.

We develop a 2-D, heterogeneous sedimentation-fluid flow model that tracks gas hydrate accumulation over geologic time scales. Simulations with a vertical fracture network, which extends through the gas hydrate stability zone and has permeability 100 times greater than the surrounding shale formation, show that focused fluid flow causes higher hydrate (25-55%) and free gas saturation (30-45%) within the fracture network compared to the surrounding, lower permeability shale. Systems with high permeability, dipping sand layers also show localized, elevated saturations of hydrate (60%) and free gas (40%) within the sand layers due to focused fluid flow.

Permeability anisotropy, with a vertical to horizontal permeability ratio on the order of 10^{-2}, enhances hydrate concentrations within high permeability conduits because anisotropy enhances transport of methane-charged fluid to high permeability conduits. Our two-dimensional (2-D), heterogeneous models quantify how focused fluid flow through high permeability zones affects local hydrate accumulation and saturation. We also show increased fluid flux and deep source methane input result in enhanced concentrations of hydrate and free gas, and also increase the flow focusing effects. From our 2-D results, we determine that the hydrate and free gas saturations can be characterized by the local Peclet number (localized, focused, advective flux relative to diffusion); which is consistent with Peclet number (net fluid flux relative to diffusion) characterization in one-dimensional (1-D) systems. This characterization suggests that even in lithologically complex systems, local hydrate and free gas saturations can be characterized by basic parameters (local flux and diffusivity). We intend to adapt our generalized model to specific field examples such as Walker Ridge in the Gulf of Mexico where enhanced hydrate saturation is noted in dipping sand layers.

(154d) Droplet Transport in near-Horizontal Low Liquid Loading Flows

Kiran Gawas, Petroleum Engineering, University of Tulsa, Tulsa, OK

Low liquid loading gas-oil-water flow is widely encountered in wet gas pipelines. Even though the pipeline is fed with single-phase gas, the condensation of the heavier components of the gas along with traces of water results in three-phase flow. In such cases top of the line corrosion is widely encountered due to condensation of water droplets containing dissolved corrosive gases. Most of the corrosion inhibitors used are in the liquid phase and need to be transported to the top of the pipe to prevent corrosion. The dominant mechanism for transport of liquid to the top, for large diameter pipes is through droplets entrained from liquid film at bottom of the pipe. However, droplet transport studies in pipes are scarce and are mostly based on experiments for small diameter (0.0025 m to 0.0076 m) pipes. Scaling up of these models leads to inconsistent results since flow distribution could be different for larger diameter pipes. Moreover there is no work on transport of droplets in three-phase flow which is encountered in wet gas pipelines.

In the current work, droplet transport in three-phase flow for horizontal and slightly inclined pipe has been studied. A new isokinetic sampling system is developed to study droplet flux for oil/gas and oil/water/gas flows in a 0.015 m (6 in.) diameter flow loop. Experiments were performed for inclination of +2° and -2° from the horizontal. The flow conditions for initiation of atomization and droplet transport to top of the pipe have also been studied.

The droplet flux was found to decrease due to presence of water in the liquid phase for the same total liquid flow rate. Water-cut in the entrained liquid was found to be lower than the input water cut. In addition, the total droplet flux and water-cut in the entrained liquid was found to be lower for downward inclined pipe as compared to upward inclined pipe.

(154e) Understanding Foam Flows - Application to Gas-Dominated Flow Systems

Haijing Gao, Chemical Engineering, Purdue University, Purdue, IN

Foam flows play significant roles in the recovery and production of oil and gas, for example, cutting and transporting with foam during drilling and flow from reservoirs during oil/gas recovery. Therefore, successful modeling of foam flows can benefit these processes involved in the upstream energy industry. For instance, a mechanistic understanding of foam flows in horizontal and/or vertical pipes can be applied for reliable prediction of the rate of gas and condensate production. Since foams exhibit strong elasticity, reliable measurements and predictions of the mechanical/rheological properties of foams are invaluable for developing models, which are crucial in predicting the flow response of foams. In this talk, a modern experimental technique, "Stokes experiment with foam", is shown to be a reliable technique for measuring the elastic modulus of liquid foams. Moreover, this study shows that wall effect must be taken into account for accurately modeling the foam flows near pipeline walls due to a reduction of the elastic modulus for foam flows near walls. The conclusions drawn from this study are also equivalently valid for dense emulsion flows, e.g. oil/water, and systems of even three fluids, e.g. oil, water, and gas.

AUTHOR INDEX

Adkins, Stephanie S. 324
Akkutlu, I. Yucel 368
Anderson, Ross 373
Appanapalli, Sridhar 377
Appathurai, Santosh 304
Arachchilage, Gayani P. 324
Baez, J. 80
Bailey, William 3
Baled, Hseen 143
Bamgbade, Babatunde 143
Barrufet, Maria 256
Bates, Martin 122
Biswal, Sibani Lisa 312
Bou-Mikael, Saade 82
Boyle, Katie L. 289
Brannon, Harold 287
Brege, Jonathan 50
Bringedal, Bjorn 79
Britton, Christopher 324
Brower, David 64
Bruce, Duncan 122
Buckley, Jill 86
Burgass, Rod 373
Burgess, Ward A. 143
Calçada, L. A. 7
Cañas, Wilson A. 220
Chapman, Walter G. 163
Chapoy, Antonin 373
Chatterjee, Sayantan 385
Chenevert, Martin 83
Creek, Jefferson 371
Cremaschi, Selen 4
Daigle, Timothy P. 63
Danielson, Tom 2
Dannenhauer, C. E. 7
De Andrade, Alex Rodrigues 7
Deo, Milind D. 360
Djamali, Essmaiil 308
Drexler, Santiago 80
Eaton, Michael 372
Economides, Michael J. 287
Ector, Daniel 113
Enick, Robert M. 143
Faria, Jimmy A. 80
Fisher, Kevin 290
Fogler, H. Scott 85
Forgiarini, Ana 50, 91, 309
Freeman, Craig M. 289
Friedheim, James 83
Gao, Haijing 387
Gawas, Kiran 386
Gomaa, Ahmed M. 31
Gonzalez, Doris L. 163, 220
Goodwin Jr., David 292
Guo, Xuhong 306
Gupta, Robin 345
Hall, Kenneth R. 369
Hantz, Savanna 63
Harwell, Jeffrey 62, 80, 106
Hatton, Gregory 375
Hedengren, John D. 64
Hewitt, Geoffrey F. 1
Hirasaki, George J. 312
Holcomb, David 113
Holditch, Stephen A. 286
Holste, James 369
Honarpor, Matt 370
Jang, Chung-Min 83
Janjua, Rafique 63
Johnson, Jeffrey N. 289
Johnson, Stephen 353
Joshi, Nikhil 220
Joshi, Sanjeev 376
Kan, Amy T. 308
Kashou, Sam 71
Kilpatrick, Peter 310
Kim, Do Hong 324
Kondapi, Phaneendra B. 66
Kouba, G. 4
Krejbjerg, Kristian 131
Krishnamoorti, Ramanan 81
Lachance, Jason W. 372
Leekumjorn, Sukit 131
Liang, Jenn-Tai 353
Lim, Frank 220
Logan, Alaistair J. 122
Loureiro, S. A. 7
Lu, Haiping 308
Lu, Jun 324
Ma, Kun 312
Martins, Andre Leibsohn 7
Maxey, Jason 18
McElfresh, Paul M. 113
McHugh, Mark A. 143
Mohan, Kshitij 345
Mohanty, Kishore K. 288, 345
Monteiro, Othon R. 122
Moreno, R. Z. 7
Moridis, George J. 289
Morreale, Bryan 143
Mullins, Oliver 90
Nagarajan, Nagi 370
Nasr-El-Din, Hisham A. 344
Nesic, Srdjan 311
Neville, Anne 307
Nikitopoulos, Dimitris 82
Nikolaou, M. 291
Nydal, Ole J. 6
Osorio, Raul 220
Pan, Lehua 289

AUTHOR INDEX

Panachareonsawad, Ekarit 87
Panuganti, Sai R. 163
Park, Daniel S. 82
Pedrosa, Nuno 270
Pennington, Jonathan A. 308
Pereyra, Eduardo 5
Phillips, Brian 63
Phirani, Jyoti 288
Pietrangeli, Gianna 50
Pope, Gary A. 324
Qu, Qi 31
Quintero, Lirio 50, 91, 122
Reagan, Matthew T. 289
Resasco, Daniel E. 80
Roberts, Ian 78
Rojas, Mario R. 106
Ronningsen, Hans P. 89
Ruiz, M. Pilar 80
Rydahl, Allan 70
Sabatini, David 62
Salager, Jean-Louis 50, 91, 309
Sambath, Krishnaraj 305
Sarica, Cem 5, 87
Scheid, C. M. 7
Sengebusch, Anita 66
Sharma, Gaurav 345
Sharma, Mukul M. 83
Shatto, Don P. 372
Shea, Richard 70
Shi, Wei 308
Shiau, Bor-Jier 80, 106
Shippen, Mack 3
Slupphaug, Olav 79
Smirnova, Antonina I. 122
Smuk, Stefan 78
Solairaj, Sriram 324
Soong, Yee 143
Souza, Elessandre 7
Stewart, Norm 72
Storkaas, Espen 79
Subramani, Hariprasad J. 4
Sum, Amadeu K. 374
Sun, Hong 31
Szczepanski, Richard 270
Talley, Larry D. 372
Tapriyal, Deepak 143
Terwad, Manjunath 377
Thompson, Karsten E. 82
Tohidi, Bahman 373
Tomson, Mason B. 84, 308
Tour, James M. 84
Turner, Doug J. 372
Valbuena, Ernesto 256
Vargas, Francisco M. 163
Waldmann, Alex Tadeu A. 7
Wang, Lu 308
Wang, Wei 308
Wang, Xiuli 287
Warpinski, Norman 290
Weerasooriya, Upali 324
Wiehe, Irv 88
Wilcock, Elizabeth 122
Willson, Clinton S. 82
Wong, Michael S. 84
Work, Sarah N. 308
Wu, Yue 143
Xu, Jun 306
Yan, Chao 308
Yang, Jinhai 373
Yang, Tao 306
Zhang, Hong-Quan 5
Zhang, Nan 308
Zhang, Xiaohong 270
Zhou, Jing 31